RODMAN E. SNEAD
Professor of Geography
University of New Mexico
Albuquerque, New Mexico

WORLD ATLAS OF

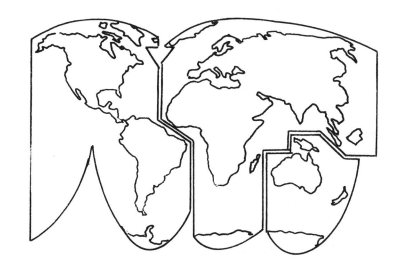

GEOMORPHIC FEATURES

Robert E. Krieger Publishing Company, Inc.
Huntington, New York

Van Nostrand Reinhold Company
New York   Cincinnati   Toronto   Melbourne

Co-Published by
Robert E. Krieger Publishing Company, Inc.
645 New York Avenue, Huntington, NY 11743, U.S.A.
and Van Nostrand Reinhold Company
A division of Litton Educational Publishing, Inc.
135 West 50th Street, New York, NY 10020, U.S.A.

10   9   8   7   6   5   4   3   2   1

Library of Congress Cataloging in Publication Data

Snead, Rodman E.
    World atlas of geomorphic features.

    1. Geomorphology—Maps.   2. Landforms—Maps.
I. Title.
[G1046.C2S6 1980]   912'.1'5514   77-28009
ISBN 0-88275-272-3

PREFACE

For many years there has been a great need for an atlas that shows the distribution of particular land features. World and regional maps have been compiled for many types of landforms, but few atlases compile this data into a single unified work. The first edition of the ATLAS OF WORLD PHYSICAL FEATURES was published by John Wiley and Sons in 1972. The original work took five years to compile and contained a great deal of new information. Now a new revised edition has been prepared with notable changes such as the enlargement of the maps to fill the entire page and a more complete numbering and reference system. New maps depict the spectacular findings on continental drift and plate tectonics as well as more detailed revised earthquake and seismological data. There are new maps on the distribution of soils, drainage patterns, and sediment transport, as well as world wave environments, mobile coasts, and the highest elevations in the United States and around the world. There is also a new map on the world distribution of arid lands.

Revised maps include a listing of many more world volcanoes, earthquakes, caves and caverns, waterfalls, meteorite craters, and large artificial lakes. There is a revision of glacial features which gives a clearer presentation of ice formed landforms on a North American base map.

The problems of cartographic presentation of physical features on world maps of small scale exists with the second edition as it does the first. For many of the landforms it is very difficult to show great detail; on some nearly impossible. Most of the dots and shaded areas show regions instead of actual numbers. For example, it would be impossible on a world scale to depict every drumlin in Boston Bay. Even on the United States map, which is a larger scale, this is a difficult task. Instead, a dot shows the approximate location of drumlin swarms. Thus, the purpose of this atlas is to indicate those areas where physical features are particularly numerous rather than to give exact numbers, size, and shape. Most of the symbols locate areas where a feature stands out noticeably on the earth's surface. no man-made landforms are shown with the exception of the maps depicting large artificial lakes. The topics selected and the outline followed were compiled mainly from the chapter headings in Arthur Strahler's and William Thornbury's physical geography and geomorphology college textbooks. However, information on a world scale is unavailable for several aspects of physical geography that I wanted to include. For instance, it would be useful, if it were possible, to show the extent of large desert alluvial fans and pediment slopes. The author would have liked to include a map of domes, other than salt domes, but not enough information exists to accurately plot these features on a world-wide basis. The new edition does, however, outline the stages of crustal deformation, and the morpho-genetic landform regions of the world.

The scale of the world map makes it difficult to plot the exact location of each feature. This is especially true for the United States and Europe, where there have been detailed investigations of most physical features. For this reason, on a number of maps, North America, the United States, and Europe are shown in this larger scale. This is also true in several cases for the Caribbean. There is no indication on the maps of the relative importance or degree of development of one feature compared with another, but most examples are so-called "classic examples" in the literature. The maps are mainly concerned with overall distribution. However, in the text that accompanies each map, the more important regions are indicated.

The brief text has several purposes. One is to define the landform for those individuals who are unfamiliar with its name or makeup. Another is to indicate the major characteristics of the feature, and a third is to indicate reasons for its occurrence and distribution. The new edition uses more numbers on the maps and then based on the numbers, lists the actual features in the text. This method requires less lettering and the maps are easier to read.

This is an introductory work. More than nine years have been spent in compiling and plotting the data, but it represents just a beginning. A work of this kind is a lifetime enterprise. Large sections of South America, Africa, and Asia have not yet been studied in sufficient detail to provide any degree of certainty about the exact location of many physical forms. I have plotted only those locations that I could find in the literature with the exception of the features of which I have first-hand knowledge. In future editions, it should be possible to show much more accurately and in greater detail the location of particular landforms. However, the new edition does plot in much greater detail a number of features that could not be found or did not exist in the literature around 1972.

The bibliography at the end of this atlas lists some of the major references used in compiling this work. It is divided into seven major sections that correspond to the sectional divisions of the atlas. In some of the descriptive texts I have included the authors and the dates of publications from which particularly significant information has been obtained.

Rodman E. Snead

iv

ACKNOWLEDGMENTS

I thank the many people who have helped in the preparation of this atlas. They include student assistants who spent long hours doing library research and compiling information on the maps; C. Managaran, graduate assistant, who spent a year doing research at the Clark University Library; Karen Hock and Gary Whiteford who helped prepare several of the maps at Clark University; David Robertson and Nancy Bott who assisted in the research and compiling of the material at Rutgers University; and Steve Anderson, Roy Goettling, William Hodges, Scott Griffiths, and William Couzins who did library research and helped draft the maps at the University of New Mexico. I am grateful to Susan Louise Courtney for drafting the base map and am particularly indebted to Judith Bateman for the long hours spent in drafting all of the other maps. Special mention must be made of the assistance given by the staff of the American Geographical Society Library in New York. Dr. Norbert P. Psuty, Dr. Robert D. Campbell, Dr. Richard E. Murphy, and Dr. Barbara Zakrezewska were most helpful in reviewing the atlas while it was being prepared.

The author wishes sincerely to thank and acknowledge the following persons, publishing firms, and professional societies for their generous permission to translate and reprint copyrighted materials: American Geographical Society; Aldine Publishing Company; Edward Arnold (Publishers) Ltd., London; Prof. Hubert A. Bauer; Prof. Nicholas Bariss; Prof. Julius Budel; Prof. Karl W. Butzer; Prof. Jack L. Davies; R. S. Dietz; Prof. R. H. Fleming; R. F. Flint; Prof. D. G. Frey; Geographische Runschau; Prof. Edwin H. Hammond; D. H. J. Harrington; Hutchison Publishing Group Ltd., London; Prof. M. W. Johnson; D. W. Junk N. V. Publishers, The Hague, Netherlands; Prof. Melvin Marcus; Marinus Nijhoff Publishers, The Hague, Netherlands; McGraw-Hill Book Company; Prof. J. V. Mieghem; Oliver and Boyd (Publishers), Edinburgh; Pergamon Publishing Company; Prentice-Hall Book Publishers; Princeton University Press; Dr. Erwin Raisz; Mr. Elliot A. Riggs; Prof. Arthus H. Robinson; Prof. Arthur N. Strahler; Prof. N. M. Strakhov; Dr. David R. Stoddart; Prof. H. V. Sverdrup; Prof. Glen T. Trewartha; Prof. P. Van Oye; Prof. Joseph E. Van Riper; Prof. Kenneth Walton; and Prof. H. W. Wright.

When one or two sources were used in compiling a map, these have been indicated at the end of the text. Where a number of sources were used, they have not been listed but are mentioned by last name and date throughout the text. Some maps were compiled from such a wide variety of sources, as well as from the author's experience, it would be impossible and misleading to list all of them. All of the major sources consulted are listed in the bibliography.

THIS BOOK IS
DEDICATED TO
MY MOTHER AND FATHER

CONTENTS

CONTENTS

Section One

General

Geography

MAPS
1-1
1-2
1-3

A WORLD WIDE SYSTEM OF LANDFORM CLASSIFICATION

A classification system that is based in part upon the genesis (origin) of landforms and in part on the actual configuration of the land surface is both appropriate and meaningful in terms of an explanatory-descriptive system of landform analysis. One such classification has been devised by Richard E. Murphy using three levels, or categories, of information in successive application to identify a landform type. These categories are: its geologic origin and rock composition (structural regions), the nature of the geomorphic process by which it has been shaped (erosional or depositional landscapes), and the configuration of its surface (topography). The threefold basis of the genetic approach to landform study - structure, process, and stage - is included in the first and second categories of the classification system. A particularly valuable attribute of the genetic approach is that a highly trained and experienced geomorphologist can use his knowledge to predict or anticipate many characteristic details of the landforms not stated in the definitions of the various classes. The empirical ingredient of the Murphy system is found in the third level of classification, topography. Here subdivisions of geometrical properties of the land surface follow strict numerical definitions. Elevations above sea level and local relief (difference in elevation between highest and lowest points in adjacent locations) form the basis for defining classes. The empirical approach to description of the topography lends an element of useful and unambiguous information to the classification system.

Source: Murphy R. E., Figures 18.13, 18.14, and 18.15 in Strahler, Arthur N., *Introduction to Physical Geography,* 2nd ed., John Wiley and Sons, Inc., New York, 1970.

MAP 1-1   STRUCTURAL REGIONS OF THE WORLD

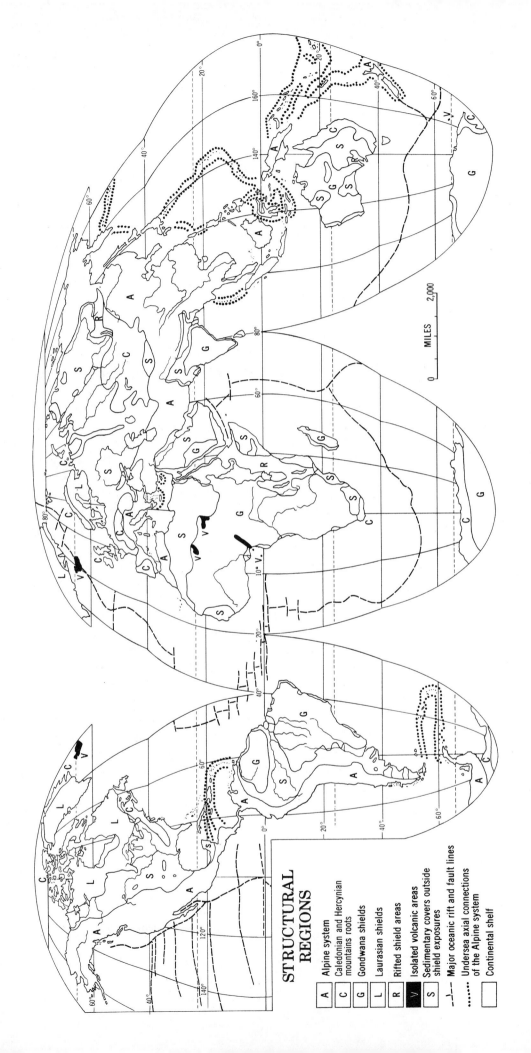

STRUCTURAL REGIONS

A	Alpine system
C	Caledonian and Hercynian mountains roots
G	Gondwana shields
L	Laurasian shields
R	Rifted shield areas
V	Isolated volcanic areas
S	Sedimentary covers outside shield exposures
-	Major oceanic rift and fault lines
· · ·	Undersea axial connections of the Alpine system
☐	Continental shelf

World structural regions (From R.E. Murphy, 1968, Annals, A.A.G., Map Supplement No. 9. Based on Goode Base Map. Cartography by J.P. Tremblay, courtesy of John Wiley and Sons, Inc.)

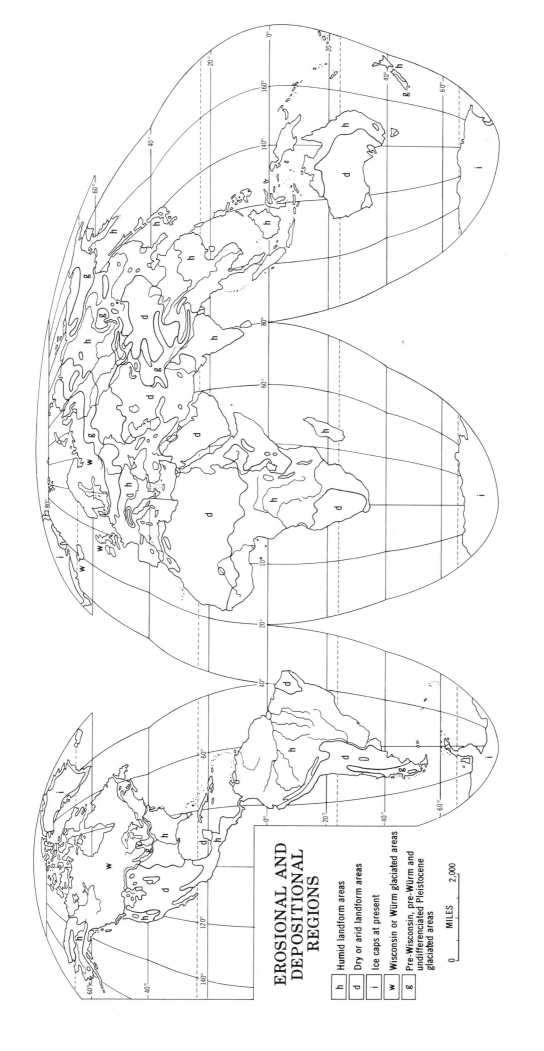

EROSIONAL AND DEPOSITIONAL REGIONS

h	Humid landform areas
d	Dry or arid landform areas
i	Ice caps at present
w	Wisconsin or Würm glaciated areas
g	Pre-Wisconsin, pre-Würm and undifferenciated Pleistocene glaciated areas

MILES

0   2,000

World distribution of erosional and depositional landforms (From R.E. Murphy, 1968, Annals, A.A.G., Map Supplement No. 9. Based on Goode Base Map. Cartography by J.P. Tremblay, courtesy of John Wiley and Sons, Inc.)

MAP 1-2   EROSIONAL AND DEPOSITIONAL REGIONS OF THE WORLD

MAP 1-3   TOPOGRAPHICAL REGIONS OF THE WORLD

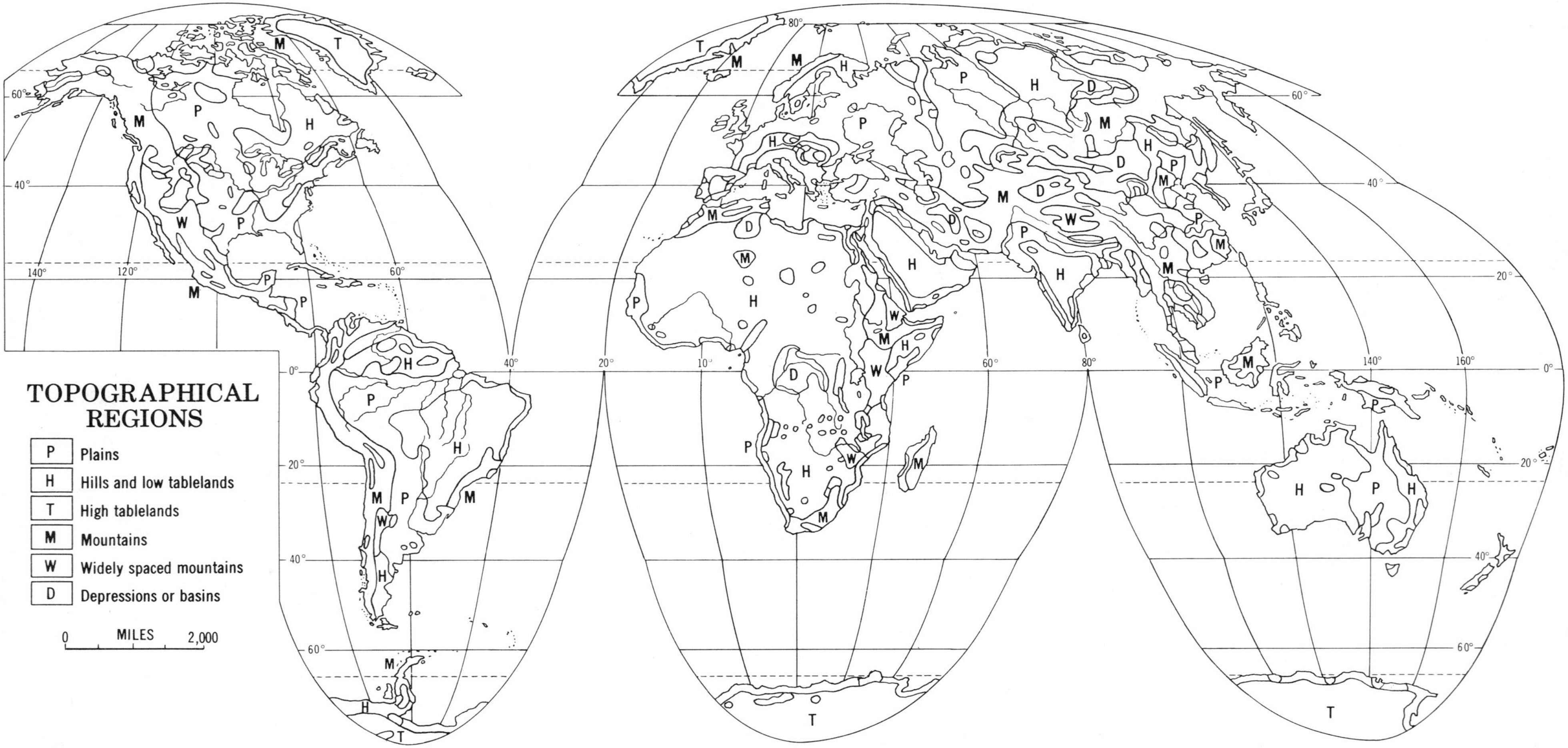

TOPOGRAPHICAL REGIONS

P — Plains
H — Hills and low tablelands
T — High tablelands
M — Mountains
W — Widely spaced mountains
D — Depressions or basins

MILES
0   2,000

World topographical regions (From R.E. Murphy, 1968, Annals, A.A.G. Map Supplement No. 9. Based on Goode Base Map. Cartography by J.P. Tremblay, courtesy of John Wiley and Sons, Inc.)

MAP
1-8

HIGH MOUNTAIN PEAKS OF THE WORLD

This is an interesting map because it shows that most of the high mountains on the earth are in only a few locations. By far, the greatest concentrations are in the Himalayan, Karakorum, Pamir Knot region of Asia. Here are found 77 peaks over 20,000 feet (6,096 meters) high and even this listing may not be complete because a number of peaks have not been correctly indentified or measured. Correct identification and plotting is also difficult in the Andes of South America. The high peaks of Peru, Chile, and Argentina have different names and heights depending on the different maps and atlases consulted. It is noteworthy that there are ten peaks in Antarctica between 14,000 and 16,860 feet (4,267 and 5,139 meters) concentrated in two main locations — on the Antarctica Peninsula, and between Victoria Land and the Queen Maud Mountains. Other interesting findings include the listing of six peaks in the Caucasus over 16,000 feet (4,877 meters) making these mountains the highest in Europe. The United States has three areas where peaks are over 14,000 feet (4,267 meters) — Alaska, California, and Colorado. In Alaska, several of the high peaks of the St. Elias Range are half in Alaska and half in Canada. In Colorado, there are a total of 30 peaks between 14,000 and 14,433 feet (4,267 and 4,399 meters).

If all peaks over 12,000 feet (3,658 meters) were given, there would be a very long list. Therefore, in the Himalayan region only peaks over 20,000 (6,096 meters) feet are presented, for Europe peaks over 10,000 (3.048 meters), in North America peaks over 14,000 feet (4,267 meters) and in South America peaks over 19,000 feet (5,791 meters). Exceptions to these elevations are Kerintji, in Sumatra, 12,467 feet (3,800 meters); Mt. Cook in New Zealand, 12,349 feet (3,764 meters); and Mt. Kosciusko in Australia, 7,310 feet (2,228 meters) Plate 1 (center-fold) is an excellent Landsat space photograph depicting the main Himalayan Ranges in the vicinity of central Nepal. The view extends from the Ganges Plain in India north to the crest of the main or greater Himalayas in northern Nepal and southern Tibet.

Source: Various sources including *The World Almanac and Book of Facts,* The Albuquerque Tribune, 1976, P. 574-575, and *The Associated Press Almanac,* Hammond Almanac, Inc., Maplewood, N. J., 1975, p. 15.

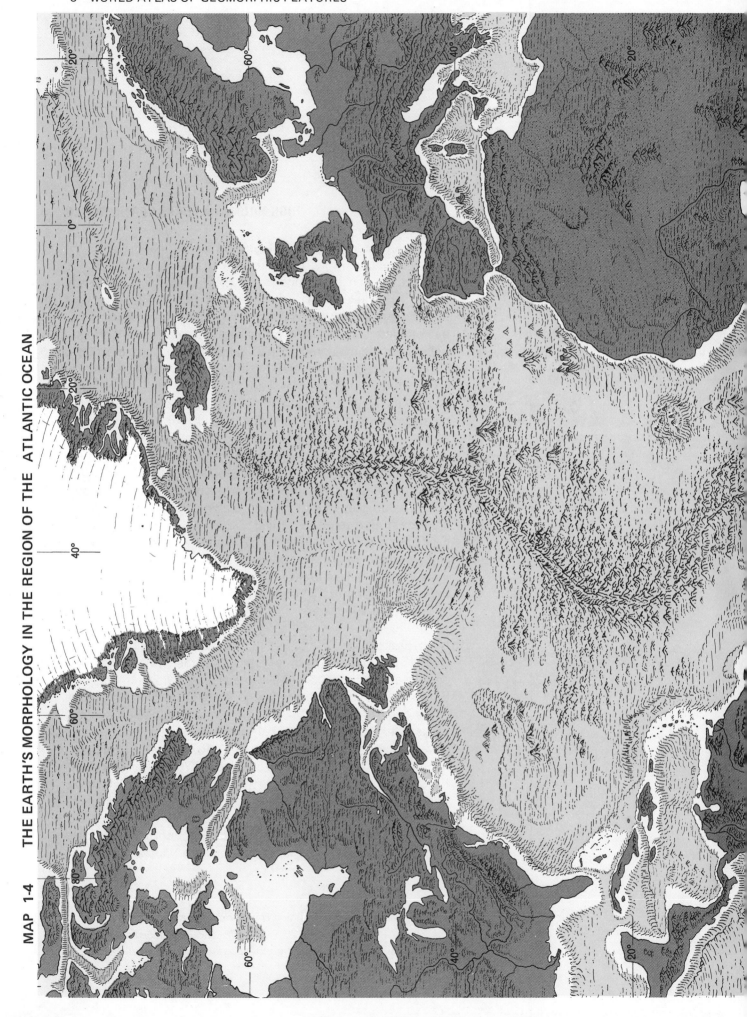

MAP 1-4   THE EARTH'S MORPHOLOGY IN THE REGION OF THE   ATLANTIC OCEAN

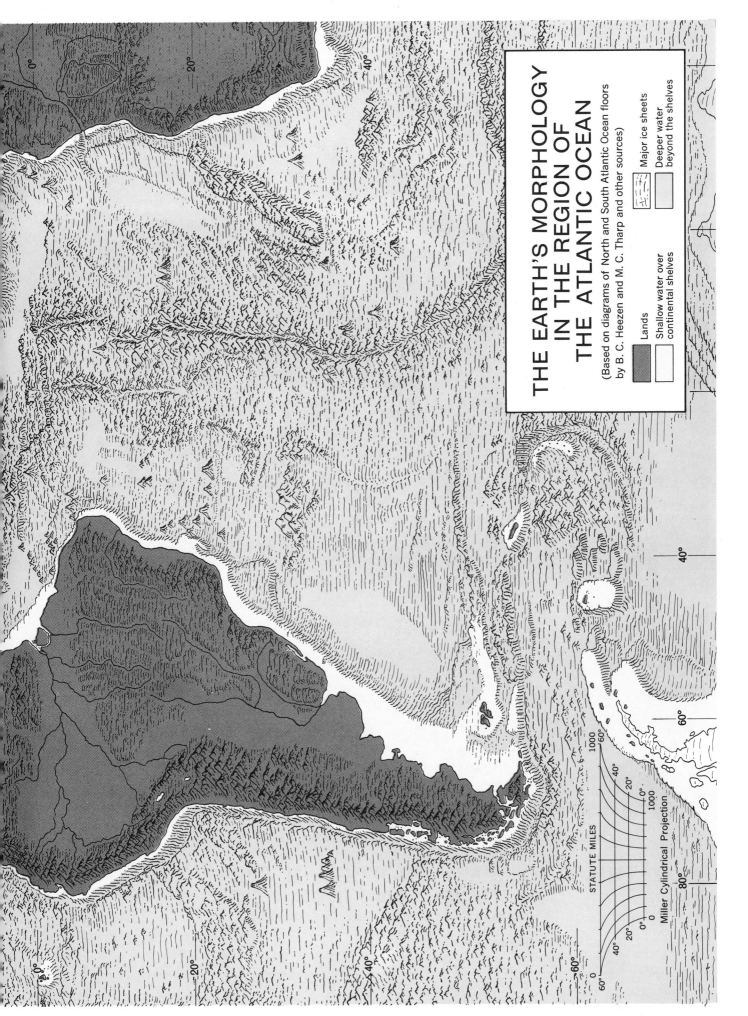

THE EARTH'S MORPHOLOGY
IN THE REGION OF
THE ATLANTIC OCEAN

(Based on diagrams of North and South Atlantic Ocean floors
by B. C. Heezen and M. C. Tharp and other sources)

Lands

Shallow water over
continental shelves

Major ice sheets

Deeper water
beyond the shelves

STATUTE MILES

Miller Cylindrical Projection

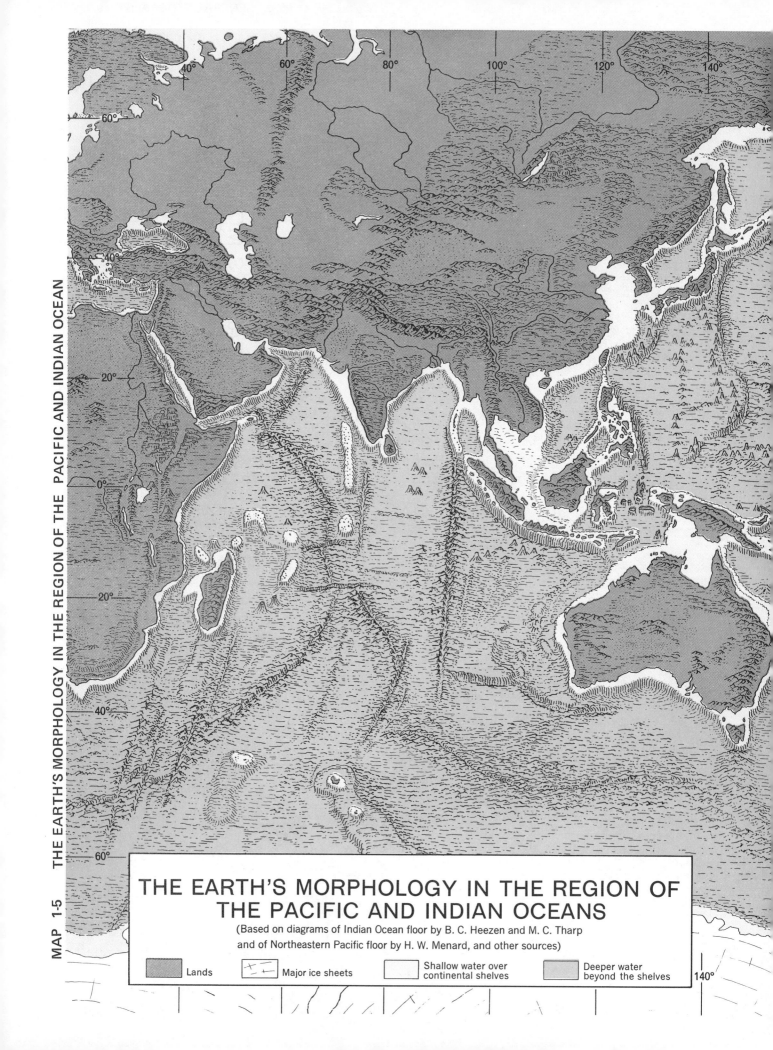

40°   60°   80°   100°   120°   140°

60°

40°

20°

0°

20°

40°

60°

THE EARTH'S MORPHOLOGY IN THE REGION OF THE PACIFIC AND INDIAN OCEANS

(Based on diagrams of Indian Ocean floor by B. C. Heezen and M. C. Tharp
and of Northeastern Pacific floor by H. W. Menard, and other sources)

Lands   Major ice sheets   Shallow water over continental shelves   Deeper water beyond the shelves

140°

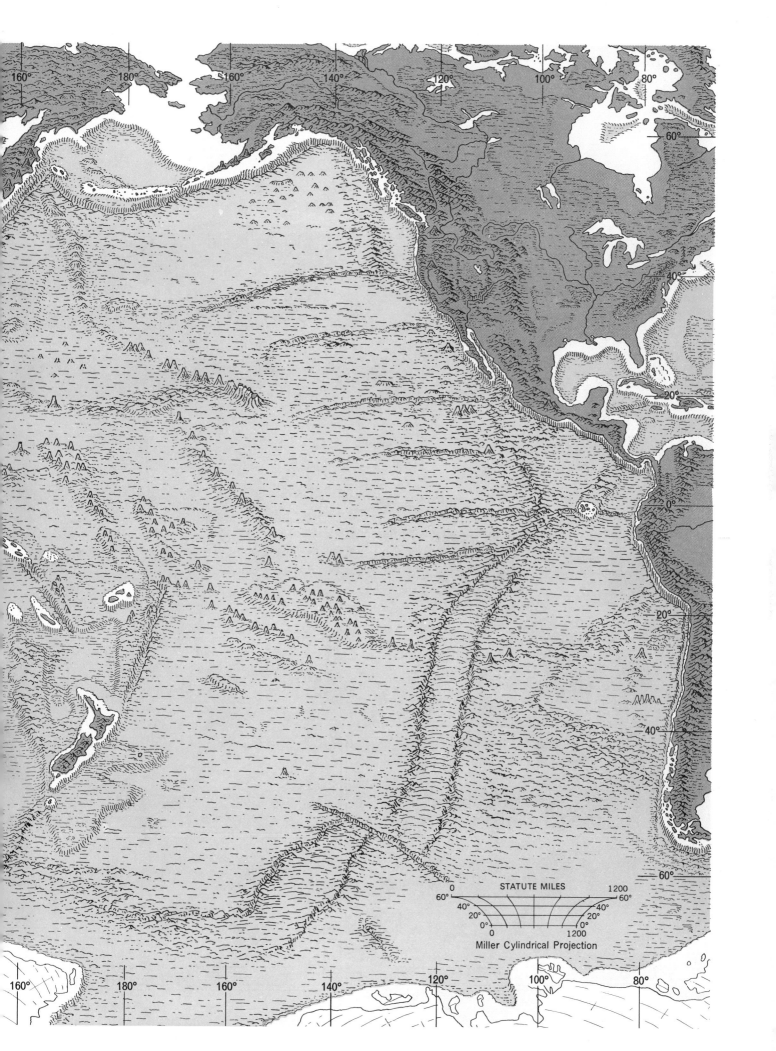

STATUTE MILES

0   1200

60°   60°

40°   40°

20°   20°

0°   0°

0   1200

Miller Cylindrical Projection

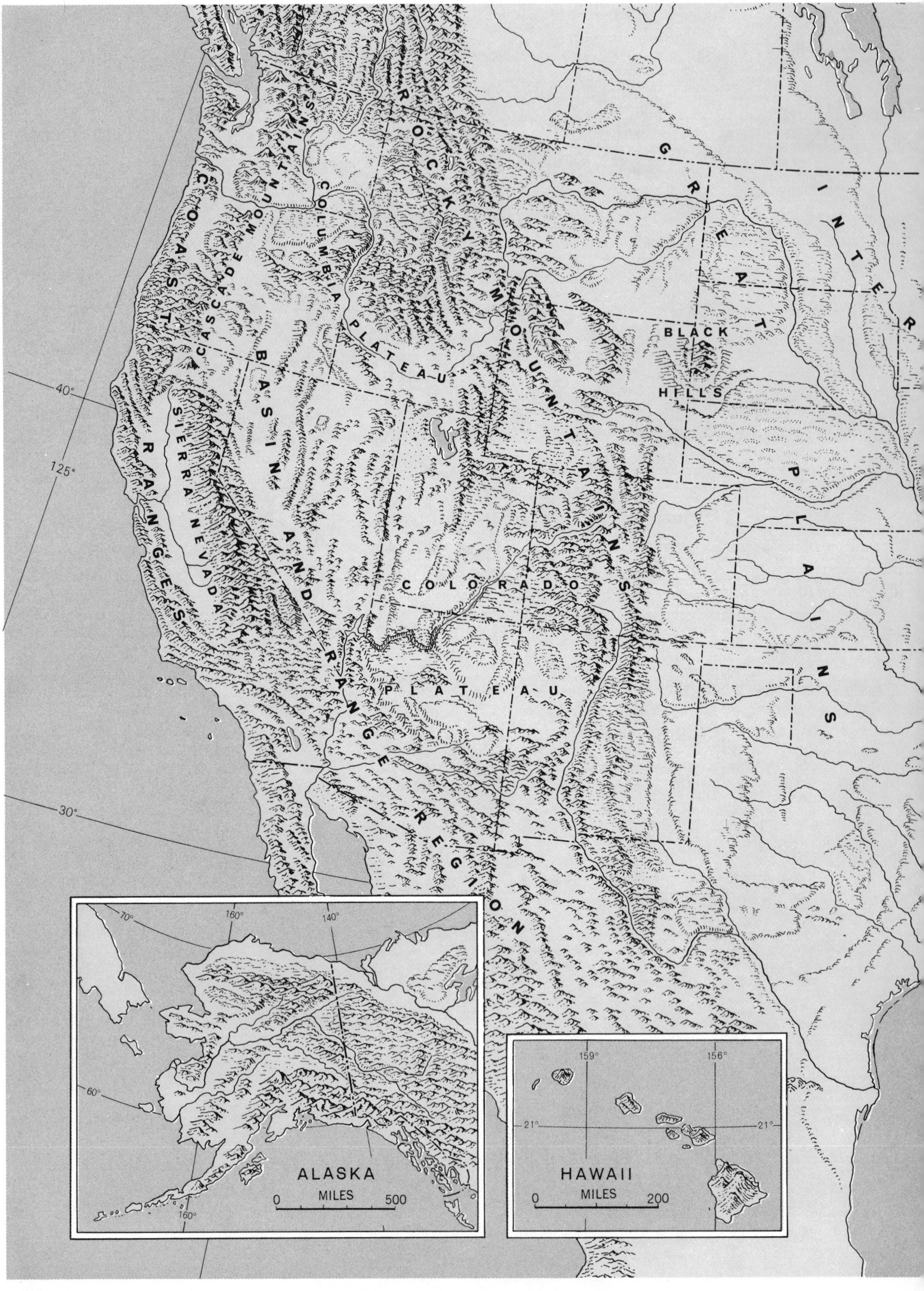

MAP 1-6   MORPHOLOGY OF THE UNITED STATES AND ADJACENT   PARTS OF CANADA AND MEXICO

ROCKY   MOUNTAINS

GREAT   PLAINS

INTER   PLAINS

BLACK   HILLS

COLUMBIA   PLATEAU

CASCADE   MOUNTAINS

COAST   RANGES

SIERRA   NEVADA

BASIN   AND   RANGE   REGION

COLORADO

PLATEAU

40°

125°

30°

ALASKA

70°

160°

140°

60°

160°

0   MILES   500

HAWAII

159°

156°

21°   21°

0   MILES   200

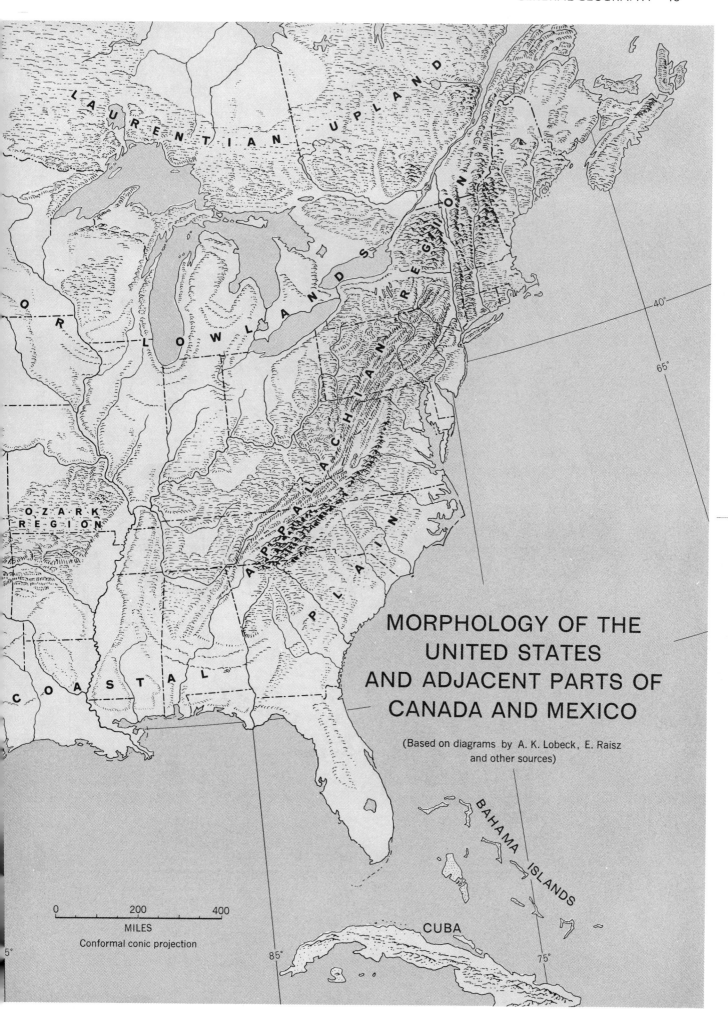

LAURENTIAN UPLAND

APPALACHIAN REGION

OR

LOWLANDS

OZARK
REGION

APPALACHIAN

COASTAL

PLAIN

MORPHOLOGY OF THE
UNITED STATES
AND ADJACENT PARTS OF
CANADA AND MEXICO

(Based on diagrams by A. K. Lobeck, E. Raisz
and other sources)

BAHAMA ISLANDS

CUBA

| 0 | 200 | 400 |
MILES
Conformal conic projection

MAP 1-7   MORPHOLOGY OF ANTARCTICA

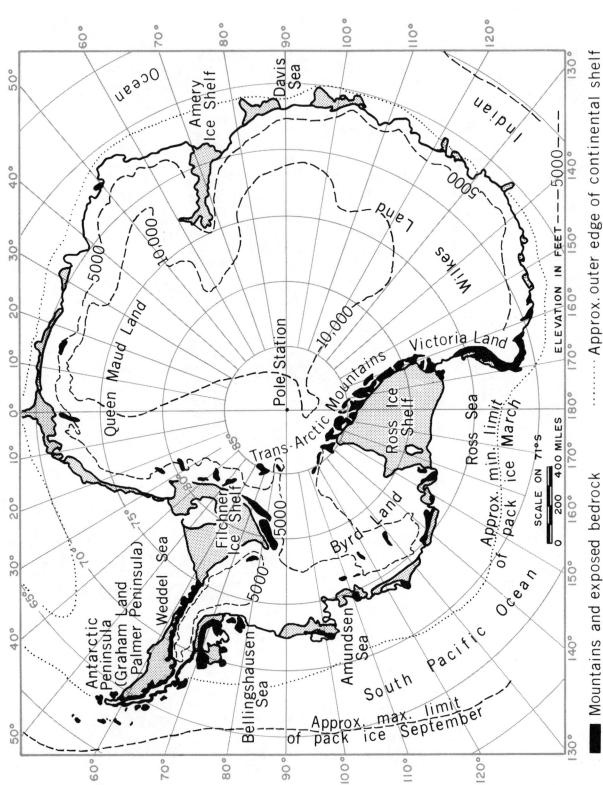

Mountains and exposed bedrock

········· Approx. outer edge of continental shelf

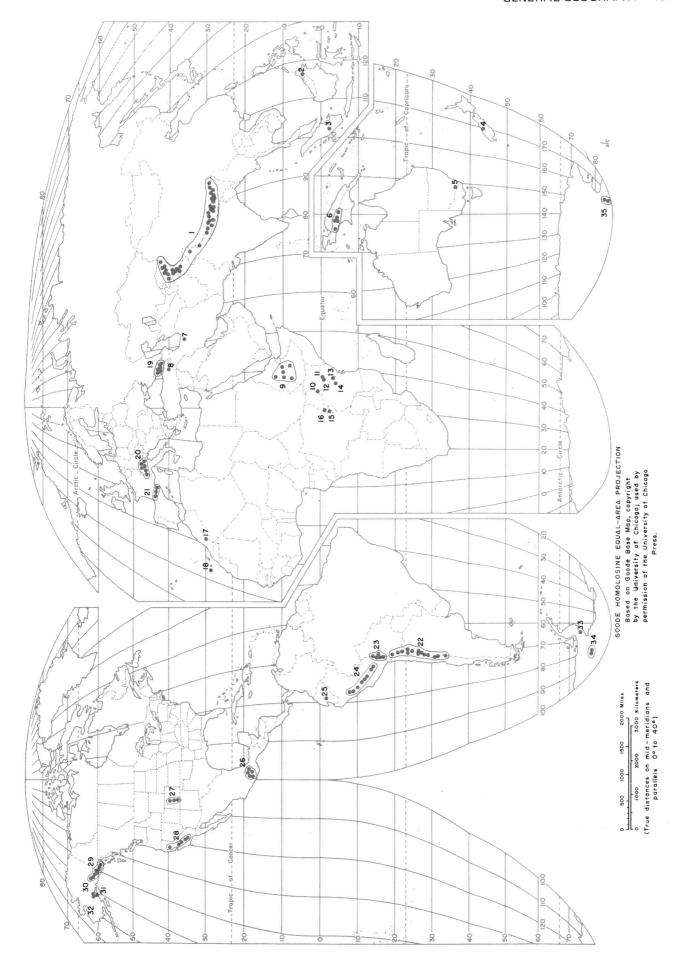

GOODE HOMOLOSINE EQUAL–AREA PROJECTION
Based on Goode Base Map, copyright
by the University of Chicago; used by
permission of University of Chicago
Press.

MAP 1-8   HIGH MOUNTAIN PEAKS OF THE WORLD

ASIA, OCEANIA AND AUSTRALIA

NAME	COUNTRY	HEIGHT IN FEET	HEIGHT IN METERS	NAME	COUNTRY	HEIGHT IN FEET	HEIGHT IN METERS
Everest	Nepal-Tibet	29,028	8,848	Pobedy Peak	Sinkiang-USSR	24,206	7,378
K2 (Godwin Austen)	Kashmir	28,250	8,611	Sia Kangri	Kashmir	24,350	7,422
Kanchenjunga	Nepal-Sikkim	28,208	8,598	Haramosh Peak	Pakistan	24,270	7,398
Lhotse (Everest)	Nepal-Tibet	27,923	8,511	Istoro Nal	Pakistan	24,240	7,388
Makalu 1	Nepal-Tibet	27,824	8,481	Tent Peak	Nepal-Sikkim	24,165	7,365
Lhotse 11 (Everest)	Nepal-Tibet	27,560	8,400	Chomo Lhari	Tibet-Bhutan	24,040	7,327
Dhaulagiri	Nepal	26,810	8,172	Chamlang	Nepal	24,012	7,319
Manaslu 1	Nepal	26,760	8,156	Kabru	Nepal-Sikkim	24,002	7,316
Cho Oyu	Nepal-Tibet	26,750	8,153	Alung Gangri	Tibet	24,000	7,315
Nanga Parbat	Kashmir	26,660	8,126	Baltoro Kangri	Kashmir	23,990	7,312
Annapurna	Nepal	26,504	8,078	Mussu Shan	Sinkiang	23,890	7,282
Gasherbrum	Kashmir	26,470	8,068	Mana	India	23,860	7,273
Broad	Kashmir	26,400	8,047	Baruntse	Nepal	23,688	7,220
Gasainthan	Tibet	26,287	8,012	Nepal Peak	Nepal-Sikkim	23,500	7,163
Annapurna 11	Nepal	26,041	7,937	Amne Machin	China	23,490	7,160
Gyachung Kang	Nepal-Tibet	25,910	7,897	Gauri Sankar	Nepal-Tibet	23,440	7,145
Disteghil Sar	Kashmir	25,868	7,885	Badrinath	India	23,420	7,138
Himalchuli	Nepal	25,801	7,864	Nunkun	Kashmir	23,410	7,135
Nuptse (Everest)	Nepal-Tibet	25,726	7,841	Lenina Peak	USSR	23,405	7,134
Masherbrum	Kashmir	25,660	7,821	Api	Nepal	23,299	7,102
Nanda Devi	India	25,645	7,817	Pauhunri	Sikkim-Tibet	23,385	7,128
Chomo Lonzo	Nepal-Tibet	25,640	7,815	Trisul	India	23,360	7,120
Rakaposhi	Kashmir	25,550	7,788	Kangto	India-Tibet	23,260	7,090
Kamet	India-Tibet	25,447	7,756	Nyenchhen Thanglha	Tibet	23,255	7,088
Namcha Barwa	Tibet	25,445	7,756	Tirsuli	India	23,210	7,074
Gurla Mandhata	Tibet	25,355	7,728	Pumori	Nepal-Tibet	23,190	7,068
Ulugh Muz Tagh	Tibet-Sinkiang	25,340	7,724	Dunagiri	India	23,184	7,066
Kungur	Sinkiang	25,325	7,719	Lombo Kangra	Tibet	23,165	7,061
Tirich Mir	Pakistan	25,230	7,690	Saipal	Nepal	23,100	7,041
Makalu 11	Nepal-Tibet	25,120	7,657	Macha Pucchare	Nepal	22,958	6,998
Minya Konka	China	24,900	7,590	Numbar	Nepal	22,817	6,955
Kula Gangri	Tibet-Bhutan	24,784	7,554	Kanjiroba	Nepal	22,580	6,882
Changtse (Everest)	Nepal-Tibet	24,780	7,553	Pyramid	Nepal-Sikkim	22,430	6,837
Muz Tagh Ata	Sinkiang	24,757	7,546	Ama Dablam	Nepal	22,350	6,812
Skyang Kangri	Kashmir	24,750	7,544	Cho Polu	Nepal	22,093	6,734
Communism Peak	USSR	24,590	7,495				
Jongsong Peak	Nepal-Sikkim	24,472	7,459				

ASIA, OCEANIA AND AUSTRALIA CONTINUED

	NAME	COUNTRY	HEIGHT IN FEET	HEIGHT IN METERS
1	Lingtren	Nepal-Tibet	21,972	6,697
	Khumbutse	Nepal-Tibet	21,785	6,640
	Hlako Gangri	Tibet	21,266	6,482
	Mt. Grosvenor	China	21,190	6,459
	Thagchhab Gangri	Tibet	20,970	6,392
2	Kinabalu	Malaysia	13,455	4,101
3	Kerintji	Sumatra	12,467	3,800
4	Cook	New Zealand	12,349	3,764
5	Kosciusko	Australia	7,310	2,228
6	Djaja	New Guinea	16,500	5,029
	Pilmsit	New Guinea	15,748	4,800
	Trikora	New Guinea	15,585	4,750
	Mandala	New Guinea	15,420	4,700
	Wilhelm	New Guinea	14,793	4,509
7	Damavand	Iran	18,934	5,771
8	Ararat	Turkey	16,946	5,165
9	Ras Dashan	Ethiopia	15,158	4,620
	Bodda	Ethiopia	13,556	4,132
	Batu	Ethiopia	14,131	4,307
	Guna	Ethiopia	13,881	4,231
	Tala	Ethiopia	13,451	4,100
	Gughe	Ethiopia	13,780	4,200
10	Elgon	Kenya-Uganda	14,178	4,321
11	Kenya	Kenya	17,058	5,199
12	Lesatima (Satimina)	Kenya	13,104	3,994
13	Kibo (Kilimanjaro)	Tanzania	19,340	5,895
14	Meru	Tanzania	14,979	4,566
15	Karisimbi (Kirunja)	Zaire-Rwanda	14,787	4,507
16	Margherita Peak	Uganda-Zaire	16,763	5,109
17	Toubkal	Morocco	13,665	4,165
18	Teide	Canary Islands	12,198	3,718

CAUCASUS (EUROPE—ASIA)

	NAME	HEIGHT IN FEET	HEIGHT IN METERS
19	El'brus	18,510	5,642
	Shkara	17,064	5,201
	Dykh Tau	17,054	5,198

CAUCASUS (EUROPE-ASIA) CONTINUED

NAME	HEIGHT IN FEET	HEIGHT IN METERS	
Kashtan Tau	16,877	5,144	19
Dzhangi Tau	16,565	5,049	
Kazbek	16,558	5,047	

EUROPE—ALPS

NAME	HEIGHT IN FEET	HEIGHT IN METERS	
Mont Blanc	15,771	4,807	
Mont Rosa (Highest peak of the group)	15,203	4,634	
Dom	14,911	4,545	
Liskamm	14,852	4,527	
Weisshorn	14,780	4,505	
Taschhorn	14,733	4,491	
Matterhorn	14,690	4,478	
Dent Blanhe	14,293	5,357	
Nadelhorn	14,196	4,327	
Grand Combin	14,154	4,314	
Lenzpitze	14,088	4,294	
Finsteraarhorn	14,022	4,274	
Castor			
Zinalrothorn	13,849	4,221	20
Hohberghorn	13,842	4,219	
Alphubel	13,799	4,206	
Rimpfischhorn	13,776	4,199	
Aletschorn	13,763	4,195	
Strahlhorn	13,747	4,190	
Dent d'Herens	13,686	4,171	
Breithorn	13,665	4,165	
Bishorn	13,645	4,159	
Jungfrau	13,642	4,158	
Ecrins	13,461	4,103	
Monch	13,448	4,099	
Pollux	13,422	4,091	
Schreckhorn	13,379	4,078	
Ober Gabelhorn	13,330	4,063	
Gran Paradiso	13,323	4,061	
Bernina	13,284	4,049	
Fiescherhorn	13,283	4,049	
Grunhorn	13,266	4,043	
Lauteraarhorn	13,261	4,042	

EUROPE-ALPS CONTINUED

NAME	HEIGHT IN FEET	HEIGHT IN METERS
Durrenhorn	13,238	4,035
Allalinhorn	13,213	4,027
Weissmies	13,199	4,023
Lagginhorn	13,156	4,010
Zupo	13,120	3,999
Fletschhorn	13,110	3,996
Adlerhorn	13,081	3,987
Gletscherhorn	13,068	3,983
Schalihorn	13,040	3,975
Scerscen	13,028	3,971
Eiger	13,025	3,970
Jagerhorn	13,024	3,970
Rottalhorn	13,022	3,969

20

EUROPE—PYRENEES

NAME	HEIGHT IN FEET	HEIGHT IN METERS
Aneto	11,168	3,404
Posets	11,073	3,375
Perdido	11,007	3,355
Maladeta	10,866	3,312
Vignemale	10,820	3,298
Long	10,479	3,194
Estats	10,304	3,141
Montcalm	10,105	3.080

21

SOUTH AMERICA—ARGENTINA/CHILE

NAME	COUNTRY	HEIGHT IN FEET	HEIGHT IN METERS
Aconcagua	Argentina	22,834	6,960
Bonete	Argentina	22,546	6,872
Ojos del Salado	Argentine-Chile	22,539	6,870
Tupungato	Argentina-Chile	22,310	6,800
Pissis	Argentina	22,241	6,779
Mercedario	Argentina	22,211	6,770
Tocorpuri	Argentina-Chile	22,195	6,765
Llullaillaco	Argentina-Chile	22,051	6,721
El Libertador	Argentina	22,047	6,720
Cachi	Argentina	22,047	6,720
Galan	Argentina	21,654	6,600
El Muerto	Argentina-Chile	21,457	6,540

22

SOUTH AMERICA-ARGENTINE/CHILE CONTINUED

NAME	COUNTRY	HEIGHT IN FEET	HEIGHT IN METERS
Nacimiento	Argentina	21,302	6,493
Laudo	Argentina	20,997	6,400
Toro	Argentina-Chile	20,932	6,380
Tres Cruces	Argentina-Chile	20,853	6,356
Tortolas	Argentina-Chile	20,745	6,323
Condor	Argentina	20,669	6,300
Gen. Manuel Belgrano	Argentina	20,505	6,250
Solo	Argentina	20,492	6,246
Polleras	Argentina	20,456	6,235
Chani	Argentina	20,341	6,200
Aucanquilcha	Chile	20,295	6,186
Juncal	Argentina	20,276	6,180
Negro	Argentina	20,184	6,152
Quela	Argentina	20,184	6,152
Palermo	Argentina	20,079	6,120
San Juan	Argentina-Chile	20,049	6,111
Sierra Nevada	Argentina-Chile	20,023	6,103
Antofalla	Argentina	20,013	6,100
Marmolejo	Argentina-Chile	20,013	6,100
Licancabur	Argentina-Chile	19,425	5,921

22

SOUTH AMERICA—BOLIVIA

NAME	HEIGHT IN FEET	HEIGHT IN METERS
Sajama	21,391	6,520
Illimani	21,201	6,462
Ancohuma	20,958	6,388
Illampu	20,873	6,362
Parinacota	20,768	6,330
Condoriri	20,095	6,125

23

SOUTH AMERICA—PERU

NAME	HEIGHT IN FEET	HEIGHT IN METERS
Huascaran	22,205	6,768
Yerupaja	21,765	6,634
Coropuna	21,079	6,425
Ausangate	20,945	6,384
Huandoy	20,852	6,356
Ampato	20,702	6,310
Salcantay	20,574	6,271
Huancarhuas	20,531	6,258
Pumasillo	20,492	6,246
Solimana	20,068	6,108

24

ECUADOR

	NAME	HEIGHT IN FEET	HEIGHT IN METERS
25	Chimborazo	20,561	6,267

NORTH AMERICA—MEXICO

	NAME	HEIGHT IN FEET	HEIGHT IN METERS
26	Orizaba (Citlatepec)	18,700	5,700
	Popocatepetl	17,887	5,452
	Iztaccihuatl	17,343	5,286
	Zinantecatl (Toluca)	15,016	4,577
	Matlalcueyetl	14,636	4,461
	Nauhcamptepetl (Cofre de Perote)	14,049	4,282
	Colima	14,003	4,268

NORTH AMERICA—UNITED STATES—COLORADO

	NAME	HEIGHT IN FEET	HEIGHT IN METERS
27	Elbert	14,433	4,399
	Massive	14,421	4,396
	Harvard	14,420	4,395
	Blanca	14,345	4,372
	La Plata	14,336	4,369
	Uncompahgre	14,309	4,361
	Crestone	14,294	4,357
	Lincoln	14,286	4,354
	Grays	14,270	4,350
	Antero	14,269	4,349
	Torreys	14,267	4,348
	Castle	14,265	4,348
	Quandary	14,265	4,348
	Evans	14,264	4,348
	Longs	14,255	4,345
	Mt. Wilson	14,246	4,342
	Shavano	14,229	4,337
	Belford	14,197	4,327
	Princeton	14,197	4,327
	Crestone Needle	14,197	4,327
	Yale	14,196	4,327
	Bross	14,172	4,320
	Kit Carson	14,165	4,317
	El Diente	14,159	4,316
	Maroon	14,156	4,315
	Tabeguache	14,155	4,314

COLORADO CONTINUED

NAME	HEIGHT IN FEET	HEIGHT IN METERS	
Oxford	14,153	4,314	27
Sneffels	14,150	4,313	
Democrat	14,148	4,312	
Capitol	14,130	4,307	
Pikes Peak	14,110	4,301	
Snowmass	14,092	4,295	
Windorn	14,087	4,294	
Eolus	14,084	4,293	
Columbia	14,073	4,289	
Missouri	14,067	4,288	
Humboldt	14,064	4,287	
Bierstadt	14,060	4,285	
Sunlight	14,059	4,285	
Handies	14,048	4,282	
Culebra	14,047	4,282	
Lindsey	14,042	4,280	
Little Bear	14,037	4,278	
Sherman	14,036	4,278	
Redcloud	14,034	4,278	
Conundrum	14,022	4,274	
Pyramid	14,018	4,273	
Wilson Peak	14,017	4,273	
Wetterhorn	14,015	4,272	
North Maroon	14,014	4,272	
San Luis	14,014	4,272	
Huron	14,005	4,269	
Holy Cross	14,005	4,269	
Sunshine	14,001	4,266	
Grizzly	14,000	4,266	

CALIFORNIA

NAME	HEIGHT IN FEET	HEIGHT IN METERS	
Whitney	14,494	4,418	28
Williamson	14,375	4,382	
White	14,242	4,341	
North Palisade	14,242	4,341	
Shasta	14,162	4,317	

CALIFORNIA CONTINUED

NAME	HEIGHT IN FEET	HEIGHT IN METERS
Sill	14,162	*4,317*
Russell	14,086	*4,293*
Split	14,058	*4,285*
Middle Palisade	14,040	*4,279*
Langley	14,028	*4,276*
Tyndall	14,018	*4,273*

(bracket labeled 28)

NORTH AMERICA—ALASKA—CANADA

(ALASKA, YUKON AND BRITISH COLUMBIA)

NAME	COUNTRY	HEIGHT IN FEET	HEIGHT IN METERS
Logan	Alaska	19,850	*6,050*
St Elias	Alaska-Canada	18,008	*5,489*
Lucania	Canada	17,147	*5,226*
King	Canada	16,971	*5,173*
Bona	Alaska	16,550	*5,044*
Blackburn	Alaska	16,523	*5,036*
Kennedy	Alaska	16,286	*4,964*
Sanford	Alaska	16,237	*4,949*
South Buttress	Alaska	15,885	*4,842*
Wood	Canada	15,885	*4,842*
Vancouver	Alaska-Canada	15,700	*4,785*
Churchill	Alaska	15,638	*4,766*
Fairweather	Alaska-Canada	15,300	*4,663*
Hubbard	Alaska-Canada	15,015	*4,577*
Walsh	Canada	14,780	*4,505*
Alverstone	Alaska-Canada	14,565	*4,439*
Browne Tower	Alaska	14,530	*4,429*
McArthur	Canada	14,253	*4,344*
Wrangell	Alaska	14,163	*4,317*
Augusta	Alaska-Canada	14,070	*4,289*

(bracket labeled 29)

NORTH AMERICA-ALASKA-CANADA CONTINUED

NAME	COUNTRY	HEIGHT IN FEET	HEIGHT IN METERS	
McKinley	Alaska Range	20,320	*6,194*	30
Hunter	Alaska Range	14,580	*4,444*	31
Foraker	Alaska Range	17,400	*5,304*	32

ANTARCTICA

NAME	HEIGHT IN FEET	HEIGHT IN METERS	
Vinson Massif	16,860	*5,139*	
Tyree	16,290	*4,965*	
Shinn	15,750	*4,801*	
Gardner	15,375	*4,686*	
Epperly	15,100	*4,602*	33., 34 and 35
Kirkpatrick	14,855	*4,528*	
Elizabeth	14,698	*4,480*	
Markham	14,290	*4,356*	
Bell	14,117	*4,303*	
Mackellar	14,098	*4,297*	

MAP
1-9

HIGHEST ELEVATIONS IN EACH STATE OF THE UNITED STATES

This map presents the great range of elevations in the United States from Mt. McKinley in Alaska — 20,320 feet (6,194 meters) to a point in western Florida at Section 30, Township 6N, Range 20W, which is only 345 feet (105 meters) above sea level. East of the Mississippi two peaks stand out; Mt. Mitchell in North Carolina with an elevation of 6,684 feet (2,037 meters) and Mt. Washington in New Hampshire, a little lower at 6,288 feet (1,917 meters). In the western United States, it is surprising how many peaks are 13,000 to 14,500 feet (3,962 to 4,420 meters) high. Mt. Whitney in California is 14,494 feet (4,418 meters), in Colorado, Mt. Elbert is only 61 feet (19 meters) lower — 14,433 feet (4,399 meters) and in Washington State, Mt. Rainier is 23 feet (7 meters) lower being 14,410 feet (4,392 meters). There are over 70 peaks in the western United States between 14,000 and 15,000 feet (4,267 and 4,572 meters).

Source: The Associated Press Almanac, Inc., 1975, Maplewood , New Jersey, p.24.

MAP 1-9   HIGHEST ELEVATIONS IN EACH STATE OF THE UNITED STATES

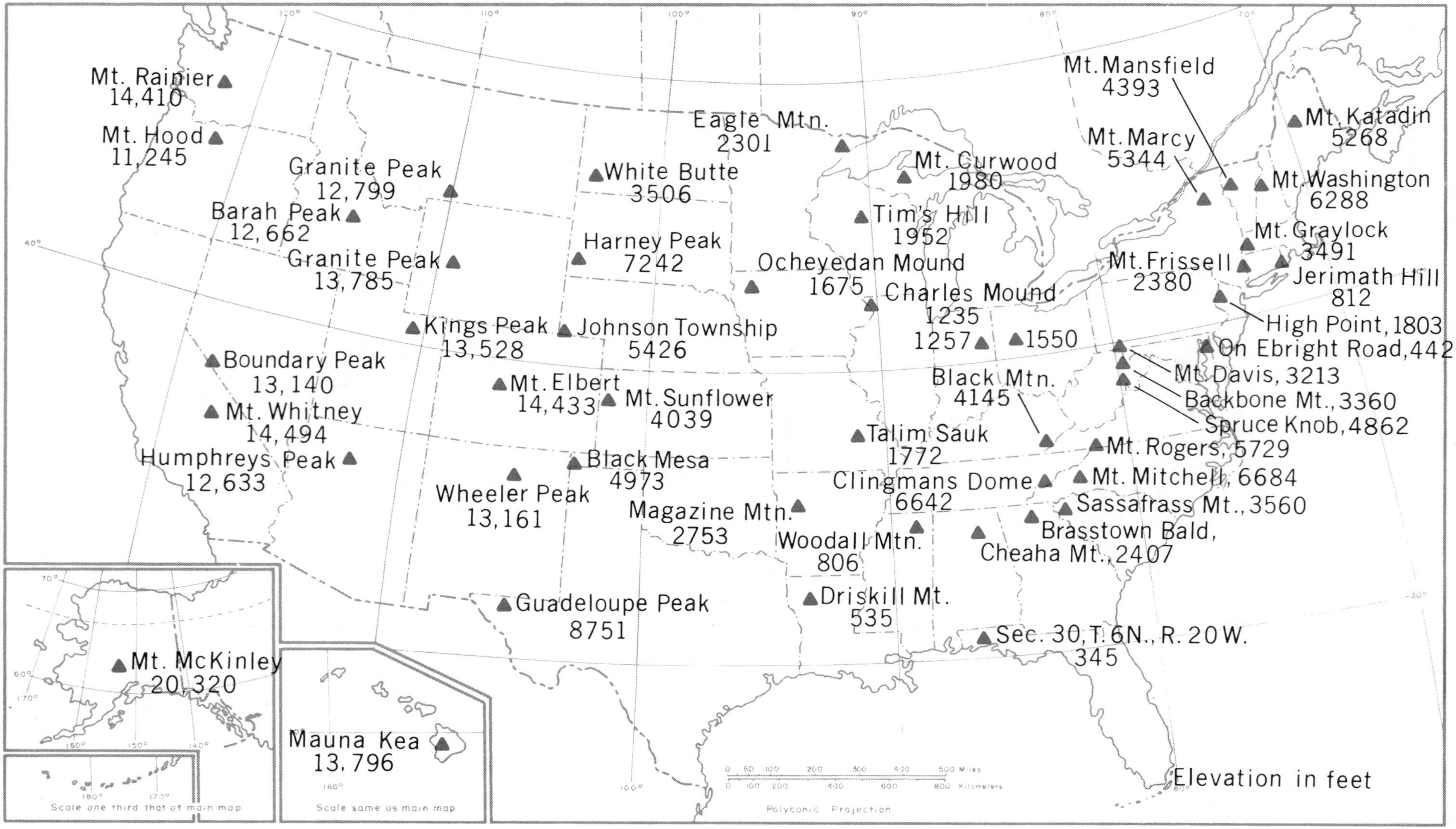

Mt. Katadin 5268
Mt. Washington 6288
Mt. Graylock 3491
Jerimath Hill 812
High Point, 1803
On Ebright Road, 442
Mt. Davis, 3213
Backbone Mt., 3360
Spruce Knob, 4862
Mt. Rogers 5729
Mt. Mitchell, 6684
Sassafrass Mt., 3560
Brasstown Bald, 4784
Cheaha Mt., 2407
Sec. 30, T. 6 N., R. 20 W. 345

Mt. Mansfield 4393
Mt. Marcy 5344
Mt. Frissell 2380
Mt. Curwood 1980
Tim's Hill 1952
Ocheyedan Mound 1675
Charles Mound 1235
1257
1550
Black Mtn. 4145
Talimi Sauk 1772
Clingmans Dome 6642
Woodall Mtn. 806
Driskill Mt. 535

Eagle Mtn. 2301
White Butte 3506
Harney Peak 7242
Johnson Township 1235
Kings Peak 13,528
Mt. Elbert 14,433
Mt. Sunflower 4039
Black Mesa 4973
Magazine Mtn. 2753
Guadeloupe Peak 8751

Mt. Rainier 14,410
Mt. Hood 11,245
Granite Peak 12,799
Barah Peak 12,662
Granite Peak 13,785
Boundary Peak 13,140
Mt. Whitney 14,494
Humphreys Peak 12,633
Wheeler Peak 13,161

Mt. McKinley 20,320

Mauna Kea 13,796

Scale same as main map
Scale one third that of main map

Polyconic Projection
0 50 100 200 300 400 500 Miles
0 100 200 400 600 800 Kilometers

Elevation in feet

TEMPERATURE AND PRECIPITATION PARAMETERS
OF GEOMORPHIC IMPORTANCE

A new dimension of the study of physical features is the finding of climatic and vegetative features that influence the development of landform assemblages. R. Common attempted to delineate geomorphically important temperature and precipitation characteristics. The problem of defining climatic morphologic regions is analogous to the problem of climatic regionalization, except that whereas climatologists, following Köppen, have generally selected climatic limits of biotic significance for their class intervals, the problem presented to Common was to set limits of geomorphic significance. The resulting maps emphasize the intricacy of the climatic mosaic with which we must deal, even when it is studied on so small a scale. In general, little work has been done on the relative importance of different climatic parameters, mainly because of the detailed interrelationships of climate and landforms. It is very difficult to employ the needed sophisticated techniques. Most schemes are highly generalized and qualitative in nature, but they do present different types of climatic data of interest to students studying physical features.

For students and geomorphologists the temperature and precipitation boundaries can be used in association with landforms. For example, on Map 1-10, where average actual temperature below 32°F (0°C) occur all year no plants can grow and this becomes a frozen land of ice and snow. On the other hand, those areas where the average actual temperatures occur above 70°F (21°C) all year, life is never restricted by lack of heat and if sufficient moisture occurs luxuriant vegetation will grow. Land areas where the diurnal range of temperature is greater than the annual range refers to equatorial regions where no true seasons exist and the coolest temperatures occur at night. It is interesting to note how in the Northern Hemisphere the length of the frost period and the range of average actual temperatures is morphologically significant over considerable areas while the inter-tropical area, where diurnal ranges are important, extends over much of the land area in the Southern Hemisphere. The extensive microthermal zone in the Northern Hemisphere corresponds to the broad zone of "D" climates which stretches across wide areas of central Canada and the Soviet Union. Because of the large amount of water area in the Southern Hemisphere, which modifies temperature extremes, the lack of large land areas, no microthermal climates are found.

MAP 1-10 TEMPERATURE PARAMETERS OF GEOMORPHIC IMPORTANCE

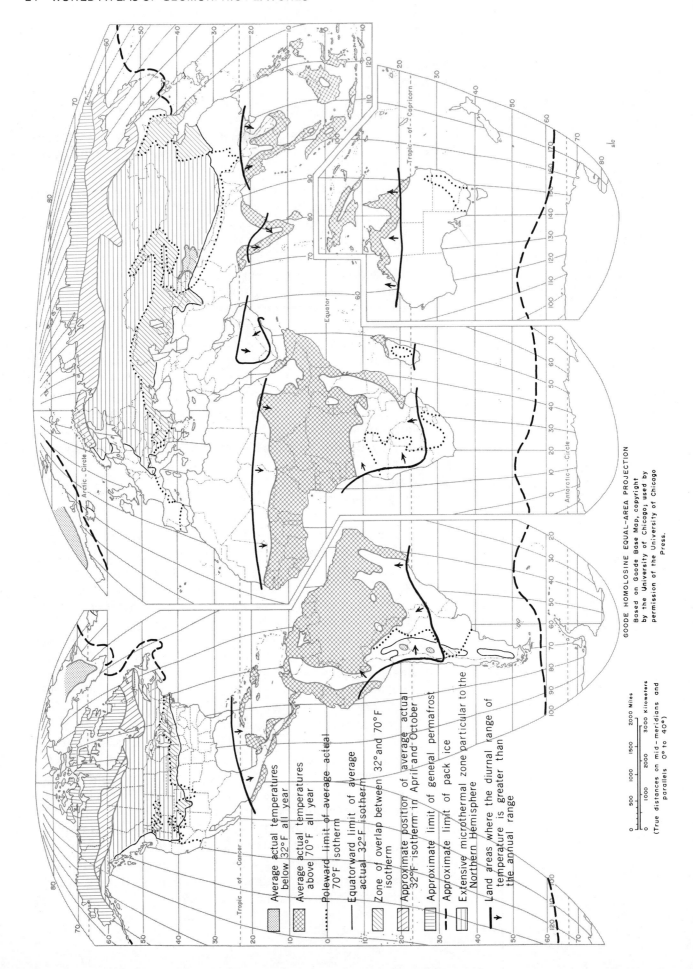

Average actual temperatures below 32°F all year

Average actual temperatures above 70°F all year

Poleward limit of average actual 70°F isotherm

Equatorward limit of average actual 32°F isotherm

Zone of overlap between 32° and 70°F isotherm

Approximate position of average actual 32°F isotherm in April and October

Approximate limit of general permafrost

Approximate limit of pack ice

Extensive microthermal zone particular to the Northern Hemisphere

Land areas where the diurnal range of temperature is greater than the annual range

GOODE HOMOLOSINE EQUAL-AREA PROJECTION
Based on Goode Base Map, copyright by the University of Chicago; used by permission of the University of Chicago Press.

Tropic of Cancer

Equator

Tropic of Capricorn

Arctic Circle

Antarctic Circle

0   500   1000   1500   2000 Miles
0   1000   2000   3000 Kilometers
(True distances on mid-meridians and parallels 0° to 40°)

Map 1-11. The legend on this map needs some explanation. Areas essentially outside temperate and tropical storm tracks are equatorial, continental, and arctic regions that are away from such storms as the severe winter blizzards in the middle latitudes and tropical hurricanes in the subtropical regions. For much of South America and Africa the most severe storms are the afternoon convective thunderstorms. Areas with at least two inches of precipitation in January, April, July and October are essentially rainless areas lacking sufficient moisture for a good vegetative cover. On the other hand, those areas with less than five inches of precipitation per year are true desert regions. Important thunderstorm zones are those areas where a large number of thunderstorms occur during part or all of the year. The Midwest and Southeast United States is a high thunderstorm zone as well as the Rocky Mountain region of New Mexico and central Mexico. Areas of true monsoon climate (six months wet and six months dry) include much of India, China, Japan, and northern Australia. The general equatorial limit of snowfall is difficult to plot because where mountains come close to the sea as the Andes do in South America, the limit of snowfall follows the crest of the higher mountain peaks close to the equator. The interior hills and plateaus of Iran and Pakistan have snowfall, but it is largely absent from the Makran costal regions.

Sources: Common, R., "Slope failure and morphogenetic regions." In Dury, G.H. (ed.), *Essays in Geomorphology.* London: Heinemann, 1966, pp. 53-81; and Peltier, L.C., "The geographical cycle in periglacial regions as it is related to climatic geomorphology," *Annals of the Association of American Geographers,* Vol. 40, 1950, pp. 214-236.

MAP 1-11   PRECIPITATION PARAMETERS OF GEOMORPHIC IMPORTANCE

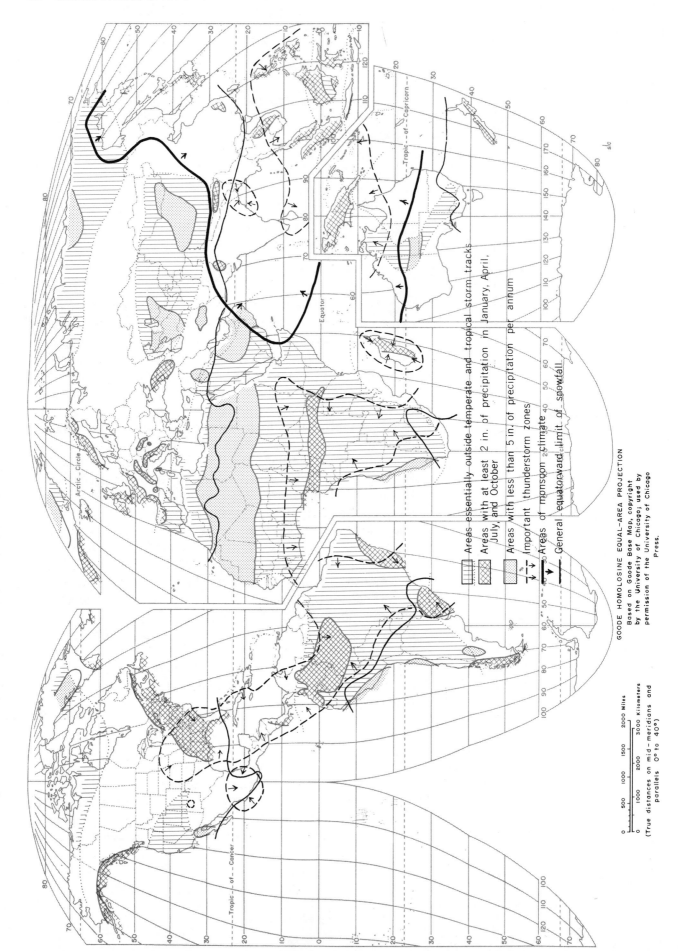

Areas essentially outside temperate and tropical storm tracks

Areas with at least 2 in. of precipitation in January, April, July, and October

Areas with less than 5 in. of precipitation per annum

Important thunderstorm zones

Areas of monsoon climate

General equatorward limit of snowfall

GOODE HOMOLOSINE EQUAL-AREA PROJECTION
Based on Goode Base Map, copyright
by the University of Chicago; used by
permission of the University of Chicago
Press.

0   500   1000   1500   2000 Miles

0   1000   2000   3000 Kilometers

(True distances on mid-meridians and
parallels 0° to 40°)

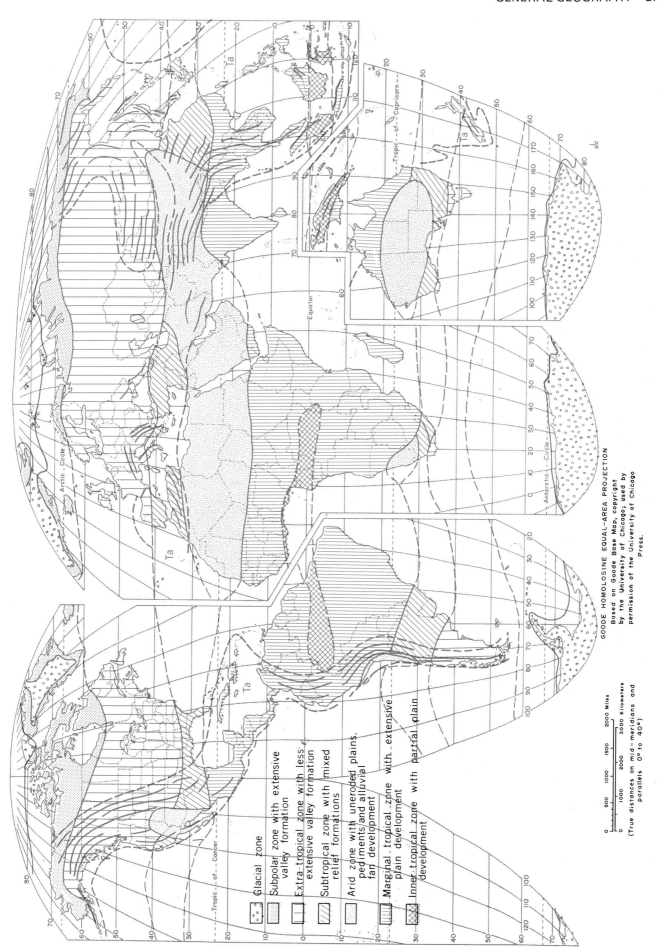

GOODE HOMOLOSINE EQUAL-AREA PROJECTION
Based on Goode Base Map, copyright
by the University of Chicago; used by
permission of the University of Chicago
Press.

0   500 1000 1500 2000 Miles
0 1000 2000 3000 Kilometers
(True distances on mid-meridians and
parallels 0° to 40°)

Glacial zone

Subpolar zone with extensive
valley formation

Extra-tropical zone with less
extensive valley formation

Subtropical zone with mixed
relief formations

Arid zone with uneroded plains,
pediments, and alluvial
fan development

Marginal tropical zone with extensive
plain development

Inner tropical zone with partial plain
development

MAP 1-12   CLIMATOLOGICAL—MORPHOLOGICAL ZONES (without high mountain ranges)

**MAP
1-12**

CLIMATOLOGICAL — MORPHOLOGICAL ZONES

This map presents Julius Budel's (1969) concept of developing landform assemblages, which may result from either present or past climates. A system of this kind is highly generalized and subjective, but it does indicate interesting and often useful relationships. It is an alternative to Louis Peltier's (1950 and 1962) inductive approach, in which he presented the geographic cycle in periglacial regions as it is related to climatic geomorphology.

The legend on this map ties climate and landform features together. For example, the glacial zone is one with glaciers or permanently frozen ground, while the subpolar zone has extensive valley formations due to the work of past glaciers and melting snows. The extra-tropical zone lies to the north of the tropical regions in the middle latitudes and has less extensive valley formation. The subtropical zone, outside the tropics has a variety of landforms while the arid zones have a great deal of interior fill with extensive plain, pediment, and alluvial fan development. On the margins of the true tropical zone in mainly the wet - dry climates there is extensive plain development. The author is here thinking of the wide expansive plains and plateaus of Africa. This is the homeland of the large herds of range animals. The inter-tropical zone is close to the equator in the hot basin regions of the Amazon and Congo. Budel is using the term plain in its broadest sense, meaning the wide expanses of nearly level land which cover much of the earth's land surface. There are many problems with this map because of its very general nature. Mountain areas are left out as well as large valleys and drainage basins but it does present one of the recent attempts to correlate climate and landform associations relationships.

Source: Budel, Julius, Abb. 1 in "Das System der Klimagenetischen Geomorphologie," *Erkunde Archiv fur Wissenschaftliche Geographie,* Band XII, Lfg. 3, Bonn, 1969, Peltier, L.C., "The Geographic Cycle in Periglacial Regions as it is Related to Climatic Geomorphology," *Annals, Association of American Geographers,* Vol. 40, 1950, pp. 214-236 and "Area Sampling for Terrain Analysis," *Professional Geographers,* Vol. 14, No. 2, 1962, pp. 24-28.

MORPHOCLIMATIC REGIONS

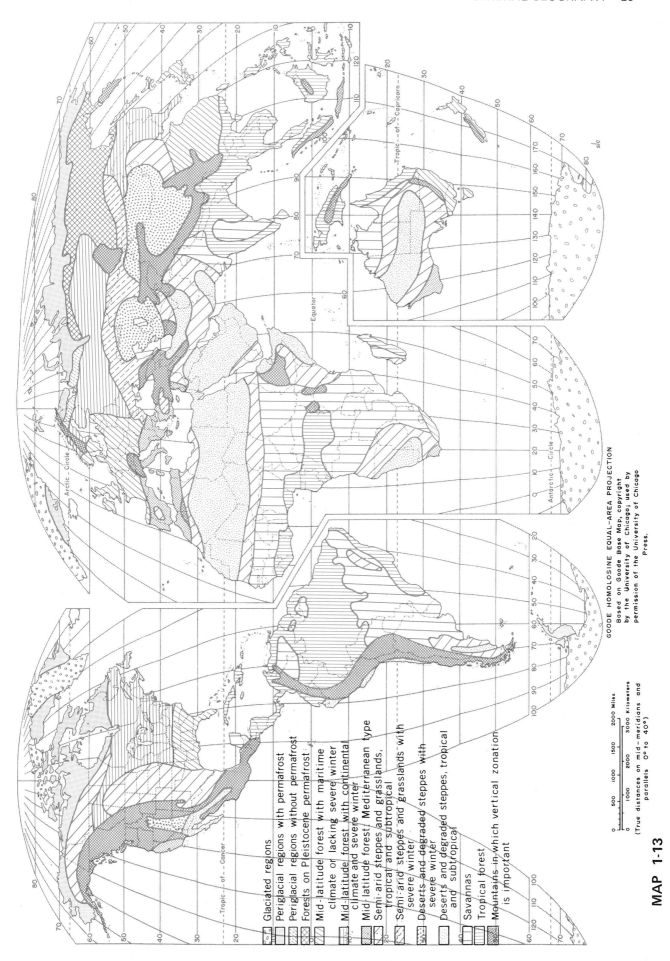

Glaciated regions

Periglacial regions with permafrost

Periglacial regions without permafrost

Forests on Pleistocene permafrost

Mid-latitude forest with maritime climate or lacking severe winter

Mid-latitude forest with continental climate and severe winter

Mid-latitude forest, Mediterranean type

Semi-arid steppes and grasslands, tropical and subtropical

Semi-arid steppes and grasslands with severe winter

Deserts and degraded steppes with severe winter

Deserts and degraded steppes, tropical and subtropical

Savannas

Tropical forest

Mountains in which vertical zonation is important

GOODE HOMOLOSINE EQUAL-AREA PROJECTION
Based on Goode Base Map, copyright
by the University of Chicago; used by
permission of the University of Chicago
Press.

(True distances on mid-meridians and
parallels 0° to 40°)

0   500   1000   1500   2000 Miles

0   1000   2000   3000 Kilometers

MAP 1-13

**MAP
1-13**

MORPHOCLIMATIC REGIONS

Budel's climatological-morphological zones (Map 1-12) can be compared with this map depicting morphoclimatic regions in which vegetation, as well as climate, is considered as a factor controlling landform development. This system, compiled by Tricart and Cailleux, is not concerned with the landforms themselves. The map shows the distribution of morphoclimatic zones based on major climatic and vegetative distributions, subdivided by climatic, paleoclimatic, and biogeographic criteria. Strakov (1967) has approached the same problem by mapping types of weathering, largely from an examination of soil characteristics (see Map 1-14). His categories are based on a zonal interpretation of the efficiency of chemical weathering, with maxima in the taiga-podzol zone and the tropical forest zone.

Although most of the legend is self explanatory a few designations need clarification. Forests on Pleistocene permafrost refer to the great Tiaga forests of Siberia where the ground is solidly frozen during the summer. Trees found on such thin soils are usually stunted. The Mediterranean type of forest includes sclerophyll vegetation consisting of low trees with small, hard, leathery leaves. Such vegetation is found in areas where there is a severe summer drought. The term savanna is usually given to the tropical grasslands which are found throughout large areas of Africa, South America, and Australia. In Africa, this is the homeland of large grazing animals. Intertropical forests (within the tropics) include the dense evergreen rainforest, called selva, found in the Congo and Amazon basins. An Azonal mountainous region is one where there is considerable slope movement taking place both of soils and vegetation. Such high mountainous regions have numerous landslides with talus slope accumulation. Large areas of this region have been modified by man's over-cutting of the forests.

Sources: Tricart, J., and A. Cailleux, Plate III in *Introduction to Climatic Geomorphology,* Longman Group Limited, London, 1972, 295 p. This map appears in modified from in Board, C., R.J. Chorley, P. Haggett, and Dr. R. Stoddart, Figure 8 in *Progress in Geography,* Vol 1, Edward Arnold (Publishers) Ltd., London, 1969, and in R.J. Chorley, *Water, Earth, and Man,* Methuen & Co., Ltd., 1969, Fig. 10.11.4, p. 480.

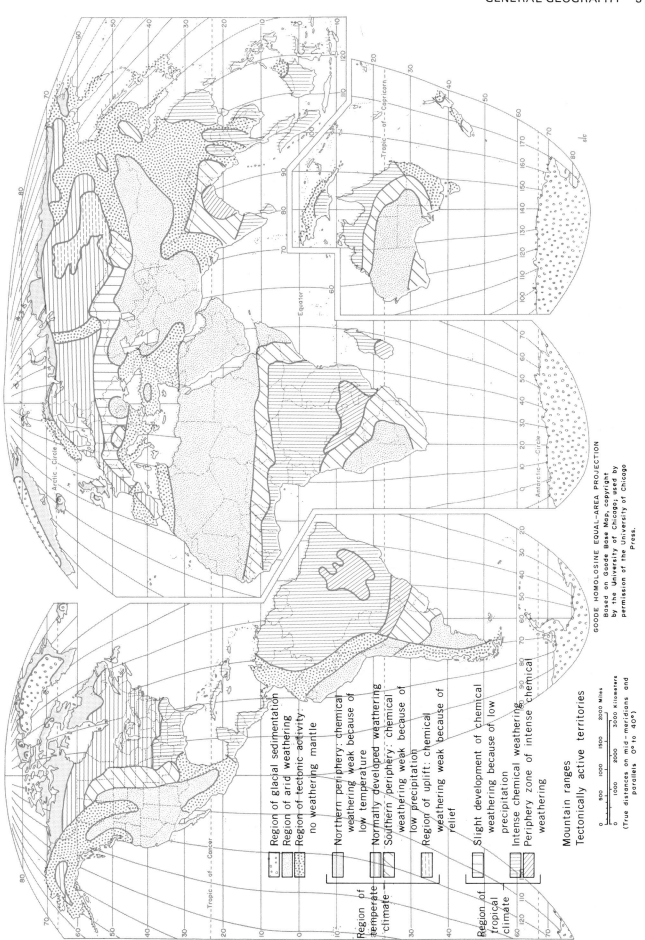

Region of glacial sedimentation

Region of arid weathering

Region of tectonic activity;
no weathering mantle

Region of
temperate
climate
— Northern periphery: chemical
weathering weak because of
low temperature

Normally developed weathering

— Southern periphery: chemical
weathering weak because of
low precipitation

Region of uplift: chemical
weathering weak because of
relief

Region of
tropical
climate
— Slight development of chemical
weathering because of low
precipitation

Intense chemical weathering

— Periphery zone of intense chemical
weathering

Mountain ranges

Tectonically active territories

GOODE HOMOLOSINE EQUAL-AREA PROJECTION
Based on Goode Base Map, copyright
by the University of Chicago; used by
permission of University of Chicago
Press.

0   500   1000   1500   2000 Miles
0   1000   2000   3000 Kilometers
(True distances on mid-meridians and
parallels 0° to 40°)

MAP 1-14   DISTRIBUTION OF TYPES OF WEATHERING

**MAP
1-14**

DISTRIBUTION OF TYPES OF WEATHERING

Following our consideration of the effects of climate and vegetation on landforms, we now examine a map showing the distribution of types of weathering found around the world. This map is concerned mainly with the effects of temperature, precipitation, relief, and uplift on different types of weathering, mainly chemical. The differences in mechanical denudation (the physical wearing away of the earth's surface) on present-day continents are shown on Map 1-15 — "Distribution of Erosion." The map of weathering not only points out climatic peculiarities but also indicates that the relief of a region has a great effect on the development of chemical weathering, particularly in various parts of moist zones. But relief by itself is not sufficient; weathering also depends on epeirogenic movements. Where rapid epeirogenic uplift gives rise to high, mountainous relief, chemical weathering is suppressed. When epeirogeny is sluggish, scarcely perceptible, the landscape consists of plains, which favor the most active chemical weathering.

The author has modified Strakhov's map in several places. "Region of arid sedimentation" has been changed to "Region of arid weathering" because the term sedimentation does not fit in some desert locations. The word "moist" has been left off the broad categories titled "Region of temperate climate" and "Region of tropical climate," to use the word moist for all areas under this category is not correct. Southwest Australia has been changed from a region of tropical climate with intense chemical weathering to a region of temperate climate with weak chemical weathering because of low precipitation.

Sources: Strakhov, N.M., Figure 12 in *Principles of Lithogenesis,* Vol. 1, Tomeieff, S.I. and Hemingway, J.E., editors, Edinburgh and London: Oliver and Boyd, Edinburgh, 1967. This map appears in modified form in Board, C., R.J. Chorley, P. Haggett, and D.R. Stoddar, Figure 9 in *Progress in Geography,* Vol. 1, New York, St. Martin's Press, pp. 174-175.

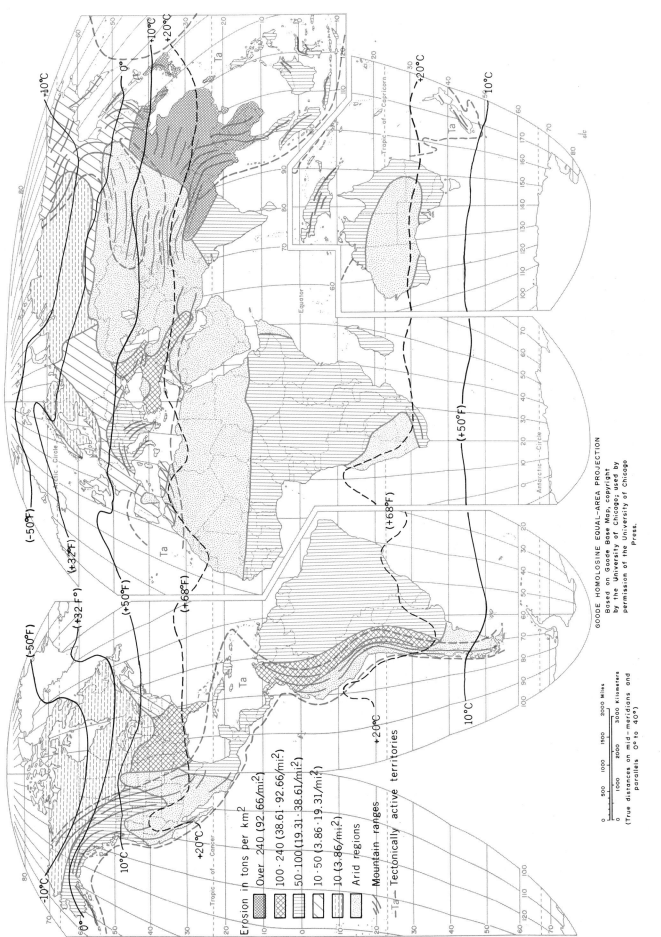

Erosion in tons per km²

	Over 240 (92.66/mi.²)
	100 - 240 (38.61 - 92.66/mi.²)
	50 - 100 (19.31 - 38.61/mi.²)
	10 - 50 (3.86 - 19.31/mi.²)
	10 (3.86/mi.²)
	Arid regions
	Mountain ranges
— Ta —	Tectonically active territories

GOODE HOMOLOSINE EQUAL-AREA PROJECTION
Based on Goode Base Map, copyright
by the University of Chicago; used by
permission of the University of Chicago
Press.

(True distances on mid-meridians and
parallels 0° to 40°)

0   500   1000   1500   2000 Miles
0   1000   2000   3000 Kilometers

MAP 1-15   DISTRIBUTION OF EROSION

**MAP
1-15**

DISTRIBUTION OF EROSION

Complementary to his studies of the distribution of types of weathering, N.M. Strakhov (1967) has also studied the distribution of various intensities of mechanical denudation, and he has prepared a map showing erosion in tons per square kilometer for broad regions of the world. (The term erosion can be defined as the group of processes whereby earth or rock material is loosened or dissolved and removed from any part of the earth's surface. Mechanical denudation is just one of these processes.) One of the factors the author fails to take into consideration is human agricultural activities and all of its related effects. Thus, this map appears to be much more of a revelation of how man has changed the erosion pattern than the effects of nature alone. On the revised form of the map reproduced here, one can pick out two broad parallel zones, where fundamentally different indices can be recognized. The first is the temperate moist belt in the Northern Hemisphere, with 16-24 in. (400-600 mm) annual precipitation in North America, Europe, and Asia. It's southern boundary is the annual 50°F (10°C) isotherm. The general intensity of mechanical denudation is small here; most commonly 10 tons/km2), rarely 10-15 tons/km2 (3.86-5.79/mi2), except in the western part of North America and a small section of China which reaches over 240 tons/km2 (92.66/mi2). The second zone includes parts of North and South America, Africa, and Southeastern Asia, adjoining the subtropical and tropical moist belt. This corresponds almost exactly to the region between the 50°F (10°C) isotherm in the Northern Hemisphere and the same isotherm in the Southern Hemisphere; over a great part of this region the annual temperature does not fall below 68°F (20°C), and it generally holds at 71.6-73.4°F (22-23°C); rainfall is high, 47-118 in. (1200-3000mm) per year. The intensity of mechanical denudation is markedly greater than in the North Temperate Zone; most commonly it holds at a level of 50-100 tons/km2 (19.31-38.61/mi2), but in a number of places it rises to

100-240 tons/km2 (38.61-92.66/mi2), and in southeastern Asia it averages 390 tons/km2 (150.58/mi2). In the basins of the Indus, Ganges, and Brahmaputra, the value is even greater - 1000 tons/km2 (386.10/mi2) and more. Strakov finds also that the intensity of chemical denudation varies zonally (latitudinally) with that of mechanical denudation. From an analysis of the individual records, he finds two laws: (1) mechanical denudation in the area of a humid climate is by nature zonal, and (2) within both temperate and tropical climate zones the energy of mechanical erosion increase with the increase in tectonic activity of the region. Plate 2.(centerfold) is an example of very rapid erosion of the high plains of northeastern New Mexico. The blown-up space photograph was taken southwest of Raton, New Mexico where the Cimarron Creek is cutting headward.

Sources: Strakhov, N.M. Figure 5 in *Principles of Lithogenesis.* Vol. 1. Tomkeieff, S.I. and Hemingway, J.E., editors, Edinburgh and London: Oliver and Boyd, Edinburgh, 1967. This map appears in modified form in Board, C., R.J. Chorley, P. Haggett, and D.R. Stoddart, Figure 16 in *Progress in Geography,* Edward Arnold (publishers) Ltd., London, 1969, and in Chorley, R.J., Figure 1.111.3, p. 48 in *Water, Earth and Man,* Methuen and Company, Ltd., London, 1969.

CLASSES OF ROCK (LITHIC REGIONS)

This map shows the main lithic regions around the world. Although numerous classifications could be used, this one divides the world into seven major categories. On such a small scale map it is difficult to show the complexities of rock form. For example, the Colorado Plateau in the western United States is shown as being composed of well-consolidated sedimentary rocks, and, although in very generalized form this is true, there are large areas with ancient metamorphic rocks and associated intrusive igneous rocks. There are even some areas of fine-grained, ashy or glassy extrusive igneous rocks. The volcanic areas near Flagstaff, Arizona and Grants, New Mexico are thus exceptions to the overall lithology of this region.

Sources: Trewartha, G.T., A.H. Robinson, and E.H. Hammond, Plate 4 in *Physical Elements of Geography,* 5th ed., McGraw-Hill Book Company, New York, 1967.

4 LITHIC REGIONS

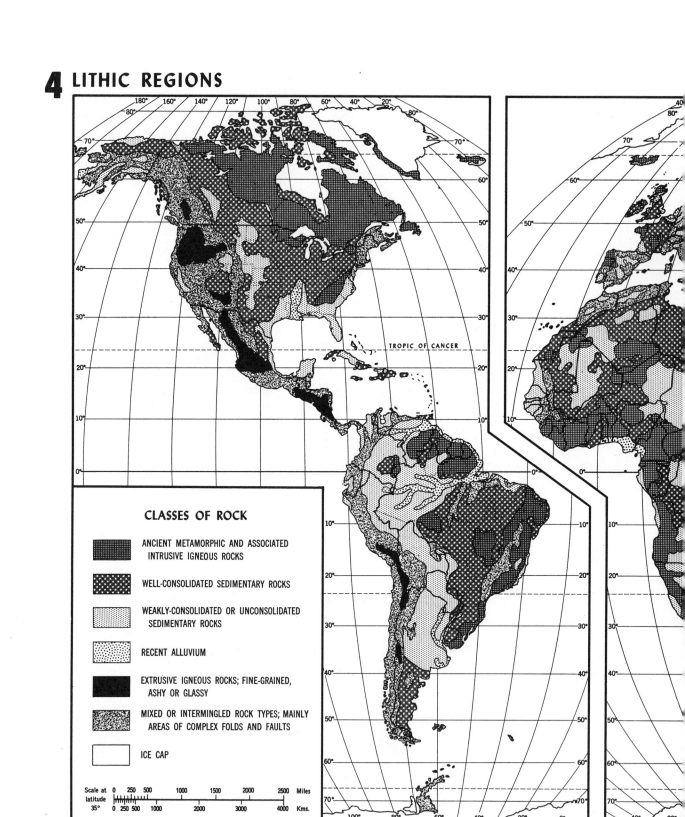

CLASSES OF ROCK

ANCIENT METAMORPHIC AND ASSOCIATED INTRUSIVE IGNEOUS ROCKS

WELL-CONSOLIDATED SEDIMENTARY ROCKS

WEAKLY-CONSOLIDATED OR UNCONSOLIDATED SEDIMENTARY ROCKS

RECENT ALLUVIUM

EXTRUSIVE IGNEOUS ROCKS; FINE-GRAINED, ASHY OR GLASSY

MIXED OR INTERMINGLED ROCK TYPES; MAINLY AREAS OF COMPLEX FOLDS AND FAULTS

ICE CAP

TROPIC OF CANCER

Scale at latitude 35°
0 250 500 1000 1500 2000 2500 Miles
0 250 500 1000 2000 3000 4000 Kms.

MAP 1-16   CLASSES OR ROCK (LITHIC REGIONS)

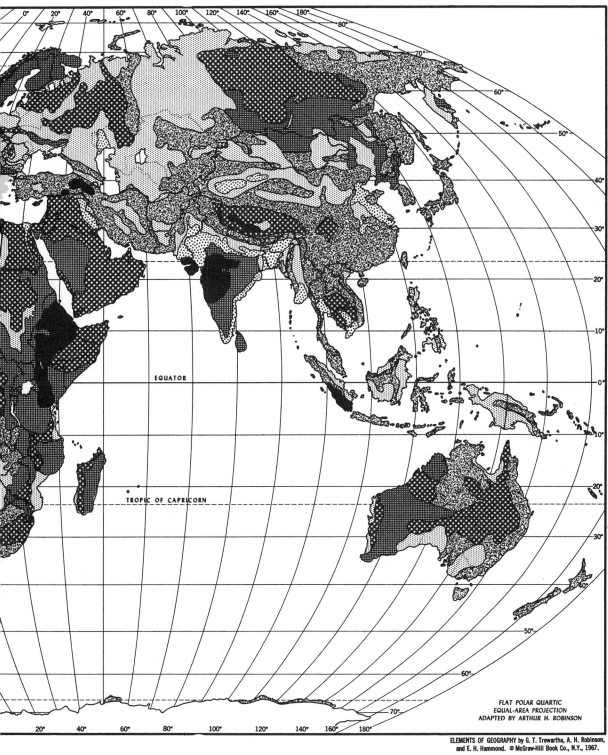

EQUATOR

TROPIC OF CAPRICORN

FLAT POLAR QUARTIC
EQUAL-AREA PROJECTION
ADAPTED BY ARTHUR H. ROBINSON

MAP 1-17   SOILS OF THE WORLD: DISTRIBUTION OF ORDERS AND PRINCIPAL SUBORDERS

U. S. DEPARTMENT OF AGRICULTURE

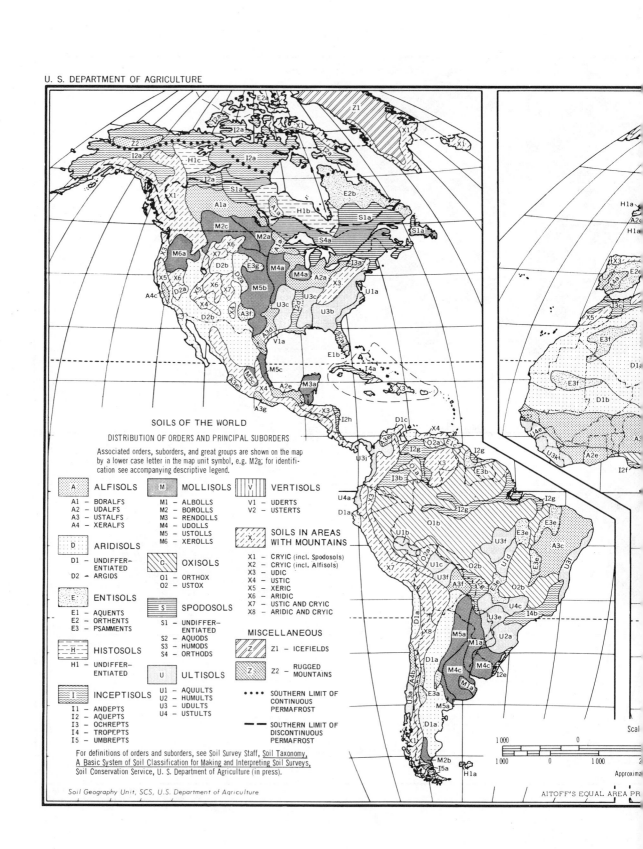

SOILS OF THE WORLD

DISTRIBUTION OF ORDERS AND PRINCIPAL SUBORDERS

Associated orders, suborders, and great groups are shown on the map by a lower case letter in the map unit symbol, e.g. M2a; for identification see accompanying descriptive legend.

A	ALFISOLS
A1	– BORALFS
A2	– UDALFS
A3	– USTALFS
A4	– XERALFS

D	ARIDISOLS
D1	– UNDIFFER-ENTIATED
D2	– ARGIDS

E	ENTISOLS
E1	– AQUENTS
E2	– ORTHENTS
E3	– PSAMMENTS

H	HISTOSOLS
H1	– UNDIFFER-ENTIATED

I	INCEPTISOLS
I1	– ANDEPTS
I2	– AQUEPTS
I3	– OCHREPTS
I4	– TROPEPTS
I5	– UMBREPTS

M	MOLLISOLS
M1	– ALBOLLS
M2	– BOROLLS
M3	– RENDOLLS
M4	– UDOLLS
M5	– USTOLLS
M6	– XEROLLS

O	OXISOLS
O1	– ORTHOX
O2	– USTOX

S	SPODOSOLS
S1	– UNDIFFER-ENTIATED
S2	– AQUODS
S3	– HUMODS
S4	– ORTHODS

U	ULTISOLS
U1	– AQUULTS
U2	– HUMULTS
U3	– UDULTS
U4	– USTULTS

V	VERTISOLS
V1	– UDERTS
V2	– USTERTS

X	SOILS IN AREAS WITH MOUNTAINS
X1	– CRYIC (incl. Spodosols)
X2	– CRYIC (incl. Alfisols)
X3	– UDIC
X4	– USTIC
X5	– XERIC
X6	– ARIDIC
X7	– USTIC AND CRYIC
X8	– ARIDIC AND CRYIC

MISCELLANEOUS

Z	
Z1	– ICEFIELDS
Z2	– RUGGED MOUNTAINS

•••• SOUTHERN LIMIT OF CONTINUOUS PERMAFROST

—— SOUTHERN LIMIT OF DISCONTINUOUS PERMAFROST

For definitions of orders and suborders, see Soil Survey Staff, Soil Taxonomy, A Basic System of Soil Classification for Making and Interpreting Soil Surveys, Soil Conservation Service, U. S. Department of Agriculture (in press).

Soil Geography Unit, SCS, U.S. Department of Agriculture

AITOFF'S EQUAL AREA PR

Scal

1000   0

1000   0   1000

Approxima

SOIL CONSERVATION SERVICE

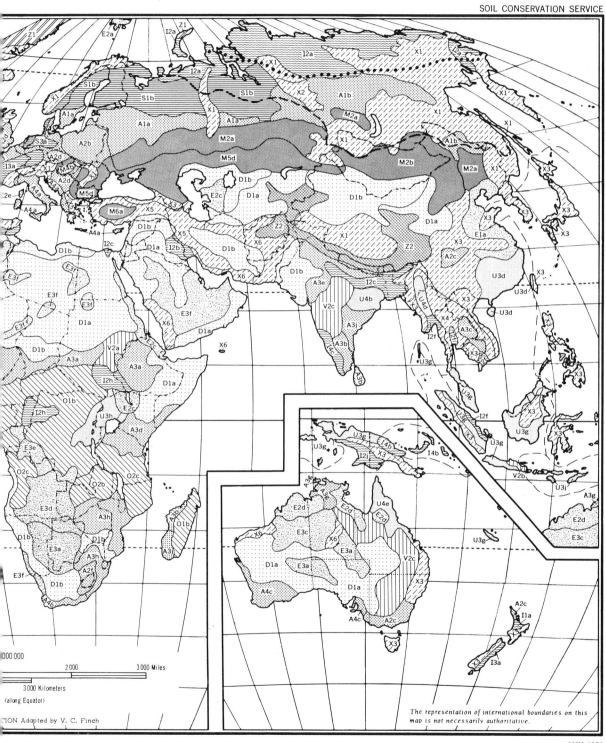

000 000

2 000   3 000 Miles

3 000 Kilometers

(along Equator)

:ION Adapted by V. C. Finch

The representation of international boundaries on this map is not necessarily authoritative.

MAY 1972
USDA-SCS-HYATTSVILLE. MD. 1972

SOILS OF THE WORLD

DISTRIBUTION OF ORDERS
AND PRINCIPAL SUBORDERS
AND GREAT GROUPS

A **ALFISOLS** — Soils with subsurface horizons of clay accumulation and medium to high base supply; either usually moist or moist for 90 consecutive days during a period when temperature is suitable for plant growth.

A1— **Boralfs** — cold.

    A1a— with Histosols, cryic temperature regimes common
    A1b— with Spodosols, cryic temperature regimes

A2— **Udalfs** — temperate to hot, usually moist.

    A2a— with Aqualfs
    A2b— with Aquolls
    A2c— with Hapludults
    A2d— with Ochrepts
    A2e— with Troporthents
    A2f— with Udorthents

A3 **Ustalfs** — temperate to hot, dry more than 90 cumulative days during periods when temperature is suitable for plant growth.

    A3a— with Tropepts
    A3b— with Troporthents
    A3c— with Tropustults
    A3d— with Usterts
    A3e— with Ustochrepts
    A3f— with Ustolls
    A3g— with Ustorthents
    A3h— with Ustox
    A3j— Plinthustalfs with Ustorthents

A4— **Xeralfs** — temperate or warm, moist in winter and dry more than 45 consecutive days in summer.

    A4a— with Xerochrepts
    A4b— with Xerorthents
    A4c— with Xerults

D **ARIDISOLS** — Soils with pedogenic horizons, usually dry in all horizons and never moist as long as 90 consecutive days during a period when temperature is suitable for plant growth.

D1— **Aridisols** — undifferentiated.

    D1a— with Orthents
    D1b— with Psamments
    D1c— with Ustalfs

D2— **Argids** — with horizons of clay accumulation.

    D2a— with Fluvents
    D2b— with Torriorthents

E **ENTISOLS** — Soils without pedogenic horizons; either usually wet, usually moist, or usually dry.

E1— **Aquents** — seasonally or perennially wet.

    E1a— Haplaquents with Udifluvents
    E1b— Psammaquents with Haplaquents
    E1c— Tropaquents with Hydraquents

E2— **Orthents** — Loamy or clayey textures, many shallow to rock

    E2a— Cryorthents
    E2b— Cryorthents with Orthods
    E2c— Torriorthents with Aridisols
    E2d— Torriorthents with Ustalfs
    E2e— Xerorthents with Keralfs

E3— **Psamments** — sand or loamy sand textures.

    E3a— with Aridisols
    E3b— with Orthox
    E3c— with Torriorthents
    E3d— with Ustalfs
    E3e— with Ustox
    E3f— shifting sands
    E3g— Ustipsamments with Ustolls

H **HISTOSOLS** — Organic soils.

H1— **Histosols** — undifferentiated.

    H1a— with Aquods
    H1b— with Boralfs
    H1c— with Cryaquepts

I **INCEPTISOLS** — Soils with pedogenic horizons of alteration or concentration but without accumulations of translocated materials other than carbonates or silica; usually moist or moist for 90 consecutive days during a period when temperature is suitable for plant growth.

I1— **Andepts** — amorphous clay or vitric volcanic ash or pumice.

    I1a— Dystrandepts with Ochrepts

I2— **Aquepts** — seasonally wet.

    I2a— Cryaquepts with Orthents
    I2b— Halaquepts with Salorthids
    I2c— Haplaquepts with Humaquepts
    I2d— Haplaquepts with Ochraqualfs
    I2e— Humaquepts with Psamments
    I2f— Tropaquepts with Hydraquents
    I2g— Tropaquepts with Plinthaquults
    I2h— Tropaquepts with Tropaquents
    I2J— Tropaquepts with Tropudults

I3— **Ochrepts** — thin, light-colored surface horizons and little organic matter.

    I3a— Dystochrepts with Fragiochrepts
    I3b— Dystrochrepts with Orthox
    I3c— Xerochrepts with Xerolls

I4— **Tropepts** — continuously warm or hot.

    I4a— with Ustalfs
    I4b— with Tropudults
    I4c— with Ustox

I5— **Umbrepts** — dark-colored surface horizons with medium to low base supply.

    I5a— with Aqualfs

M **MOLLISOLS** — Soils with nearly black, organic-rich surface horizons and high base supply; either usually moist or usually dry.

M1— **Albolls** — light grey subsurface horizon over slowly permeable horizon; seasonally wet.

    M1a— with Aquepts

M2— **Borolls** — cold.

    M2a— with Aquolls
    M2b— with Orthids
    M2c— with Torriorthents

SOILS OF THE WORLD CONTINUED

M3— **Rendolls** — subsurface horizons have much calcium carbonate but no accumulation of clay.

    M3a— with Usterts

M4— **Udolls** — temperate or warm, usually moist.

    M4a— with Aquolls
    M4b— with Eutrochrepts
    M4c— with Humaquepts

M5— **Ustolls** — temperate to hot, dry more than 90 cumulative days in year.

    M5a— with Argialbolls
    M5b— with Ustalfs
    M5c— with Usterts
    M5d— with Ustochrepts

M6— **Xerolls** — cool to warm, moist in winter and dry more than 45 consecutive days in summer.

    M6a— with Xerorthents

O   **OXISOLS** — Soils with pedogenic horizons that are mixtures principally of kaolin, hydrated oxides, and quartz, and are low in weatherable minerals.

O1— **Orthox** — hot, nearly always moist.

    O1a— with Plinthaquults
    O1b— with Tropudults

O2— **Ustox** — warm or hot, dry for long periods but moist more than 90 consecutive days in the year.

    O2a— with Plinthaquults
    O2b— with Tropustults
    O2c— with Ustalfs

S   **SPODOSOLS** — Soils with accumulation of amorphous materials in subsurface horizons; usually moist or wet.

S1— **Spodosols** — undifferentiated.

    S1a— cryic temperature regimes; with Boralfs
    S1b— cryic temperature regimes; with Histosols

S2— **Aquods** — seasonally wet.

    S2a— Haplaquods with Quartzipsamments

S3— **Humods** — with accumulations of organic matter in subsurface horizons.

    S3a— with Hapludalfs

S4— **Orthods** — with accumulations of organic matter, iron, and aluminum in subsurface horizons.

    S4a— Haplorthods with Boralfs

U   **ULTISOLS** — Soils with subsurface horizons of clay accumulation and low base supply; usually moist or moist for 90 consecutive days during a period when temperature is suitable for plant growth.

U1— **Aquults** — seasonally wet.

    U1a— Ochraquults with Udults
    U1b— Plinthaquults with Orthox
    U1c— Plinthaquults with Plinthaquox
    U1d— Plinthaquults with Tropaquepts

U2— **Humults** — temperate or warm moist all of year; high content of organic matter.

    U2a— with Umbrepts

U3— **Udults** — temperate to hot; never dry more than 90 consecutive days in the year.

    U3a— with Andepts
    U3b— with Dystrochrepts
    U3c— with Udalfs
    U3d— Hapludults with Dystrochrepts
    U3e— Rhodudults with Udalfs
    U3f— Tropudults with Aquults
    U3g— Tropudults with Hydraquents
    U3h— Tropudults with Orthox
    U3j— Tropudults with Tropepts
    U3k— Tropudults with Tropudalfs

U4— **Ustults** — warm or hot; dry more than 90 cumulative days in the year.

    U4a— with Ustochrepts
    U4b— Plinthustults with Ustorthents
    U4c— Rhodustults with Ustalfs
    U4d— Tropustults with Tropaquepts
    U4e— Tropustults with Ustalfs

V   **VERTISOLS** — Soils with high content of swelling clays; deep, wide cracks develop during dry periods.

V1— **Uderts** — usually moist in some part in most years; cracks open less than 90 cumulative days in the year.

    V1a— with Usterts

V2— **Usterts** — cracks open more than 90 cumulative days in the year.

    V2a— with Tropaquepts
    V2b— with Tropofluvents
    V2c— with Ustalfs

X   **Soils in areas with mountains** — Soils with various moisture and temperature regimes; many steep slopes; relief and total elevation vary greatly from place to place. Soils vary greatly within short distances and with changes in altitude; vertical zonation common.

    X1— Cryic great groups of Entisols, Inceptisols, and Spodosols.
    X2— Boralfs and cryic great groups of Entisols and Inceptisols.
    X3— Udic great groups of Alfisols, Entisols, and Ultisols; Inceptisols.
    X4— Ustic great groups of Alfisols, Inceptisols, and Ultisols.
    X5— Xeric great groups of Alfisols, Entisols, Inceptisols, Mollisols, and Ultisols.
    X6— Torric great groups of Entisols; Aridisols.
    X7— Ustic and cryic great groups of Alfisols, Entisols, Inceptisols, and Mollisols; ustic great groups of Ultisols; cryic great groups of Spodosols.
    X8— Aridisols, torric and cryic great groups of Entisols, and cryic great groups of Spodosols and Inceptisols.

Z   **MISCELLANEOUS**

    Z1— Icefields.
    Z2— Rugged mountains — mostly devoid of soil (includes glaciers, permanent snow fields, and, in some places, areas of soil).

    Southern limit of continuous permafrost.

    Southern limit of discontinuous permafrost.

**MAP
1-17**

SOILS OF THE WORLD, DISTRIBUTION OF ORDERS
AND PRINCIPAL SUBORDERS

Although soils derive from rocks (Map 1-16) the distribution of soils does not really match the distribution of rocks. The soil classification demonstrates why this is true, because it is based on properties that are mainly the result of climatic differences. This system was chosen because climatic data are available in areas where detailed soil surveys have not been made, and also because local soil differences reflecting slope, drainage, and parent rock material cannot be generalized adequately on a global scale. The climatic soils, or zonal soils, as they are often termed, are grouped in three main classes that correspond to the three major climatic processes of profile development: (1) podzolization, the cool, humid process, (2) ferralization, the warm, humid process, also known as laterization, and (3) calcification, the process associated with dry climates. Areas of mountain soils, alluvium, and desert soils and sands are also shown on this map. These soils should not be considered zonal, but they cover a large enough area to warrant being included on the world soil map.

Sources: U.S. Department of Agriculture, Soil Conservation Service, Soil Geography Unit, SCS, U.S. Department of Agriculture, Scale 1:50,000,000, May, 1972.

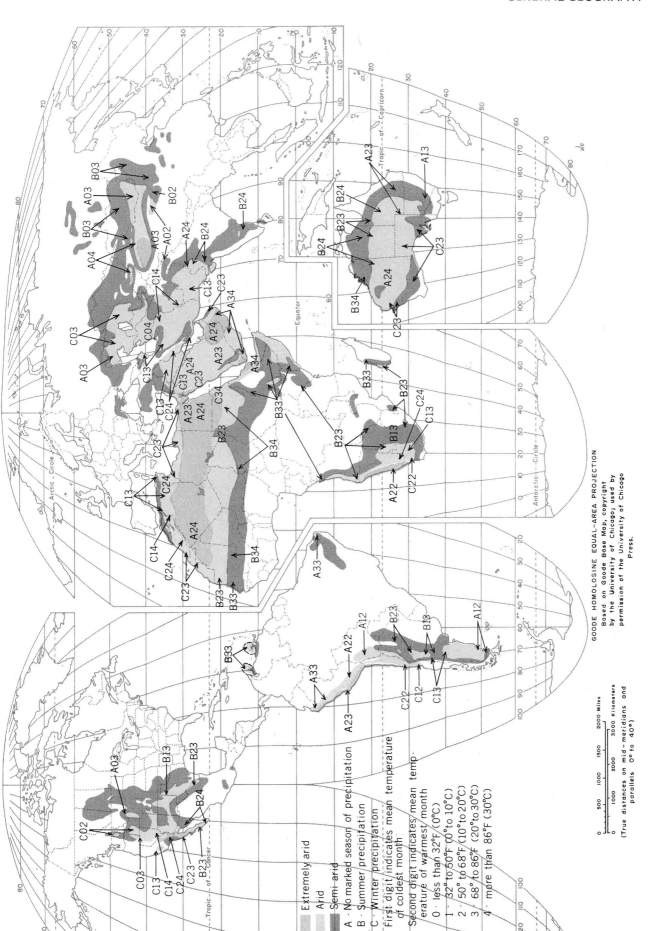

GOODE HOMOLOSINE EQUAL–AREA PROJECTION
Based on Goode Base Map, copyright
by the University of Chicago; used by
permission of the University of Chicago
Press.

Extremely arid

Arid

Semi-arid

A - No marked season of precipitation
B - Summer precipitation
C - Winter precipitation

First digit indicates mean temperature of warmest month

Second digit indicates mean temperature of coldest month

0 - less than 32°F (0°C)
1 - 32° to 50°F (0°to 10°C)
2 - 50° to 68°F (10°to 20°C)
3 - 68° to 86°F (20°to 30°C)
4 - more than 86°F (30°C)

0   500   1000   1500   2000 Miles
0 1000 2000 3000 Kilometers
(True distances on mid-meridians and
parallels 0° to 40°)

MAP 1-18   DISTRIBUTION OF ARID LANDS

MAP
1-18

WORLD DISTRIBUTION OF ARID LANDS

This map has a lot of information and is one of the new maps added to the second edition. The arid areas occupy a third of the earth's land surface. Some four per cent of the land surface is extremely arid, 15 per cent is arid, and 14.6 per cent is semi-arid. We need to define these boundaries. The information is adopted from Meigs', (1953) which is a widely accepted scheme based upon Thornthwaite's moisture index, involving a consideration of potential evaporation and water balance (Meigs, 1953; McGinnies et al, 1968). Extremely arid areas are defined as those where at least 12 consecutive months are without rainfall and where there is no regular seasonal rhythm of rainfall. The breakdown between arid and semi-arid is based upon the moisture index devised by C.W. Thornthwaite (1948). Those areas with a moisture index of less than −40 are considered arid, while areas bounded by the −20 and −40 values are designated semi-arid. In this classification, the outer limit of all dry lands is taken to be at the −20 value of the moisture index. A broad division which is not a particularly accurate one because the problem of evapotranspiration states that arid regions have an average of 0 to 10 inches (0 to 254 millimeters) of rainfall per year and semi-arid regions have an average of 10 to 20 inches (254 to 508 millimeters) of rainfall per year.

The extremely arid, and semi-arid zones are further classified according to the period of the year when precipitation occurs and the mean temperature of the warmest and coldest months. For example, most of the Patagonian desert of Southern Argentina is semi-arid with no marked season of precipitation. For this semi-arid region mean temperature of the coldest month is 32° to 50°F (0° to 10°C) and the mean temperature of the warmest month is 50° to 68°F (10° to 20°C). The largest desert in the world is the Sahara stretching from the Atlantic Ocean east to the Red Sea. Its central part is classified as extremely arid with no marked season of precipitation for most of its surface. The Sahara has a mean temperature of the coldest month between 50° to 68°F (10° to 20°C) and the mean temperature for the warmest month more than 86°F (30°C). Circumpolar and high-altitude deserts, where physiological aridity is associated with low temperatures, are excluded from this map.

Sources: Cooke, R.U. and A. Warren, Fig. 1.1 in *Geomorphology in Deserts.* University of California Press, Berkeley and Los Angeles, 1973, pp. 7-10; Thornthwaite, C.W., "An Approach Toward a Rational Classification of Climate, " *Geographical Review,* Vol. 38, 1948, pp. 55-94; Meigs, P., "World Distribution of Arid and Semi-arid Homo-climates," in *Reviews of Research on Arid Zone Hydrology,* UNESCO, Paris, 1953, pp. 203-209; and McGinnies, W.G., et al., *Deserts of the World,* University of Arizona Press, Tucson, 1968, 788 p.

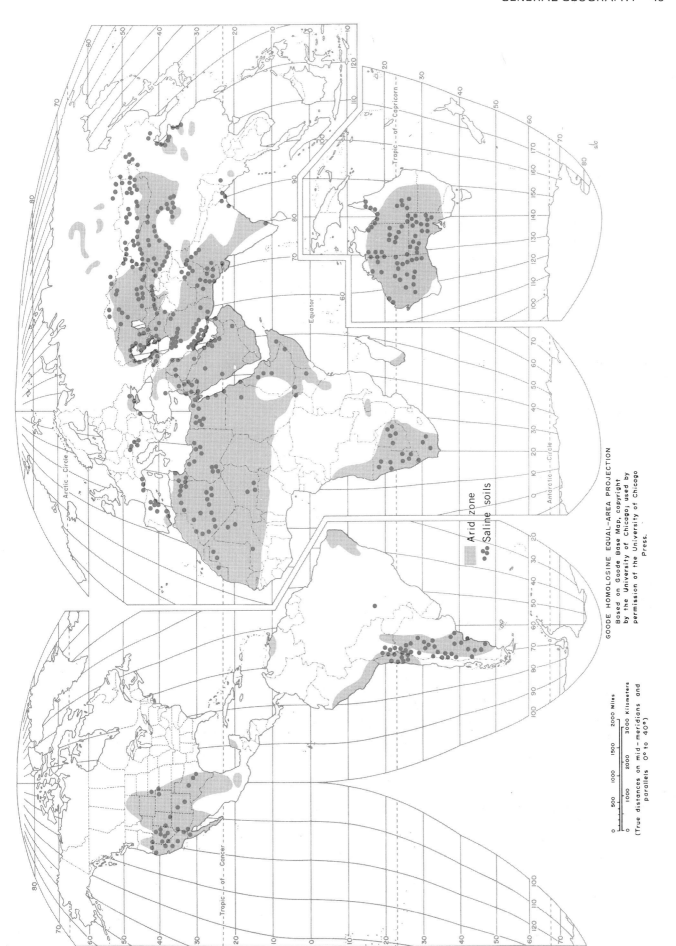

GOODE HOMOLOSINE EQUAL-AREA PROJECTION
Based on Goode Base Map, copyright
by the University of Chicago; used by
permission of the University of Chicago
Press.

Arid zone
Saline soils

0 500 1000 1500 2000 Miles
0 1000 2000 3000 Kilometers
(True distances on mid-meridians and
parallels 0° to 40°)

MAP 1-19   DISTRIBUTION OF SALINE SOILS

**MAP
1-19**

DISTRIBUTION OF SALINE SOILS

Salinity is a common characteristic of soils in desert regions. Black dots locate areas where saline soils have been reported or present a problem to agriculture. Although this map is very generalized, it does illustrate that the distribution of these soils is surprisingly widespread. Saline soils can be divided into two major categories: white alkali (solonchak), and black alkali (solonetz). In the Soviet Union alone, they cover more than 30 million acres of 3.4 percent of the land area. In the drier continental climate of central Europe, there are more than 200,000 acres (80,937 hectares) of these soils in the Hungarian plain. In China, 8 million acres (3,240,000 hectares) have high degrees of soil salinity, and many areas are similarily affected in the regions surrounding the Mediterranean Sea, the United States, Australia, Africa, India, and Pakistan. Certain regions show great concentrations as, for example, lands bordering the Persian Gulf and Caspian Sea. Southern Bolivia, the Atacama Desert of Chile, and large sections of Argentina possess extensive areas of saline soils. Several areas outside the so-called "arid zone" have a high concentration of salt. Spain has a particularly high percentage, and delta regions of several large rivers (for example, the Ganges-Brahmaputra) have saline soils that generally produce poor crop yields. Waterlogging is often accompanied by an increase in salinity — sometimes to levels that even the most salt-tolerant crops cannot withstand. Large irrigated tracts in West Pakistan are faced with this problem.

Sources: After Woldstedt, *Das Eiszeitalter,* Ferdinand Enke, Stuttgart (1954-1965), and UNESCO "Compte rendu des recherches relatives a l'hydrologie de la zone aride," 1952. This map appears in Walton, K., Figure 8, *The Arid Zones,* Aldine Publishing Company, Chicago, 1969.

Section Two

Structure
and
Tectonics

MAP
2-1

THE PLATE'S OF THE EARTH'S LITHOSPHERE

This is a new map for the atlas and is an important one because it shows the regions of the earth where active movements have been and still are taking place. The mosaic of plates forms the earth's litosphere, or outer shell and according to the recently developed theory of plate tectonics, the plates are not only rigid but also in constant relative motion. The boundaries of the plates are of three main types: ridge axes, where plates are diverging and a new oceanic floor is generated; transforms, where the plates slide past one another; and subduction zones, where plates converge and one plate dives under the leading edge of its neighbor. The triangles of the subduction zone lines indicate the leading edge of a plate. The arrows indicate the direction of plate motion and the dashed lines have been drawn where the plate boundary is uncertain (see Maps 2-2, 4-1, and 4-2). Several of the major rift systems include the Mid-Atlantic Ridge; the San Andreas Fault system; the Aleutian Trench; the Peru-Chile Trench; and the Macquarie Ridge. These ridge, trench and transform zones separate plates and indicate some of the most mobile zones of the world.

Sources: Revised from J.F. Dewey, "Plate Tectonics," *Scientific Amer-ican,* May, 1972, pp. 56-57; forepiece of F. Press and R. Siever, *Earth,* W.H. Freeman and Company, San Francisco, 1974; J. Guest (Editor) *The Earth and Its Satellite,* David McKay Co., Inc.. New York, 1971, Fig. 11, p.40.

MAP 2-1   THE PLATES OF THE EARTH'S LITHOSPHERE

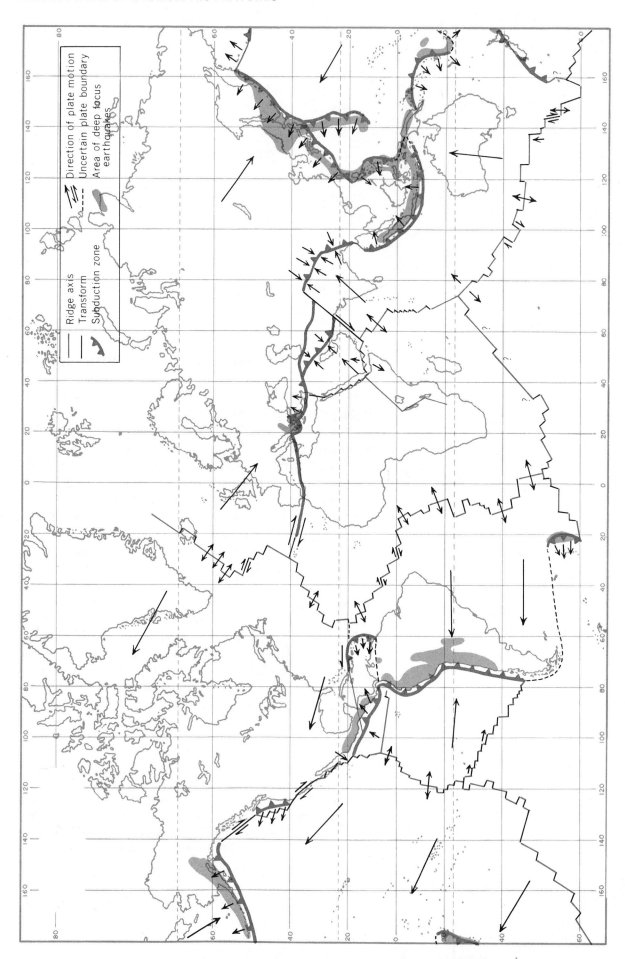

MAP
2-2

CONTINENTAL DRIFT

The group of four maps (A through D) are presented in the atlas because of the current geological ideas on the break-up of the universal land mass called Pangaea. For many years scientists were skeptical about the slow movement of one land mass away from another, but now most geologists accept these revolutionary concepts. Map A depicts the universal land mass called Pangaea as it may have looked 200 million years ago. Panthalassa was the ancestral Pacific Ocean and the Tethys Sea, the ancestral Mediterranean formed a large bay separating Africa and Eurasia. The relative positions of the continents, except for India, are based on the best fits made by a computer, using the 1000-fathom isobath to define continental boundaries. Map B shows the arrangement of the continents after 65 million years of drift, at the end of the Jurrassic period, 135 million years ago. The North Atlantic and Indian Ocean have opened considerable. The birth of the South Atlantic has been initiated by a rift. The rotation of the Eurasian land mass has begun to close the eastern end of the Tethys Sea. The Indian plate is about to pass over a thermal center (colored dot) that will pour out basalt to form the Deccan plateau. Map C is the position of the continents after 135 million years of drift. At 65 million years ago, at the end of the Cretaceous period, the South Atlantic had widened into a major ocean. A new rift has carved Madagascar away from Africa and the rift in the North Atlantic has switched from the west side to the east side of Greenland. But, Australia still remains attached to Antarctica. The position of the continents as they look today and the major rift belts are shown on Map 25. Map D shows how the earth may look 50 million years from now. An attempt has been made to extrapolate present-day plate movements to indicate how the

MAP 2-2 CONTINENTAL DRIFT

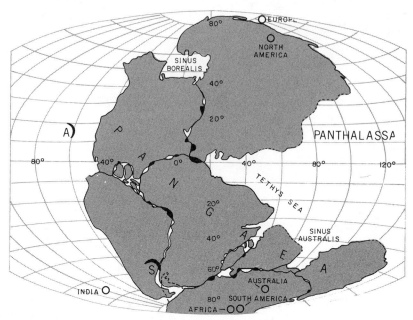

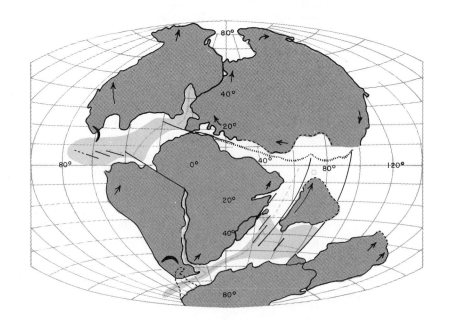

MAP 2-2 CONTINENTAL DRIFT, CONTINUED

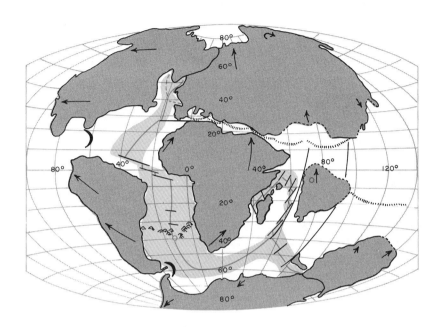

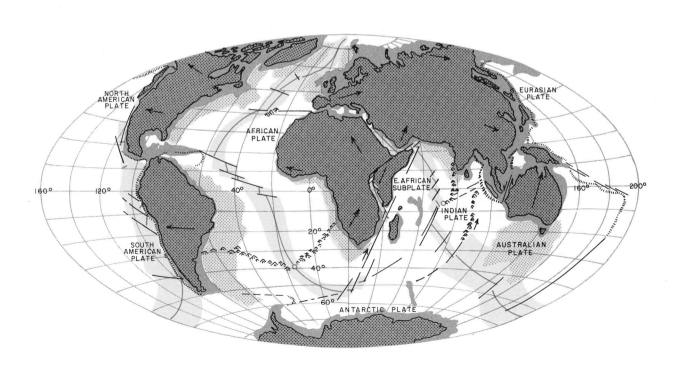

continents have drifted in what the authors call the Psychozoic era (the age of awareness). The Antarctic remains essentially fixed, but may rotate slightly clockwise. The Atlantic, particularly the South Atlantic, and the Indian Ocean continue to grow at the expense of the Pacific. Australia drifts northward and begins to rub against the Eurasian plate. The eastern portion of Africa is split off, while its northward drift closes the Bay of Biscay and virtually collapses the Mediterranean. New land area is created in the Caribbean by compressional uplift. Baja California and a sliver of California west of the San Andreas fault are severed from North America and begin drifting to the northwest. In about 10 million years Los Angeles will be abreast to San Francisco, still fixed to the mainland, but in about 60 million years, Los Angeles will start sliding into the Aleutian trench.

As more and more new material is added to the concept of continental drift, the above speculation will be added to or revised. Certainly the earth is a more mobile body than geologists realized 30 to 40 years ago. Plate 3.(page155A) is a Gemini XI space photograph of the continent of India and the island of Ceylon. Southern India, especially the Deccan Shield, is one of main areas of the world experiencing very fast continental drift. As it moves north into Asia and slides under it along a great thrust fault, the Himalayan mountains are forced up.

Sources: R.S.Dietz and J.C. Holden, ''Reconstruction of Pangaea: Breakup and Dispersion of Continents, Permian to Present,'' *Journal of Geophysical Research,* Vol. 75, No. 26, Sept. 10, 1970, pp. 4939-4956; R.S. Dietz and J.C. Holden, ''The Breakup of Pangaea,'' *Scientific American,* Oct., 1970, pp. 30-41; J.F. Dewey, ''Plate Tectonics,'' *Scientific American,* May, 1972, pp. 56-68; S.K. Runcorn, *Continental Drift,* Academic Press, Inc., 1962 and J. Tuzo Wilson, *Continents Adrift,* W.H. Freeman and Company, 1972.

MAP
2-3

PRE-CAMBRIAN SHIELDS AND OLD MASSIFS

The next 14 maps (Nos. 2-1 to 2-18) are concerned with the morphology, structure, and instability of the crust of the earth. The core of the continental land masses are the ancient crystal-line rock complexes called "shields" (because their surface is usually broadly convex in general profile) and "old massifs." Massifs are extensive, often elevated platforms, ancient geomor-phic units, in places covering hundreds of square miles. The basement of all ancient platforms is made up of Pre-Cambrian (Archean, Proterozoic, and Riphean) metamorphosed igneous and sedimentary rocks. The Riphean time period corresponds to the end of the Proterozoic or late Pre-Cambrian, the time interval of 1,500 to 600 million years ago. The granite-gneiss basement of these large rock masses often has an irregular surface, rising and out-cropping high above sea level or sub-siding below sea level under a sedimentary mantle. The large outcrops of the crystalline basement are called shields or massifs, smaller outcrops, that only approach the surface and sometimes penetrate through the sedimentary mantle are one type of anticline or arch. Deep, broad depressions in which the platform basement is greatly submerged and covered by a thick sedimentary fill, are called geosynclines. Ancient Pre-Cambrian platforms are the most stable and massive sections of the earth's crust. They represent a mosaic of blocks depicting various ages whose formation was completed on the whole by the beginning of the Cambrian. Since then some sections of platforms have subsided into synclines and other sections have been uplifted into sheilds; the platform basement has been broken up in many places, however, without undergoing major changes in its inter-nal structure.

The map is divided into three Pre-Cambrian time periods, un-differentiated, lower and middle, and upper Proterozoic crystal-line rocks. The Proterozoic is an era of geological history that includes the interval between the Archeozoic and the Paleozoic. And in turn makes up the entire Pre-Cambrian according to the United States Geological Survey. The fourth category of the legend depicts areas where the sedimentary strata overlies old (Pre-Cambrian) platforms. The boundaries of shield areas are difficult to find on the surface because only in a few areas is there a clear-cut border between shields and interior plains of the continents. Sheets of sedimentary rock mantle portions of some of the ancient shields, indicating short periods of sub-sidence and deposition of sediments. This is particularly true of the Ethiopian Shield, which covers nearly two-thirds of the African continent, The Amazonian Shield (Brazilian Uplands), and the Angara Shield of central Siberia. Around the ancient crystalline shields, on nearly all sides, are the great mountain chains (see Map 2-7). Most of the margins of the continents and the ocean basins are areas of crustal instability that have been unstable for a long period in earth history.

Sources: *Pergamon World Atlas,* Pergamon Press, New York, (PWN Poland, Polish Scientific Publishers, Warsaw, 1968, p. 22, and *Morski Atlas,* (Oceanic Atlas), edited by I.S. Isakov, Naval General Staff, Moscow, 1950.

PRECAMBRIAN SHIELDS AND MASSIFS

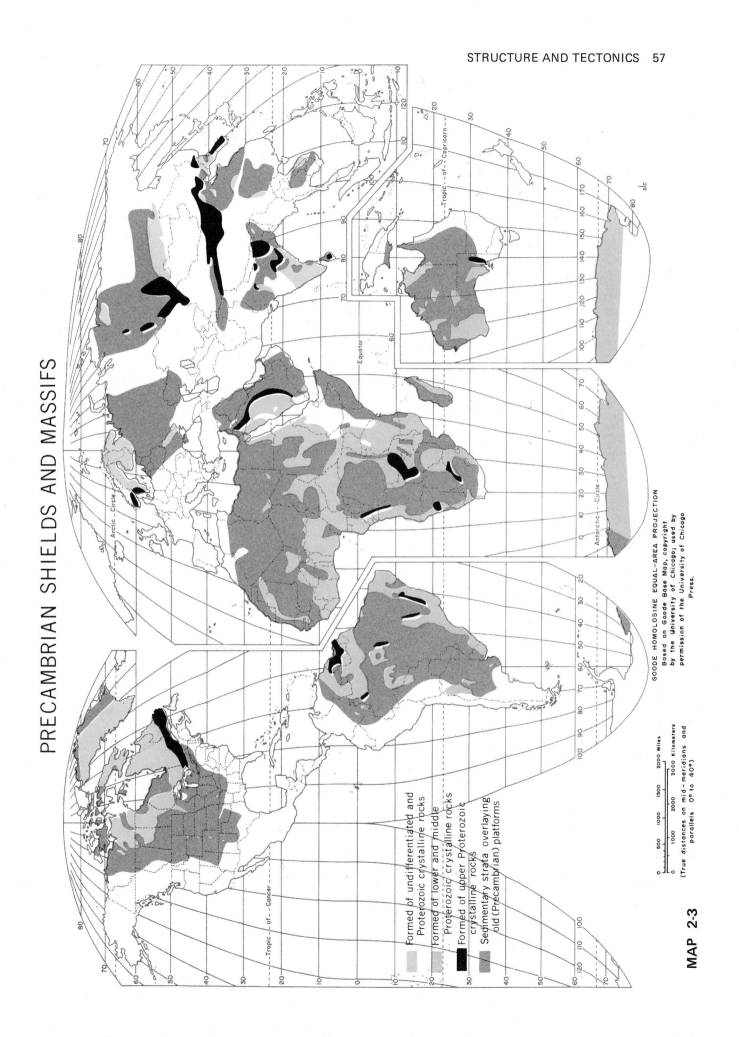

Formed of undifferentiated and Proterozoic crystalline rocks

Formed of lower and middle Proterozoic crystalline rocks

Formed of upper Proterozoic crystalline rocks

Sedimentary strata overlaying old (Precambrian) platforms

GOODE HOMOLOSINE EQUAL-AREA PROJECTION
Based on Goode Base Map, copyright by the University of Chicago; used by permission of the University of Chicago Press.

0   500   1000   1500   2000 Miles
0 1000   2000   3000 Kilometers
(True distances on mid-meridians and parallels 0° to 40°)

MAP 2-3

MAPS
2-4
2-5

SALT DOMES AND MUD VOLCANOES

A salt dome is a structure resulting from the upward movement of a salt mass, and with which oil and gas fields are frequently associated. Salt domes are of particular interest to structural geologists because they afford good examples of the plastic movement of large bodies of rock. A typical salt dome consists of a central core of rock salt and a surrounding dome of sedimentary strata. More than 300 of these features are known in the Gulf coastal region of Mississippi, Louisiana, and Texas. These structures most often form as domal or anticlinal to ridgelike diapiric folds, many of great structural magnitude with varying heights and aerial extent. Many salt domes are not expressed topographically, but the ones that are exposed generally rise from a few feet to 40 feet (12 meters) above the surrounding lowlands; in a few instances, they may rise as much as 80 to 100 feet (30 meters) or more. They generally cover an area of about a mile in diameter. The salt domes shown on the world map are the ones that are exposed topographically. In a number of cases lakes occupy depressions in the craters of domes, and frequently oil seeps and salt springs are associated with them. Often these associated features lead to their discovery (Billings, 1942). Good examples of the type topography that develops on a typical salt dome can be found at the West Point (Butler) salt dome in southeastern Freestone County, Texas (DeGotyer, 1919 and Powers, 1920) and at Avery and Jefferson Islands in southern Louisiana (Thornbury, 1969). The United States map shows a number of possible salt domes (open circles) lying beneath the continental shelf areas of Louisiana and Texas (Murray, 1961). In several other areas of the world there are great concentrations of salt domes. The Emba district of Russia on the northeast coast of the Caspian Sea has more than 100 domes (Sanders, 1939), and there are a large number in western Iran (Fisher, 1968).

Mud volcanoes qualify as unique and unusual landforms. They are not formed by igneous activity, hence, they cannot be considered true volcanoes. They are more closely related to salt domes, oil domes, or mineral springs, but here again they do not entirely "fit," so I have given them a separate map. A mud volcano is a high-pressure gas seepage that carries with it water, mud, sand, fragments of rocks, and occasionally oil. Mud volcanoes are generally confined to regions underlain by softer shales, boulders, sands, clays, and unconsolidated sediments. Most mud volcanoes, especially the larger ones, are

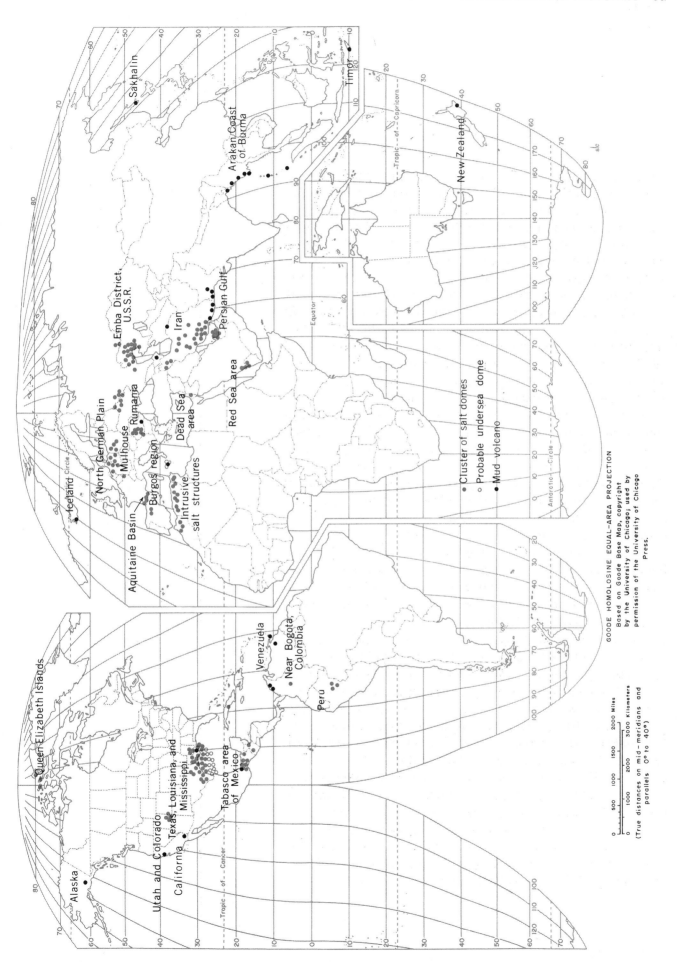

GOODE HOMOLOSINE EQUAL–AREA PROJECTION
Based on Goode Base Map, copyright
by the University of Chicago; used by
permission of the University of Chicago
Press.

MAP 2-4   SALT DOMES AND MUD VOLCANOES OF THE WORLD

associated with anticlines, faults, or diapiric folds. A mud volcano is especially likely to form on a anticline overlain by stiff, thick clay. During dry weather the clay becomes desiccated and cracked and, if the cracks cut deep enough, a little trapped gas manages to escape. As the gas rises, it mixes with the clay and groundwater to form a mud that erupts either steadily or spasmodically, depending on the local pressure conditions, and on the available amounts of gas, water and mud, and the size and shape of the opening. Mud volcanoes are most active after a long drought, probably because the desiccation cracks are then wider and penetrate deeper underground (Craig, 1912; Weeks, 1952; Levorsen, 1954). The surface expression of a mud volcano is commonly a cone of mud through which gas escapes either continuously or intermittently. Single cones, or groups of cones, may cover an area of several square miles and extend more than one thousand feet in height. However, they are more often measured in tens and hundreds of feet. Some mud volcanoes show at the surface as basinlike depressions or as level stretches of ground strewn with blocks of rock carried up from below. Mud volcanoes of this type generally occur where the rainfall is heavy, or at places on the seacoast where the tides and waves wash the soft muds away as fast as they are extruded, leaving behind erratic pebbles and boulders, many of them very large. The largest known mud volcanoes are found along the Makran coast of West Pakistan and Iran. Here individual cones resembling volcanoes reach elevations of 340 feet (103 meters). Sometimes they occur on the tops of mountain ranges, and in these locations they are more than 1,500 feet above sea level. These large features appear to be associated with faulting and result from pressures related to rapid sedimentation. During earthquakes the eruptions become very active because of the increase in pressure (Harrison, 1941; Snead, 1964, 1967). Mud volcanoes are also found along the Arakan coast of Burma, from Chittagong in East Pakistan south to the Andaman and Nicobar Islands (Chhibber, 1934; Wadia, 1953), and along a line of fissures near Paterno in eastern Sicily, at Krafla in Iceland, and on Long Island, near New York City (until apartment houses were built on top of them!) The island of Trinidad has some of the most famous mud volcanoes, although they cannot compare with the 1,000-foot-high (305-meter-high) giants near Baku in the Soviet Union or the 5,700-foot-wide (1,737-meter-wide) features on the island of Timor.

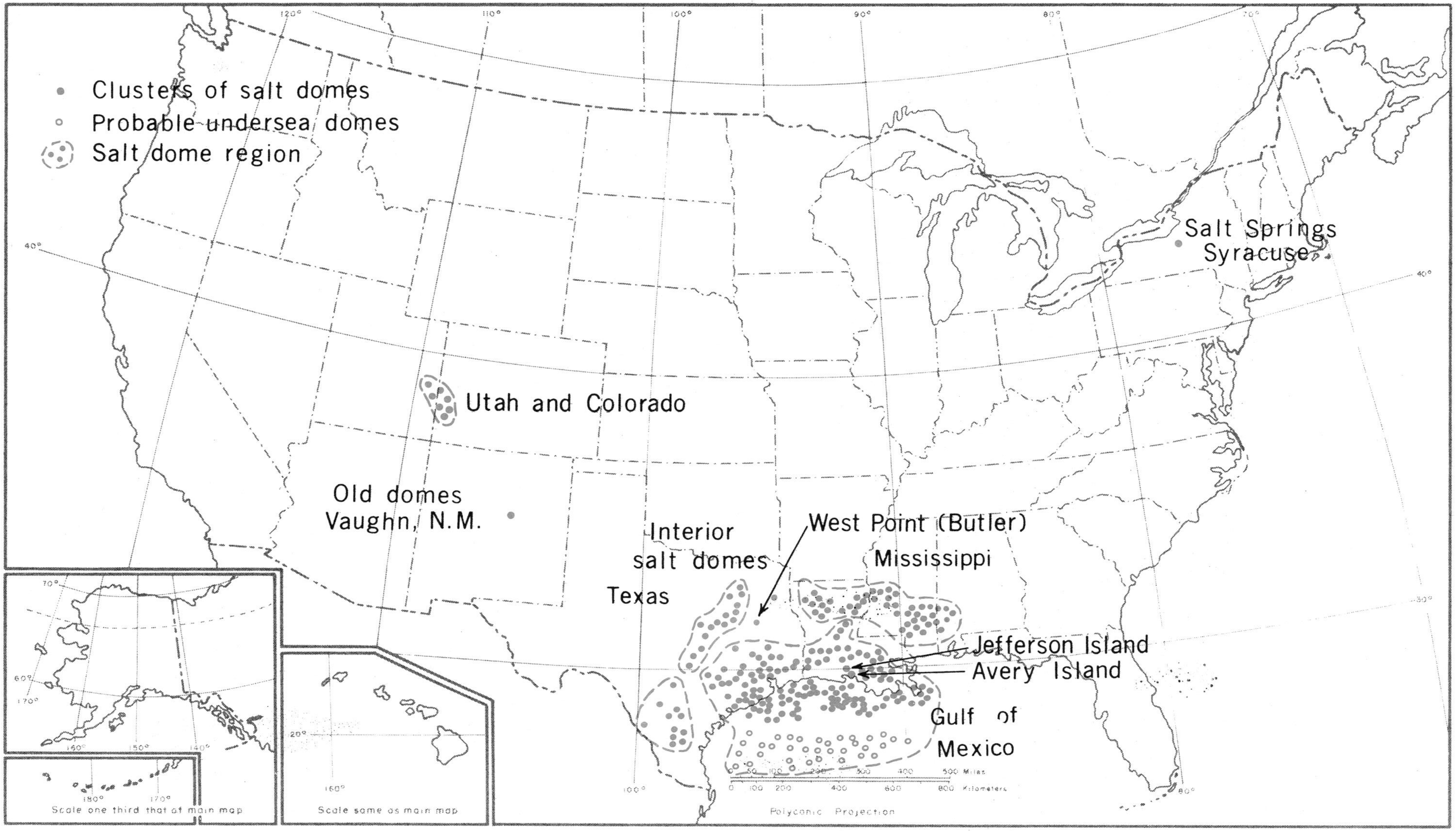

Clusters of salt domes
Probable undersea domes
Salt dome region

Salt Springs
Syracuse

West Point (Butler)
Mississippi

Jefferson Island
Avery Island

Gulf of
Mexico

Interior
salt domes
Texas

Utah and Colorado

Old domes
Vaughn, N.M.

Polyconic Projection

0 100 200 300 400 500 600 700 800 Kilometers
0   100   200   300 400 500 Miles

Scale same as main map

Scale one third that of main map

MAP 2-5   SALT DOMES AND MUD VOLCANOES OF THE UNITED STATES

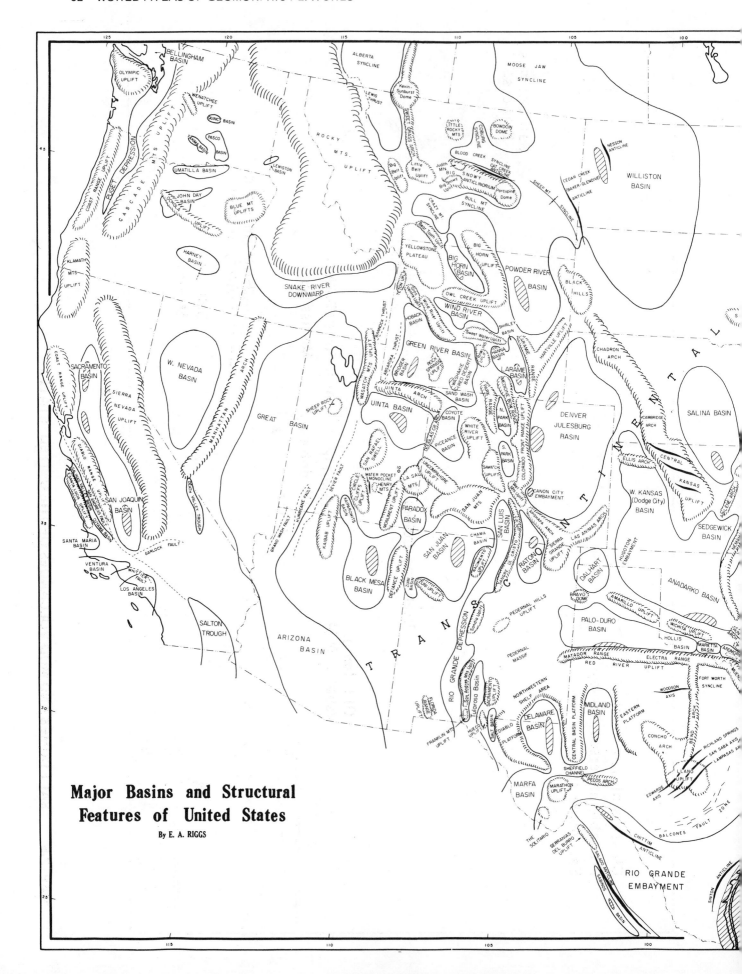

Major Basins and Structural Features of United States

By E. A. RIGGS

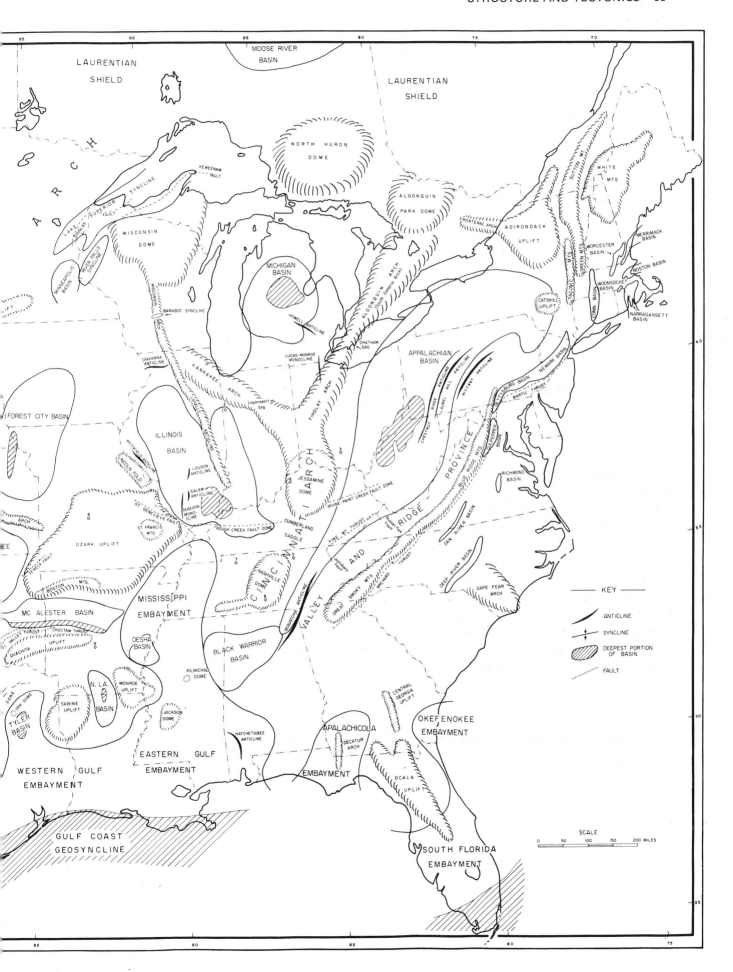

**MAP
2-6**

MAJOR BASINS AND STRUCTURAL FEATURES OF THE
UNITED STATES

When I prepared an outline for this atlas a world map showing the distribution of structural domes and basins was included. However, after much research and discussions with colleagues, I found that there was not enough data available to produce a reliable world map at this time. Domes and basins constitute a major pattern of deformation of the earth's crust, but the location and recognition of these features is difficult, and a world map would be generalized to the point of inaccuracy. However, for the United states such a map has been prepared (Riggs, 1960). Although Riggs' map is generalized and in places may be rather inaccurate and overexaggerated, still it presents an overall view of major domes and basins found in the United States.

Source: Riggs, E. A., *Major Basins and Structural Features of United States,* C.S. Hammond and Company, Maplewood, New Jersey, 1960.

MAP
2-7

FOLDED MOUNTAIN RANGES

One of the major ways that crustal deformation takes place is through folding. In large areas of the earth, especially around the borders of the major continentel blocks, stresses have been applied to the crust so slowly, or else the rocks have been so confined by the weight of overlying materials, that instead of breaking they have bent or buckled. Some materials act much more like plastic than others depending upon temperature and pressure constraints. Broad belts of the crust have been thrown into systems of folds that range from small wrinkles to great wave-like structures measuring many miles from crest to crest and thousands of feet from top to base. An anticline is the arch or crest of a simple fold while the trough is a syncline. A one-sided simple fold is a monocline or flexure. In some areas the folds are tightly jammed together, or even overturned and piled on top of one another in structures of remarkable complexity. Most of the great folded mountain areas (Alps, Himalayas, Zagros of Iran, Appalachains, and Rockies) are extremely complex. This map shows regions of crustal folding and mountain building covering approximately the last 600 million years of earth history. Most of the present high mountain chains were formed during the Cenozoic Folding (Umbgrove, 1947). Folded mountain ranges follow certain belts of structural instability. Notice that a belt of high folded mountains surrounds the Pacific Ocean. The main ranges include the Alaskan Ranges, the Canadian Coastal Mountains, the Cascades, the Sierra Nevada, the Sierra Madre Occidental of Mexico, the

Andes, the New Zealand Alps, and numerous ranges of Indonesia, the Philippines, Formosa, Japan, and Kamchatka. A second major system of mountains crosses the southern portion of the Eurasian continent north of the three major peninsulas of Iberia, Arabia, and India. Although the general trend of this transverse belt is roughly toward the east-southeast or east-northeast. This general belt of cordilleras (a group of mountain ranges) includes the Pyrenees, the Alps, the Balkan Ranges, the Pontus and Taurus mountains of Turkey, the Elburz, the Hindu Kush, and the Himalayas. Several ranges offset from this belt may also be considered part of it: the Atlas Mountains of North Africa, the Carpathians, the Caucasus, and the very high ranges radiating from the Pamir Knot (often called the "roof of the world") - the region where Tibet, West Pakistan, Afghanistan, and the Soviet Union come together. A third group of mountains can be considered roots of old mountain ranges formed during the period of Paleozoic folding. At one time these mountain ranges were much higher, but have since been eroded down, still, some of these ranges are high enough and rugged enough to be considered true mountains. The Great Smokies and the Blue Rige mountains of the Appalachian chain are such mountains. Other old mountain areas of Paleozoic folding are parts of the Grampian Mountains of Scotland and the rugged ranges bordering Greenland and Norway. High rugged mountain ranges are unusual in the interior of continents. Exceptions are the mountains of east-central Eurasia, the Urals, and the interior of Canada and the United States.

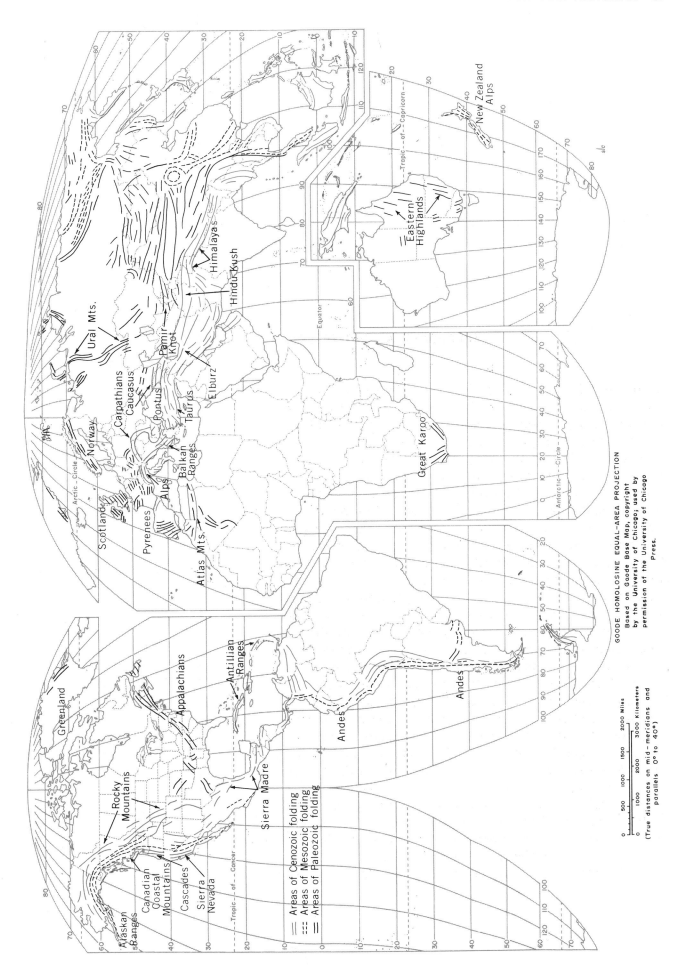

GOODE HOMOLOSINE EQUAL–AREA PROJECTION
Based on Goode Base Map, copyright
by the University of Chicago; used by
permission of the University of Chicago
Press.

MAP 2-7   FOLDED MOUNTAIN RANGES

**MAPS
2-8
2-9**

FAULT ZONES AND RIFT VALLEYS

The process of faulting is one of the major ways in which land-forms are produced. (A fault is a slipping of blocks of the earth's crust along great ruptures or cracks.) The crust of the earth is subject to a number of tensions and stresses, and some-times it breaks. This is proving to be more and more true as new evidence on the shifting of continents and ocean basins is re-vealed. Few areas of the world are completely free of the stresses taking place, and faulting is very widespread. This map shows the main fault zones. Faults are usually grouped into three broad classes - normal faults, thrust faults, and strike-slip faults. Normal faults, produced mainly by tensional stresses, are characterized by much steeper angles than are thrust faults. Thrust faults, resulting from compressional stresses, occur at very low angles, which means that one rock mass may slide many miles over another. A strike-slip fault occurs when there is horizontal relative movement between two blocks. Where rapid and relatively recent faulting has produced a great amount of vertical displacement, the resulting formation may produce a fault escarpment, a sharp slope, or cliff. Usually, the most prominent landforms are sheer cliffs that have been formed along normal faults rather than along thrust faults. The United States has excellent examples of faulting, so many, in fact, that only the largest and most famous can be mentioned. The San Andreas fault system, which runs for more than 600 miles through California, is one of the most famous strike-slip faults in the world. The Basin and Range Province of the western United States is an area where features produced by faulting dominate the topography. It is bounded by the steep eastern wall of the Sierra Nevada of California and the abrupt western edge of the Colorado Plateau in Utah and Arizona. Two major fault lines are the abrupt western border of the Wasatch Mountain near Salt Lake City, Utah, and the major fault system along the eastern face of the Northern Rocky Mountains in Glacier National Park, Montana. A very long fault is the Balcones Escarpment, which forms a steep cliff facing the coastal plain in south-central Texas.

Distinctive landforms are associated with faulting. Where a succession of tensional faults occurs, a series of large down-dropped troughs, or basins (rift valleys) may result. Rift valleys (also known as "grabens," a German word meaning a grave or ditch) may develop from two major tensional faults on each side of a down-sloped valley; or the rift may result from compressional stresses in which the depressed central block of the graben was pushed downward, and erosion may have prevented the development of an overhang along the sides. The longest and deepest rifts in the world occur in east Africa and central Asia. In east Africa, from Lake Malawi on the south through the Red Sea and into the Jordan Valley, there is a succession of rifts with walls several thousand feet high where they cross the highlands of Tanzania, Uganda, Kenya and Ethiopia. Lakes Tanganyika, Albert, and Rudolf occupy sections of this great rift system. The lakes are usually very deep, and Lake Baykal (Baikal), which occupies a rift valley in Siberia, is the deepest lake in the world (5,712 feet/1,741 meters). One of the best known grabens in North America is Death Valley in California, whose floor is 282 feet (86 meters) below sea level. Other examples of grabens include the middle Rhine Valley, between the Vosges Mountains on the west and the Schwarzwald (Black Forest) on the east, the Upper Vardar Valley in Yugoslavia, and the Dead Sea with an elevation of 1,286 feet (392 meters) below sea level.

MAP 2-8   FAULT ZONES AND RIFT VALLEYS OF THE WORLD

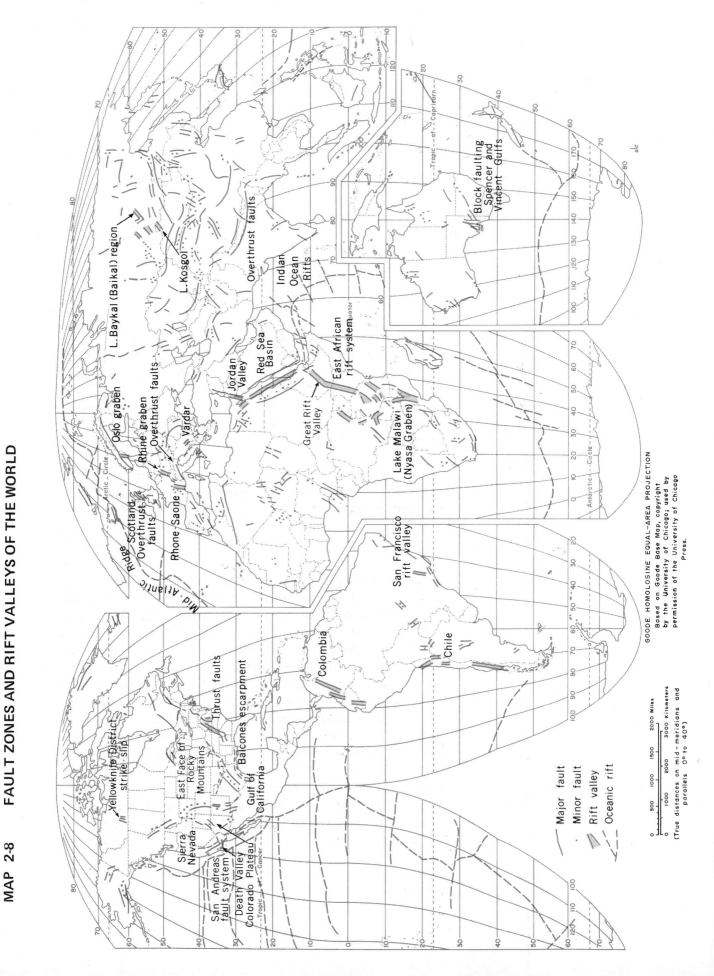

GOODE HOMOLOSINE EQUAL−AREA PROJECTION
Based on Goode Base Map, copyright
by the University of Chicago; used by
permission of the University of Chicago
Press.

Major fault
Minor fault
Rift valley
Oceanic rift

0   500   1000   1500   2000 Miles
0   1000   2000   3000 Kilometers

(True distances on mid−meridians and
parallels 0° to 40°)

Ridge
Scotland
Overthrust
faults
Oslo graben
Rhine graben
Overthrust faults
Vardar
Rhone−Saone
Mid−Atlantic
L. Baykal (Baikal) region
L. Kosgol
Overthrust faults
Jordan
Valley
Red Sea
Basin
Indian
Ocean
Rifts
East African
rift system
Great Rift
Valley
Lake Malawi
(Nyasa Graben)
Block faulting
Spencer and
Vincent Gulfs

Yellowknife District
strike−slip
East Face of
Rocky
Mountains
Thrust faults
Balcones escarpment
Gulf of
California
Colorado Plateau
Death Valley
San Andreas
fault system
Sierra
Nevada
Colombia
Chile
San Francisco
rift valley

Arctic Circle
Tropic of Cancer
Tropic of Capricorn
Equator
Antarctic Circle

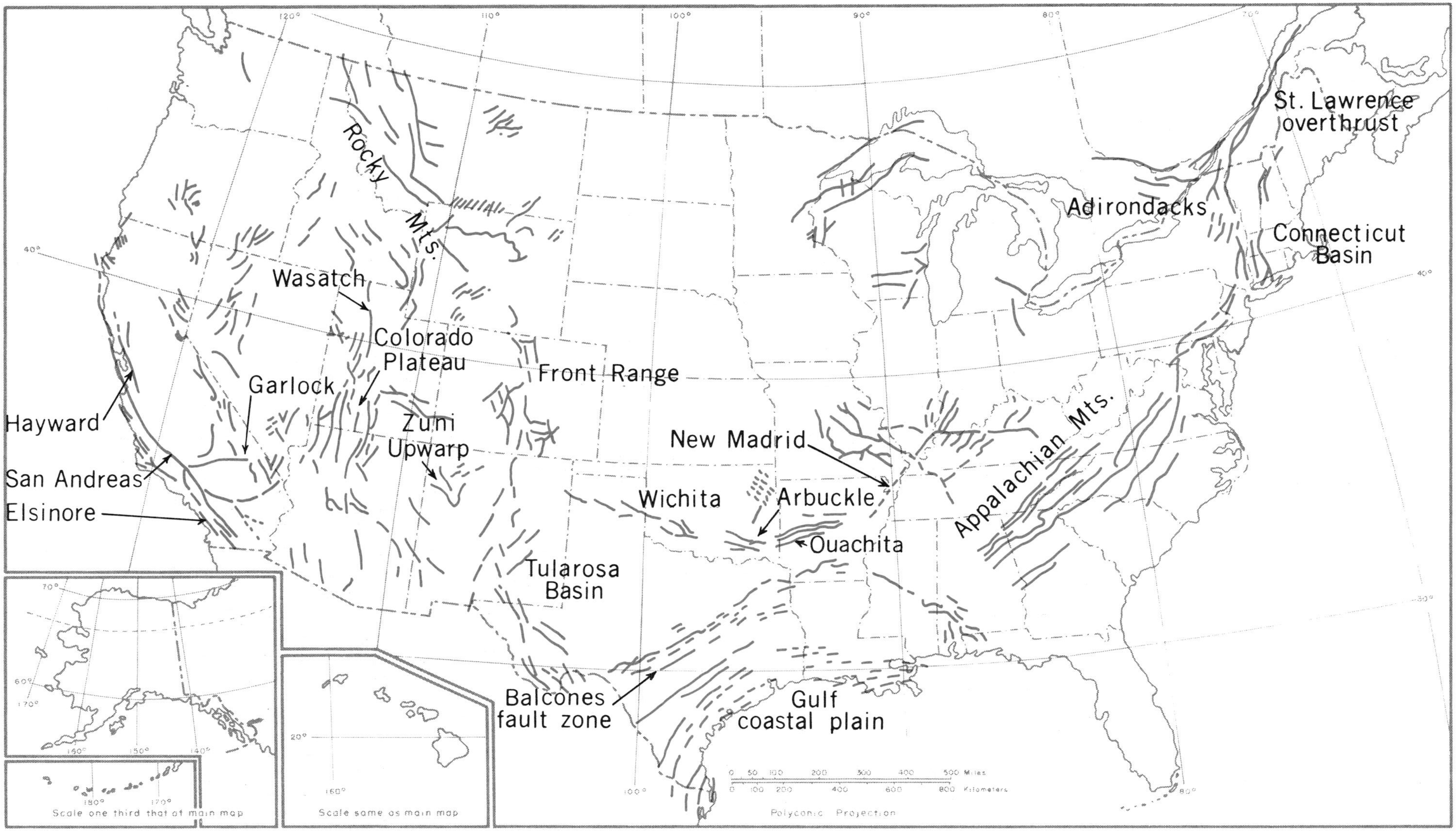

MAP 2-9   FAULT ZONES AND RIFT VALLEYS OF THE UNITED STATES

St. Lawrence overthrust

Connecticut Basin

Adirondacks

Appalachian Mts.

New Madrid

Arbuckle

Ouachita

Wichita

Gulf coastal plain

Front Range

Tularosa Basin

Balcones fault zone

Rocky Mts.

Wasatch

Colorado Plateau

Zuni Upwarp

Garlock

Hayward

San Andreas

Elsinore

Polyconic Projection

0   50 100   200   300   400   500 Miles
0 100 200   400   600   800 Kilometers

Scale same as main map

Scale one third that of main map

MAP
2-10

MAJOR EARTHQUAKES

Although earthquakes appear to be widespread, certain belts or zones exist where they are most frequent. Earthquakes occasionally occur in some of the most stable regions of the earth, such as the earth tremors in eastern Canada in 1927, but these are rare. Sections of Africa and Australia are relatively free of earthquakes, whereas Peru and Chile constitute one of the most active earthquake regions of the world. The map depicts where and when major earthquakes have occurred. The list at the end of the text gives the location, date, and number of deaths where known. The most active seismic zones are along Plate boundaries and subduction zones associated with island and mountain arcs. The island arcs of the Aleutians, the Kurile-Kamchatka arcs, and the chain of islands extending south through Japan, Okinawa, Taiwan, Philippines, and Indonesia to the small islands south of New Zealand are all subject to frequent earthquakes (Hodgson, 1964). At the times the tremors in this region are associated with volcanism, forming the so-called "ring of fire" around the Pacific. This system includes the Andean seismic region of South America with great ocean depths just offshore. In Chile, in 1960, a series of shocks lasted from May until December, produced by uplifting and down-dropping of large sections of the coast. A second type of earthquake zone is related to block faulting. Movement along the San Andreas Fault in California caused the great San Francisco earthquake of 1906, and it is highly probable that severe earthquakes will continue in this region. Block faulting creates earthquakes in the Philippines, in New Zealand, the Mediterranean region and Turkey. A third type of earthquake zone occurs along mid-ocean ridges. One such ridge follows the center of the Atlantic Ocean from north of Iceland to Antarctica. Similar ridges can be followed around Africa into the Indian Ocean, and there are also several long ridges in the Pacific, all shown by·dashed lines. Earthquakes are associated with some but not all of these ridges; this association appears to depend on how active the movements are along the ridges. The mid-Atlantic ridge is very active. Earthquakes are also quite frequent where mountains are being formed. The eastern Himalayas, especially in northern Assam, are a high seismic area, as are some of the mountains in central Asia and in West Pakistan. Earthquakes are occasionally associated with isolated volcanoes far away from active mountain-building regions. This is especially true where large amounts of material are rapidly ejected from a cone. The active volcanoes of Hawaii, the West Indies, and Italy often have created

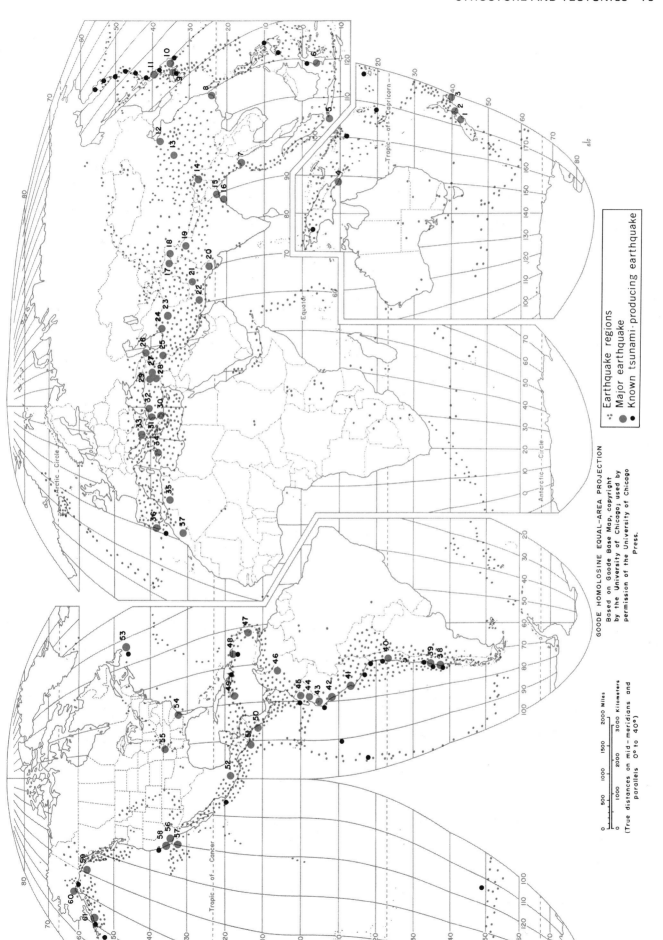

GOODE HOMOLOSINE EQUAL-AREA PROJECTION
Based on Goode Base Map, copyright
by the University of Chicago; used by
permission of the University of Chicago
Press.

Earthquake regions
Major earthquake
Known tsunami-producing earthquake

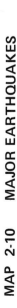

0   500   1000   1500   2000 Miles

0   500   1000   2000   3000 Kilometers

(True distances on mid-meridians and
parallels 0° to 40°)

MAP 2-10   MAJOR EARTHQUAKES

THE FOLLOWING IS A LIST OF THE MAJOR EARTHQUAKES THAT HAVE OCCURRED BETWEEN 1531 and 1976:

	LOCATION	YEAR	DEATHS WHERE KNOWN
1.	New Zealand, Murchison	June 17, 1927	---
	New Zealand, Inangahua	1968	---
2.	New Zealand, Wellington/Marlborough	1855	---
3.	New Zealand, Hawke's Bay	February 3, 1931	---
4.	Papua Territory, New Guinea	January 18-21, 1951	3,000
5.	Krakatoa, Sumatra/Java	1883	36,000
6.	Sulawesi/Celebes, Indonesia	February 23, 1964	
7.	Burma, Pegu	May 5, 1930	---
8.	China, Kansu	December 16, 1920	180,000
9.	Japan, Sanrihu (sea wave)	June 15, 1896	22,000
10.	Japan, Tokyo	September 1, 1923	143,000
11.	Japan, Mino-Owari	October 28, 1891	7,300
12.	China, T'ien-ching (Tienstsin)	July 7, 1969	3,000
	China, Tangshan	August, 1976	
13.	China, Shensi	January 24, 1556	830,000
14.	India, Assam	June 12, 1897 / August 15, 1950	20,000-30,000
15.	Indian Earthquake	January 15, 1934	---
16.	India, Calcutta	1960	---
17.	India, Rann of Cutch	June 16, 1819	---
18.	India, Kangra	April 14, 1905	20,000
19.	Pakistan/India, Kashmir	1975	---
20.	Pakistan (9 towns)	December 28, 1974	5,200
21.	Pakistan, Quetta	1935	50,000
22.	Iran, Southern	April 10, 1972	5,057
23.	Iran, Northeastern	August 31, 1968	11,588
24.	Iran, Northern	July 2, 1957	2,500
25.	Iran, Northwestern	September 1, 1962	10,000
26.	U.S.S.R. Caucasia, Shemaka	November, 1667	80,000
27.	Turkey, Eastern	August 19, 1966	2,529
28.	Turkey, Eastern, City of Lice	September 6, 1975	2,319
29.	Turkey, City of Erzingan	December 27, 1939	---
30.	Turkey, Western	March 28, 1970	1,086
31.	Turkey, Northwestern	March 18, 1953	1,200
32.	Turkey, Northwestern	April 29-30, 1968	15,000
33.	Yugoslavia, Skopje	July 26, 1963	1,100
34.	Italy, Catania	January 11, 1693	60,000
	Italy, Calabria	February 4, 1783	50,000
	Italy, Messina	December 28, 1908	100,000
	Italy, Avezzano	January 13, 1915	29,970
	Italy, Naples	1857	---
35.	Algeria, Northern	September 9-12, 1954	1,657
36.	Portugal, Lisbon	November 1, 1755 / January 26, 1531	60,000 / 30,000
37.	Morocco, Agadir	February 29, 1960	12,000
38.	Chile, South-Central	May 21-30, 1960	5,700
	Chile, Chillan	January 24, 1939	30,000
39.	Chile, Valparaiso	August 16-17, 1906	1,500
40.	Chile, North-Central	July 9, 1971	90
41.	Peru, Lima	May 16, 1868	16,000
42.	Peru, Northern	May 31, 1970	66,794
43.	Ecuador/Peru	August 13-15, 1868	25,000
44.	Ecuador, Pelileo	1949	---
45.	Ecuador, Quito	February 4, 1797	41,000
46.	Columbia/Venezuela	May 16, 1875	16,000
47.	Matinique/West Indies	May 8, 1902	40,000
48.	Puerto Rico	1918	---
49.	Jamaica	June 7, 1692 / January 14, 1907	---
50.	Nicaragua	December 23, 1972	6,000
51.	Guatemala	1976	---
52.	Mexico, Central	August 28, 1973	700
53.	Canada, Eastern Grand Banks	1929	---
54.	South Carolina, Charleston	1886	
55.	Missouri, New Madrid	December 16, 1811 / January 23, 1812 / February 7, 1812	
56.	California, Kern County	1958	---
57.	California, Long Beach	March 10, 1933	115
58.	California, San Francisco	April 18-19, 1906	452
59.	Alaska, Anchorage	March 27, 1964	131
60.	Alaska, Yakutat Bay	September 3, 1899 / October 1899	
61.	Alaska, Hegen Lake	1959	

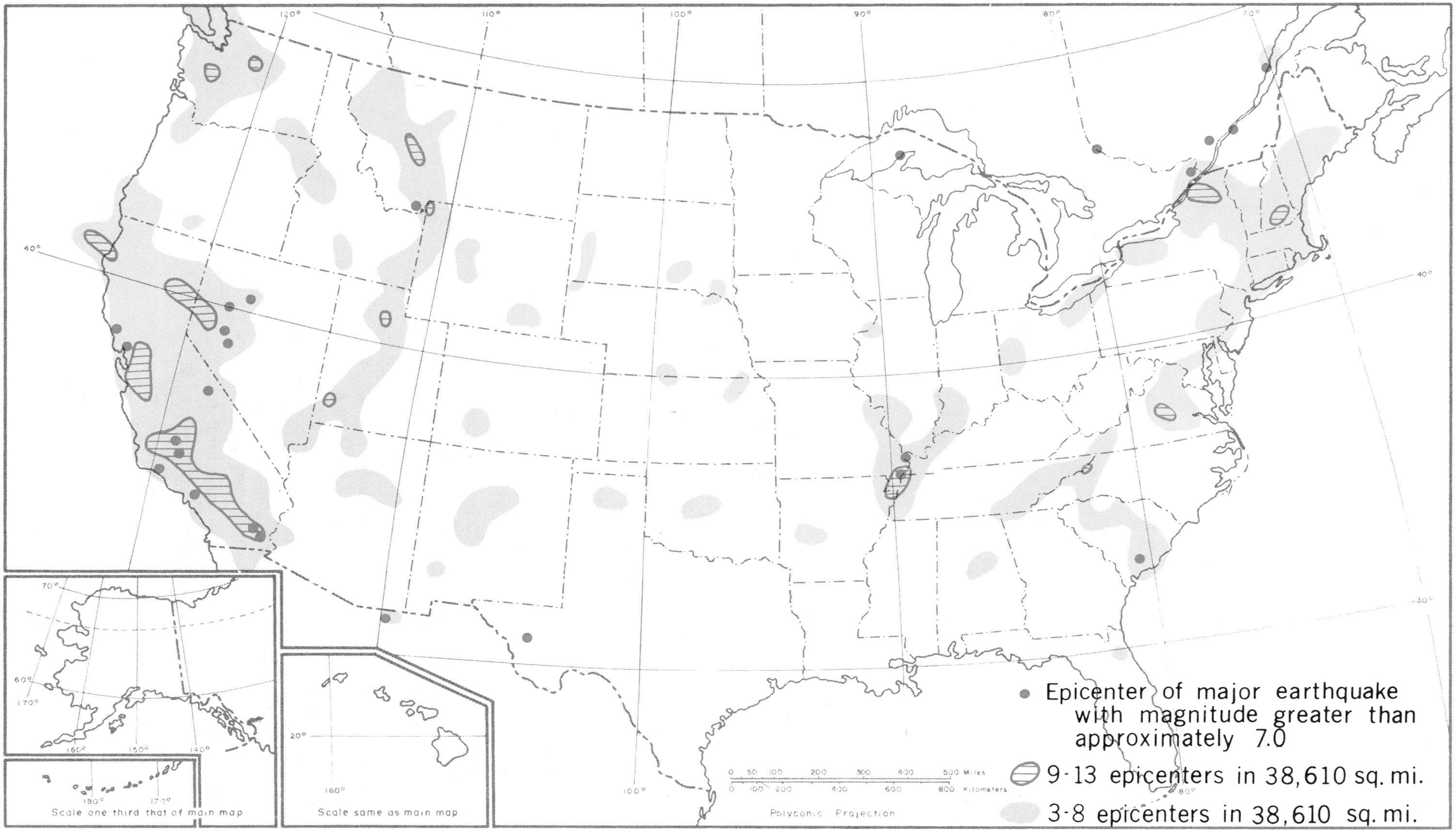
Epicenter of major earthquake with magnitude greater than approximately 7.0
9-13 epicenters in 38,610 sq. mi.
3-8 epicenters in 38,610 sq. mi.
0 50 100 200 300 400 500 Miles
0 100 200 400 600 800 Kilometers
Polyconic Projection
Scale one third that of main map
Scale same as main map
120°
110°
100°
90°
80°
70°
40°
30°
20°
160°
150°
140°
180°
170°
60°

MAP 2-11 EARTHQUAKE EPICENTERS AND REGIONS OF THE UNITED STATES tremors with their explosions. Plate 4. (centerfold) is a color infrared space photograph of southern California. Because of the numerous tectonic movements taking place, this a very active region geologically for faulting and earthquakes.

A larger red dot is used to indicate locations where 54 major earthquakes have occurred. The term major is used for the most destructive quakes in recent times.

**MAP
2-11**

EARTHQUAKE EPICENTERS AND REGIONS OF THE UNITED STATES

The United States map shows areas of tectonic activity indicated by recorded earthquake epicenters as of 1957 (Woollard, 1958; Wright and Frey. 1965). Earthquakes have occurred in four major regions. By far the largest number occur in the Pacific region adjacent to the San Andreas effects. The San Andreas fault broke over a length of 200 miles [322 kilometers] (with a maximum horizontal displacement of 21 feet [6 meters] on the shore near Tomales Bay!) The second region is the Rocky Mountains, and the third is in the Mississippi Valley, centered on New Madrid, Missouri. In this latter area there have been frequent small earthquakes over the past 150 years, with a major occurence in 1811 to 1812. Such drastic changes took place on the surface of the earth during this earthquake that it is regarded by many as being the most serious series of earth movements in United States history. The fourth region is in the eastern United States and Canada, with one epicenter near Charleston, South Carolina, and another epicenter on the Grand Banks, off the Canadian coast. The Grand Banks earthquake in 1929 occurred under the sea. Most of the earthquakes in the northeast appear to result from very slow structural rebound that goes back to the release of weight from glacial ice. Most seismic shocks of this kind are minor (Leet, 1942).

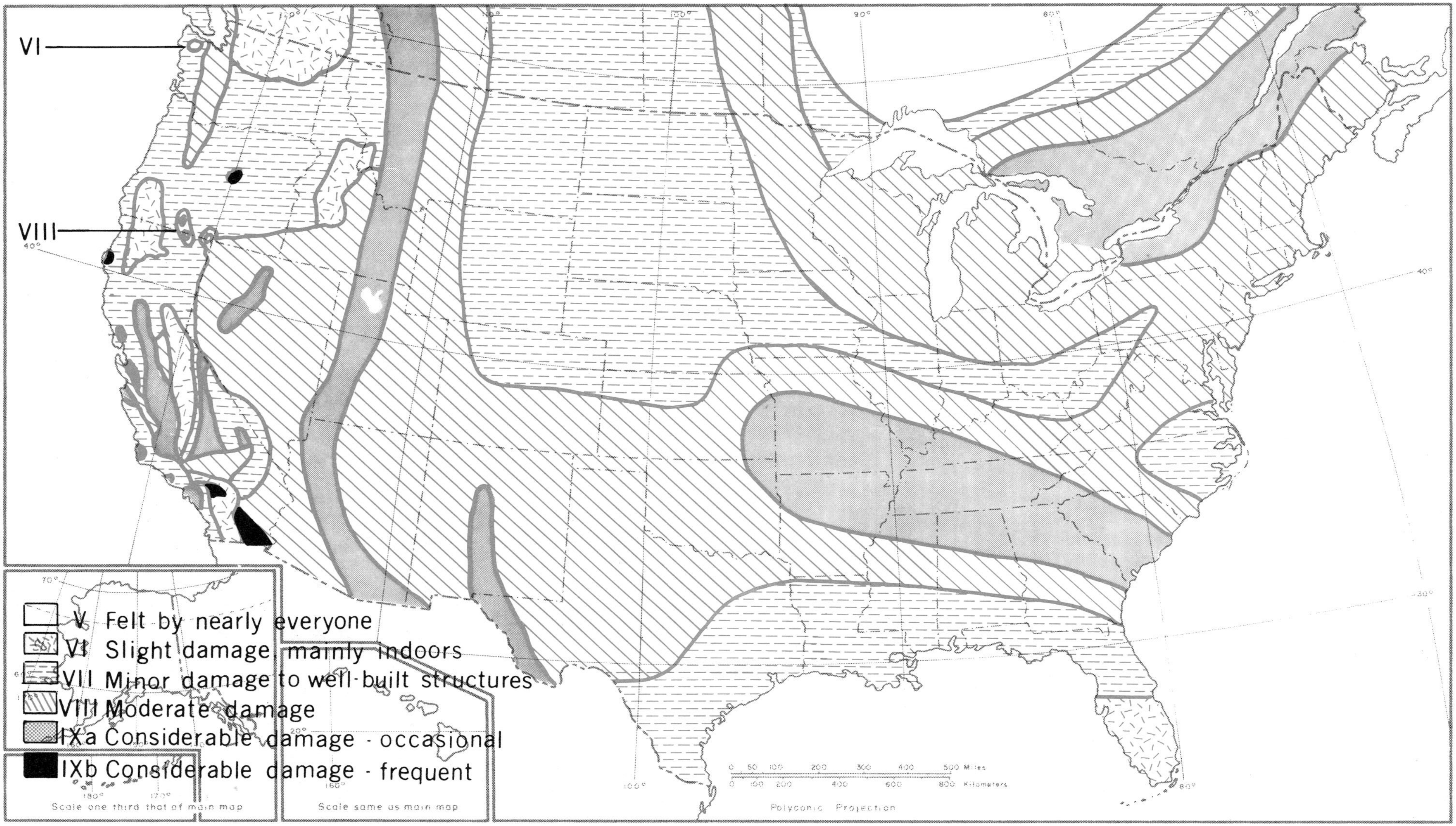

	V	Felt by nearly everyone
	VI	Slight damage mainly indoors
	VII	Minor damage to well-built structures
	VIII	Moderate damage
	IXa	Considerable damage - occasional
	IXb	Considerable damage - frequent

Scale one-third that of main map

Scale same us main map

0 50 100   200   300   400   500 Miles
0 100 200   300   400   600   800 Kilometers

Polyconic Projection

MAP 2-12   SEISMIC REGIONALIZATION OF THE UNITED STATES AND ADJACENT PARTS OF CANADA

**MAP
2-12**

SEISMIC REGIONALIZATION MAP FOR THE UNITED STATES AND ADJACENT PARTS OF CANADA

This map draws upon a modified Mercalli Scale compiled by C.F. Richter indicating the intensity of earthquakes which have occurred and could occur across the United States and adjacent parts of Canada. This map can be compared with Map 2-10 showing epicenters of major earthquakes.

One must realize the limitations of such generalized maps. The precise drawing of lines is largely a matter of guesswork and the writer agrees with Richter when he states that he "hopes no one will go to the length of enlarging the map and using it to assign different risk values to towns on the same type of ground five miles apart, simply because on the enlarged scale the line appears to pass between."

Only Scales V through IX are shown on the map because the first four are too minor to plot. The following is a complete listing of Mercalli scale as it appears in the Hodgson (1964) volume.

The Modified Mercalli Scale

I. Not felt except by a very few under especially favorable circumstances.

II. Felt only by a few persons at rest, especially on upper floors of buildings. Delicately suspended objects may swing.

III. Felt quite noticeably indoors, especially on upper floors of buildings, but many people do not recognize it as an earthquake. Standing motor cars may rock slightly. Vibration like passing truck. Duration estimated.

IV. During the day felt indoors by many, outdoors by few. At night some awakened. Dishes, windows, doors disturbed; walls make creaking sound. Sensation like heavy truck striking building. Standing motor cars rocked noticeably.

V. Felt by nearly everyone; many awakened. Some dishes, windows, etc., broken; a few instances of cracked plaster; unstable objects overturned. Disturbances of trees, poles, and other tall objects sometimes noticed. Pendulum clocks may stop.

VI. Felt by all; many frightened and run outdoors. Some heavy furniture moved; a few instances of fallen plaster or damaged chimneys. Damage slight.

VII. Everybody runs outdoors. Damage negligible in buildings of good design and construction; slight to moderate in well-built ordinary structures; considerable in poorly built or badly designed structures; some chimneys broken. Noticed by persons driving motor cars.

VIII. Damage slight in specially designed structures; considerable in ordinary substantial buildings, with partial collapse; great in poorly built structures. Panel walls thrown out of frame structures. Fall of chimneys, factory stacks, columns, monuments, walls. Heavy furniture overturned. Sand and mud ejected in small amounts. Changes in well water. Disturbs persons driving motor cars.

IX. Damage considerable in specially designed structures; well-designed frame structures thrown out of plumb; great in substantial buildings, with partial collapse. Buildings shifted off foundations. Ground cracked conspicuously. Underground pipes broken.

X. Some well-built, wooden structures destroyed; most masonry and frame structures destroyed with foundations; ground badly cracked. Rails bent. Landslides considerable from river banks and steep slopes. Shifted sand and mud. Water splashed over banks.

XI. Few, if any, (masonry) structures remain standing. Bridges destroyed. Broad fissures in ground. Underground pipelines completely out of service. Earth slumps and land slips in soft ground. Rails bent greatly.

XII. Damage total. Waves seen on ground surfaces. Lines of sight and level distorted. Objects thrown upward into the air.

Because of limited space, the legend has been condensed on the map. Note that IX is broken into two parts, occasional and frequent, because of the severity of these earthquakes.

This intensity scale is obviously of limited value. The description "many frightened" doesn't mean very much because a shock which would alarm the residents of Toronto or Cleveland would pass unnoticed in Tokyo. Also, the damage to buildings varies with soil conditions and to some extent with the standard of "good" building in the particular country involved that the intercomparison of earthquakes in different countries obviously doesn't mean very much. But when the scale is applied uniformly over an area to describe the effects of a single earthquake, it does appear to have a good deal of value.

Source: J. H. Hodgson (1964) *Earthquakes and Earth Structures,* Prentice-Hall, Inc., pp. 58-59; 132; 140.

MAP
2-13

VOLCANOES

Compare the maps of earthquakes, (Map 2-10) midoceanic ridges and island arcs, (Map 2-1) and fault zones (Map 2-8) with the map showing the distribution of major volcanoes; there are close correlations, since many volcanoes follow areas corresponding to the zones of weakness in the crust of the earth. Because volcanoes are found along cracks or fissures in the earth's crust, severe earthquakes are often associated with their eruptions. The seismograph station close to Kilauea volcano on the island of Hawaii, for example, is used to indicate when an eruption is about to take place. Since there are 516 known active volcanoes and over 289 known inactive volcanoes in the world today, it would be difficult to plot on the world map all of these features. After checking several sources the author selected 161 better known volcanoes. These are shown with a dot and a small number. The name, location, and height of each volcano is given in the list below. An active volcano is one that either is still erupting volcanic ejecta or has smoke and/or gases emitting from its cone or sides. Many volcanoes are inactive today but are not considered dead as long as there is some heat escaping from vents or fissures. Thermal springs and geysers are good evidence that a volcanic region still has the potential to erupt again. A dormant volcano is one that has been known to erupt in the past or shows evidence of activity but is temporarily devoid of external activity at the present time.

Most volcanoes occur in broad zones or belts along island arcs or major fracture zones. Broad regions where active volcanoes can be found are shown on the world map as dashed lines and a circled number represents the total active arcs to be found in that region. The major volcanic belts are indicated in red. These areas are where the most active volcanism can be found.

There have been four active volcanoes in Siberia (Continental Asia) but little is known about them. They were reported to have erupted between 1300 and 1770. An unnamed volcano in Tibet had rising smoke and stones thrown out in 1951. Two volcanoes in Antarctica can not be shown because of the map projection. Buckle Island (66 48' S; 163 15' E) 4,063 feet (1,238 meters) erupted in 1899 and Erebus (77 35' S; 167 10' E), 13,200 feet (4,023 meters) high, was a subglacial eruption on Ross Island in 1947. Also, not included on the world map are the numerous young volcanoes, many of them seamounts in the Pacific basin. Hundreds of these submarine cones erupt below the surface of the ocean and do not appear above sea level.

A major concentration of volcanoes is associated with the zone of weakness that surrounds the Pacific Ocean (the "ring of fire"), and another concentration that extends from the Caribbean, via the Azores, to the Mediterranean, then eastward through Turkey to the great mountain chains of central and southeastern Asia. Crossing this east to west chain is the mid-Atlantic ridge, along which volcanoes are found from Iceland through the Azores and Canary Islands to Tristan da Cunha in the South Atlantic Ocean, many of them being submerged under the sea. The volcanoes in east-central Africa and on several of the mid-Pacific islands occur individually and not as parts of "belts." Landforms associated with volcanic activity vary with the type of volcanic action. Three dominant types of eruptions are recognized: quiet, intermediate, and explosive. The violence of eruptions ranges from slow out-pourings to enormous, cataclysmic explosions. The fissure flow, the quiet type of volcano, results from the rise of highly fluid lavas to the surface; this kind of volcano produces large lava flows. An intermediate type is the shield or dome volcano. The third type of volcano, the explosive one, may suddenly erupt and eject several cubic miles of earth material. Most of the volcanoes shown are the explosive type, best represented by high volcanic cones such as Mount Fuji in Japan. The explosive volcanoes are built up, layer on layer, by the eruption of volcanic ash, cinders, or other fragmented material. Perfectly symmetrical cones (for example, Mount Mayon in the Philippines) are not common. More frequently, high volcanic peaks are shattered and blown apart by violent explosions, so that many volcanoes are fragments of cones or are cones within cones. Mount Vesuvius, in Italy, is an example of one crater within another (nested craters). In areas where explosive eruptions are frequent, many cones of different sizes can be found, ranging from small cinder cones about a hundred feet high to such towering volcanic peaks as Popocatepetl, Ixtaccihuatl, and Orizaba, all of which rise 17,000 feet (5,182 meters) or more above sea level on the Mexican tableland. Some cones may form extremely rapidly: Paricutin, an ash cone in west-central Mexico, appeared suddenly on February 20, 1943, in a farmer's cornfield. It grew into a mountain peak with a few weeks, and although its rate of growth is much slower today, Paricutin has reached a height of more than 1,500 feet (457 meters).

There are many volcanic cones in the United States, especially in Hawaii, Alaska, and the western states. Mounts Shasta, Pitt, Jefferson, Hood, Rainier and St. Helens are large volcanoes that are inactive but not extinct. Mount Baker has given signs of renewed activity. Mount Lassen, in northern California, erupted last in 1914 and continues to emit gases. Mount Katmai, in Alaska, has had a number of explosive eruptions, several within the last few years. Instead of building up, its cone collapsed, resulting in a huge crater (caldera) nearly three miles (5 meters) long and two miles (3 meters) wide. Another region of active volcanism is central America from Guatemala south to Panama. Irazu and Poas volcanoes are very active at the present time. Volcanic cones create some of the highest mountains in the world. The Andes are dominated by volcanic peaks of impressive height: for example, Mount Aconcagua rises to a height of 22,835 feet (6,960 meters). Several Bolivian peaks exceed 20,000 feet (6,096 meters), and some near Lake Titcaca are active. Kilimanjaro, almost on the equator in Tanzania, Africa, rises above 19,000 feet (5,791 meters) into a region of perpetual snow.

Sources: The following are several of the major references consulted. G. A. MacDonald, *Volcanoes,* Prentice-Hall, Inc., Englewood Cliffs, N. J., 1972, Fig. 14-1, pp. 429-450; Newmann van Padang, M., Richards, A.F., Machado, F., Bravo, T., Baker, P.E., and LeMaitre, R. W., *Catalogue of the Active Volcanoes of the World including Solfatara Fields,* Part 21, Atlantic Ocean, International Association of Volcanology, Naples, 128p., 1967; *The World Almanac and Book of Facts 1976,* pp. 572-573; L. S. Dillon, "Neovolcanism: A Proposed Replacement for the Concepts of Plate Tectonics and Continental Drift," *Plate Tectonics Assessments and Reassessments,* The American Association of Petroleum Geologists, Tulsa, Oklahoma, 1974, 514p.; A Rittermann, *Vulkan und ihre Tatigkeit,* Ferdinand Enke Verlag, Stuttgart, 1936, pp. 162-163; R. R. Coats, "Volcanic Activity in the Aleutian Arc," *U. S. Geological Survey Bulletin,* 974-B, 1950, pp. 35-49; and G. S. Gorshkov, *Volcanism and the Upper Mantle; Investigations in the Kurile Island Arc,* Plenum Publishing Corporation, New York, 1970, 385p.

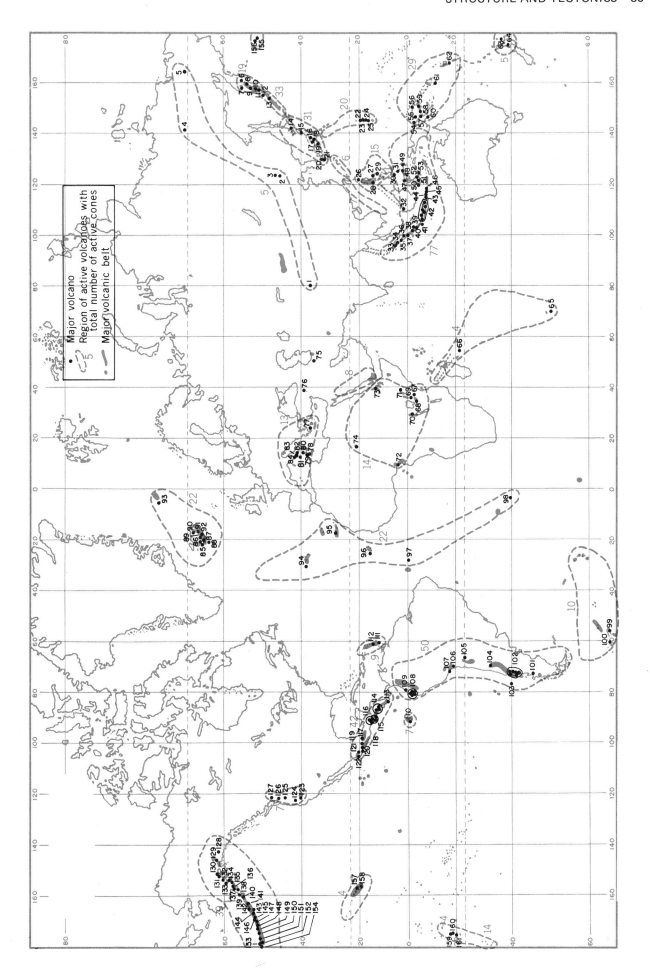

MAP 2-13 VOLCANOES

MAJOR VOLCANOES OF THE WORLD

The following is a list of 161 major volcanoes plotted on the world map. In the list below the letters next to the name of the volcano indicate: R = Rumbling; St = Steaming; E = Erupted (with year of last eruption); and D = Dormant.

CONTINENTAL ASIA

	NAME	LOCATION	HEIGHT IN FEET	HEIGHT IN METERS
1.	Tibet ("Smoke was rising and stones were thrown out") (E - 1951)	China	19,500	5,944
2.	Laoheishan (E - 1720-1722)	Siberia, USSR	1,730	527
3.	Huoshaoshan (E - 1720-1722)	Siberia, USSR	1,500	457
4.	Indigirsky (E - 1770)	Siberia, USSR		
5.	Anjuisky (E - between 1300 and 1700 A.D.)	Siberia, USSR	3,465	1,056

FAR EAST AND OCEANIA

	NAME	LOCATION	HEIGHT IN FEET	HEIGHT IN METERS
6.	Klyuchevskaya (R)	Kamchatka-Peninsula	15,584	4,750
	Plosky Tolbachik (E - 1941)	Kamchatka Peninsula	12,080	3,682
	Shiveluch (E - 1964)	Kamchatka Peninsula	10,771	3,283
	Bezymiannaya (Bezymianny) (E - 1961)	Kamchatka Peninsula	9,514	2,900
7.	Ichinskaya (E - 1901)	Kamchatka Peninsula	11,880	3,621
8.	Kronotskaya (D)	Kamchatka Peninsula	11,575	3,528
9.	Koryakskaya (E - 1957)	Kamchatka Peninsula	11,339	3,456
	Avachinskaya (St)	Kamchatka Peninsula	9,026	2,751
10.	Karymskaya (E - 1970)	Kamchatka Peninsula	5,560	1,695
11.	Zheltovskaya (E - 1972)	Kamchatka Peninsula	6,407	1,953
12.	Alaid (E - 1972)	Kuril Islands	7,662	2,335
13.	Sarychev	Kuril Islands	4,960	1,512

JAPAN

	NAME	LOCATION	HEIGHT IN FEET	HEIGHT IN METERS
14.	Me-Akan (E - 1966)	Hokkaido	4,931	1,503
15.	Bandai (E - 1888)	Honshu	5,968	1,819
16.	Asama (E - 1973)	Honshu	8,340	2,542
17.	Ontake (E - 1970)	Honshu	10,049	3,063
18.	Fugi (D)	Honshu	12,460	3,798
19.	Mihara (E - 1964)	Honshu	2,028	618
20.	Aso (E - 1970)	Kyushu	5,255	1,602

	NAME	LOCATION	HEIGHT IN FEET	HEIGHT IN METERS
21.	Kirishima (St)	Kyushu	5,610	1,710
	Sakurajima (E - 1948)	Kyushu	3,690	1,125

MARIANAS AND IZU ISLANDS

	NAME	LOCATION	HEIGHT IN FEET	HEIGHT IN METERS
22.	Farallon de Pajaros (E - 1952)		1,096	334
23.	Asuncion (St)		2,927	892
24.	Pagan (D)		1,870	570
25.	Alamagan (E - 1945)		2,441	744

PHILIPPINES

	NAME	LOCATION	HEIGHT IN FEET	HEIGHT IN METERS
26.	Didicas (E - 1952)	Didicas	900	274
27.	Mayon (E - 1968)	Luzon	8,077	2,462
28.	Taal (E - 1971)	Luzon	984	300
29.	Bulusan (E - 1966)	Luzon	5,115	1,559
30.	Apo (D)	Mindanoa	9,369	2,856

INDONESIA

	NAME	LOCATION	HEIGHT IN FEET	HEIGHT IN METERS
31.	Awu (E - 1966)	Sanjihe	6,102	1,860
	Api Siau (E - 1974)	Hulu	5,843	1,781
32.	Silawaih Agam (St)	Sumatra	5,695	1,736
33.	Burni Telong (E - 1924)	Sumatra	8,660	2,640
34.	Tindikat (St)	Sumatra	8,045	2,452
35.	Sorikmarapi (E - 1917)	Sumatra	7,080	2,158
36.	Marapi (D)	Sumatra	9,541	2,908
37.	Kerintji (St)	Sumatra	12,540	3,822
38.	Sumbing (E - 1926)	Sumatra	8,275	2,522
39.	Kaba (E - 1941)	Sumatra	6,440	1,963
40.	Dempo (St)	Sumatra	10,470	3,191
41.	Anak Krakatoa (E - 1960)	Sumatra	510	155

JAVA

No.	Volcano	Region	ft	m
42.	Semeru (E - 1968)	Java	12,130	3,697
	Slamet (E - 1967)	Java	11,325	3,452
	Raung (E - 1945)	Java	10,995	3,351
	Sundoro (D)	Java	10,397	3,169
	Tjerimai (E - 1938)	Java	10,155	3,095
	Gedeh (E - 1949)	Java	9,760	2,975
	Merapi (E - 1969)	Java	9,605	2,928
	Bromo (E - 1950)	Java	7,685	2,342
	Kelut (E - 1967)	Java	5,710	1,740
BALI				
43.	Agung (E - 1964)	Bali	10,370	3,161
	Batur (E - 1968)	Bali	5,665	1,727
LOMBOK				
44.	Rindjani (E - 1966)	Lombok	12,295	3,748
SUMBAWA				
45.	Tambora (D)	Sumbawa	9,410	2,868
46.	Sangeang Api (E - 1966)	Raba	6,430	1,960
47.	Una Una (E - 1960)	Kalolio	1,640	500
48.	Gamkonora (E - 1949)	Halmahera	5,364	1,635
49.	Ibu (Gunnuns) (D)		4,921	1,500
50.	Ija (E - 1969)		2,162	659
51.	Lewotobi Lakilaki (E - 1940)		5,217	1,590
52.	Sirung (E - 1947)		2,828	862
53.	Banda Api (D)		2,152	656
	Nila (E - 1932)		2,562	781
	Teon (D)		2,149	655
	Serua (D)		2,103	641
MELANESIA				
54.	Bam Island (D)	Bismarck Arch	1,969	600

No.	Volcano	Region	ft	m
55.	Manam (E - 1974)	Bismarch Arch	4,265	386
56.	Bagana (E - 1966)	Solomons	5,730	1,747
57.	Long Island (E - 1953)	Bismarch Arch	4,278	1,304
58.	Bamus (D)	New Britain	7,338	2,337
59.	Lolobau (D)	Bismarch Arch	3,058	932
60.	Lamington (E - 1951-56)	New Guinea	5,840	1,780
61.	Tina Hula (E - 1971)	Santa Cruz Island	3,000	914
62.	Lopoui (E - 1967)	New Hebrides	4,755	1,449
NEW ZEALAND				
63.	White Island (E - 1971)	New Zealand	1,053	321
64.	Ruapehu (E - 1974)	New Zealand	9,175	2,797
	Ngauruhoe (E - 1975)	New Zealand	7,515	2,291
	Tongariro (E - 1950)	New Zealand	6,495	1,980
INDIAN OCEAN				
65.	Big Ben (E - 1950)	Heard Island	9,007	2,745
66.	Piton de la Fournaise (E - 1973)	Reunion Island	8,680	2,646
AFRICA				
67.	Kilimanjaro (D)	Tanzania	19,340	5,895
68.	Meru (D)	Tanzania	14,979	4,566
69.	Ol Doinyo Lengai (E - 1955)	Kenya	9,442	2,878
70.	Nyiragongo (E - 1973)	Zaire	11,400	3,475
	Nyramlagira (E - 1971)	Zaire	10,028	3,057
71.	Telchi (E - 1922)	Kenya	2,120	646
72.	Cameroon Mt. (E - 1959)	Cameroon	13,350	4,069
73.	Erta Ale (E - 1973)	Ethiopia	1,650	503
74.	Tibesti (D)	Chad		
MIDDLE EAST				
75.	Demauind (D)	Iran	18,376	5,601
76.	Ararat (D)	Turkey	16,946	5,165
77.	Santorin (E - 1950)	Thira, Greece	4,316	1,316

MAJOR VOLCANOES OF THE WORLD, CONTINUED

NAME	LOCATION	HEIGHT IN FEET	HEIGHT IN METERS
EUROPE AND MEDITERRANEAN			
78. Etna (E - 1974)	Sicily, Italy	10,902	3,323
79. Vulcano (1964)	Italy	1,637	499
80. Stromboli (E - 1971)	Italy	3,038	926
81. Ischia (D)	Italy	2,588	789
82. Vesuvius (St)	Italy	4,190	1,277
83. Nuovo (D)	Italy	460	140
84. Albano	Italy		
ICELAND AND SURROUNDING ISLANDS			
85. Hehla (E - 1970)	Iceland	4,892	1,491
86. Eldyjar (E - 1926)	Iceland		
87. Katla (E - 1955)	Iceland	4,750	1,448
88. Surtsey (E - 1963-67)	Iceland	568	173
89. Laki (D)	Iceland	2,700	823
90. Kirkjufell (E - 1973)	Iceland	725	221
91. Oraefajokull (E - 1954)	Iceland	6,429	1,960
92. Askja (E - 1961)	Iceland	4,954	1,510
93. Beerenberg (E - 1971)	Jan Mayen, Norway	8,347	2,544
ATLANTIC OCEAN			
94. Fayal (E - 1957-58)	Azores	3,440	1,049
95. Teide (D)	Teneriffe, Canary Is.	12,198	3,718
96. Fogo (E - 1951)	Cape Verde Is.		
97. St. Paul Rock	Atlantic Ocean		
98. Tristan da Cunha (E - 1961)	Atlantic Ocean	6,760	2,060
ANTARCTICA			
99. Erebus (E - 1974)		12,450	3,795
100. Deception Island (E - 1970)		1,890	576

NAME	LOCATION	HEIGHT IN FEET	HEIGHT IN METERS
SOUTH AMERICA			
101. Ventisquero (E - 1971)	Chile	7,546	2,300
Puyahue (E - 1971)	Chile	7,349	2,240
Hudson (E - 1973)	Chile	8,580	2,615
Calbuco (E - 1961)	Chile	6,611	2,015
Lautaro (St)	Chile	11,090	3,380
Llaima (E - 1955)	Chile	10,239	3,121
Villarrica (E - 1972)	Chile	9,318	2,840
102. Shoshuenco (E - 1960)	Chile	7,743	2,360
Casablanca (E - 1960)	Chile	6,529	1,990
Cauye (E - 1960)	Chile	4,692	1,430
103. El Yunque (E - 1943)	Juan Fernandez Is. (Chile)	3,020	921
104. Tupungatito (E - 1964)	Chile	18,610	5,672
105. Lascar (E - 1951)	Chile	19,652	5,990
106. Guullatiri (E - 1959)	Chile	19,882	6,060
107. El Misti (D)	Peru	19,098	5,821
108. Cutopaxi (St)	Ecuador	19,347	5,897
Cuyambe (D)	Ecuador	18,996	5,790
Sangay (E - 1946)	Ecuador	17,159	5,230
Tungurahua (R)	Ecuador	16,512	5,033
Catacachi (E - 1955)	Ecuador	16,204	4,939
Guagua Pichincha (D)	Ecuador	15,696	4,784
Chimborazo	Ecuador	20,702	6,310
Reventador (E - 1958)	Ecuador	11,500	3,505
109. Purace (E - 1950)	Columbia	15,604	4,756
110. Alcedo (E - 1970)	Galapagos Is.	3,599	1,097

MAJOR VOLCANOES OF THE WORLD, CONTINUED

CENTRAL AMERICA AND CARIBBEAN

	NAME	LOCATION	HEIGHT IN FEET	HEIGHT IN METERS
111.	Soufriere (E - 1972)	St. Vincent	4,048	1,234
112.	Pelee (D)	Marinique	4,583	1,397
113.	Arenal (E - 1970)	Costa Rica	5,092	1,552
	Rincon de la Vieja (E - 1950)	Costa Rica	6,234	1,900
	Poas (St)	Costa Rica	8,930	2,722
	Irazu (E - 1964)	Costa Rica	11,260	3,432
114.	Ometepe (Concepcion) (E - 1957)	Nicaragua	5,106	1,556
	Momotombo (E - 1952)	Nicaragua	4,199	1,280
	Telica (E - 1971)	Nicaragua	3,409	1,039
	Cerro Negro (E - 1971)	Nicaragua	3,204	977
	El Viejo (E - 1971)	Nicaragua	5,840	1,780
	Coseguina (E - 1835)	Nicaragua	3,830	1,167
115.	Izalco (E - 1967)	El Salvador	7,749	2,362
	San Miguel (E - 1970)	El Salvador	6,988	2,130
	Conchagua (E - 1947)	El Salvador	4,100	1,250
116.	Tajumulco (R)	Guatemala	13,845	4,220
	Tacana (R)	Guatemala	13,428	4,093
	Acatenango (R)	Guatemala	13,070	3,984
	Fuego (E - 1974)	Guatemala	12,582	3,835
	Santa Maria (E - 1973)	Guatemala	12,762	3,890
	Atitlan (R)	Guatemala	11,565	3,525
	San Pedro (R)	Guatemala	9,921	3,024
	Pacaya (E - 1972)	Guatemala	8,346	2,544

	NAME	LOCATION	HEIGHT IN FEET	HEIGHT IN METERS
117.	Citlaltepec (D)	Mexico	18,700	5,700
118.	Popocatepetl (St)	Mexico	17,887	5,452
119.	Colima (St)	Mexico	14,003	4,268
120.	Jorullo (D)	Mexico	4,390	1,338
121.	Paricutin (D)	Mexico	7,451	2,271
122.	Barcena (E - 1952)	Mexico	1,235	376
	NORTH AMERICA			
123.	Lassen (D)	California	10,457	3,187
124.	Shasta (D)	California	14,162	4,317
125.	Hood (D)	Oregon	11,245	3,427
126.	Rainier (D)	Washington	14,408	4,392
127.	Baker (St)	Washington	10,778	3,285
128.	Wrangell (St)	Alaska	14,163	4,317
129.	Martin (E - 1960)	Alaska	6,050	1,844
130.	Yorbert (E - 1953)	Alaska	11,413	3,479
131.	Spurr (E - 1953)	Alaska	11,069	3,374
132.	Redoubt (E - 1966)	Alaska	10,197	3,108
133.	Iliama (St)	Alaska	10,016	3,053
134.	St. Augustine (E - 1976)	Alaska	3,927	1,197
135.	Douglas (St)	Alaska	7,064	2,153
136.	Katmai (E - 1962)	Alaska	6,715	2,047
	Kukak (St)	Alaska	6,700	2,042
137.	Mageik (St)	Alaska	7,250	2,210

MAJOR VOLCANOES OF THE WORLD, CONTINUED

NAME	LOCATION	HEIGHT IN FEET	HEIGHT IN METERS
138. Chiginagak (D)	Alaska	6,900	2,103
139. Aniahchak (D)	Alaska	4,400	1,341
140. Veniaminof (D)	Alaska	8,225	2,507
141. Paulof (E · 1973)	Alaska	8,261	2,518
142. Shishaldin (St)	Aleutians	9,387	2,861
143. Pogrami (E · 1964)	Aleutians	6,568	2,002
144. Bogoslov (E · 1907)	Aleutians		
145. Akutan (E · 1974)	Aleutians	4,275	1,303
146. Makushin (D)	Aleutians	6,680	2,036
Okmoko (E · 1958)	Aleutians	3,519	1,073
147. Cleveland (E · 1944)	Aleutians	5,675	1,730
Kogamil (D)	Aleutians	2,930	893
148. Yunaska (D)	Aleutians	3,133	955
149. Seguam (D)	Aleutians	3,458	1,054
150. Korouin (D)	Aleutians	4,852	1,479
151. Great Sitkin (E · 1974)	Aleutians	5,710	1,740
152. Kanaga (D)	Aleutians	4,416	1,346
153. Tanaga (D)	Aleutians	5,925	1,806
154. Gareloi (D)	Aleutians	5,334	1,626
155. Little Sitkiu (St)	Aleutians	3,897	1,188
156. Kiska (1962)	Aleutians	4,004	1,220
MID-PACIFIC			
157. Haleakala	Maui	10,025	3,056
158. Mauna Kea (D)	Hawaii	13,796	4,205
Mauna Loa (E · 1975)	Hawaii	13,680	4,170
159. Niwafo ou (E · 1946)	Tonga	853	260
160. Fonualei (E · 1939)	Tonga	600	183
161. Tofua (D)	Tonga Is.	1,660	506

MAPS
2-14
2-15

LAVA FLOWS AND REMNANT VOLCANIC FEATURES

Highly fluid lavas have risen to the surface of the earth, often between sedimentary rocks, to form fissure flows or quiet volcanoes. Although few of these flows are active today, during past geologic periods "quiet" volcanoes produced immense lava plateaus. The map depicts the distribution of large past and present lava flows. Dark red areas indicate where the lava is still intact and where one can actually walk across the surface and see the basaltic lava in all its jagged shape. Light red areas depict the distribution of older lava flows where the evidence no longer exists on the surface and, in many cases, may now be buried by a thick soil cover. Unlike volcanoes, fissure flows are not always located in areas of major crustal instability, that is, regions of compressional crustal buckling and earthquakes. Instead, they appear to be the outgrowth of tensional stresses deep within the crust and were produced as a result of compressional stresses some distance away. Across the Deccan (Dekkan) of India, the Parana Uplands of Brazil, and over the Columbia Plateau region of the northwestern United States, successive layers of dark, basaltic lava spread over thousands of square miles, in some places to depths of several thousand feet. However, individual flows rarely were more than 20 feet (6 meters) thick. Where valleys have cut into the lava uplands, very steep sides result because lava sheets tend to develop either a vertical system of cracks or joints or a fluted structure of hexagonal columns. The so-called *Ghats* (steps) of India are really the western edge of the great Deccan lava flows which have eroded, forming a series of steep columnar platforms. A most spectacular example of columnar is the Giant's Causeway on the north coast of Ireland, where marine erosion has exposed and emphasized the steep columns.

Not all of the many recently active lava flows around the world could be plotted on the map. Instead, a few areas are indicated where the lava is intact. Craters of the Moon National Monument, in south-central Idaho, is one of the most spectacular areas. In many parts of Arizona, New Mexico, and Oregon, the lava looks as if it had erupted only yesterday, when actually it has retained its shape for thousands of years. On the islands of Maui and Hawaii there are spectacular lava flows, several of which are active at present. Basaltic lava takes two main shapes: (1) *pahoehoe*, or *dermolithic* lava, cools slowly, forming a ropy, twisted, smooth surface; (2) *aa*, or *clastolithic* lava, cools much more rapidly, forming jagged surfaces with many holes where gases escape during cooling. Both types may occur together in a single flow.

The map also shows the distribution of lava tunnels and caves, mainly in the United States, but also in a few locations in other parts of the world. These features are probably found in every major lava flow, but unless they have been well documented in the literature they are not indicated on the maps.

Also shown are well-known remnant volcanic features. The late stages in the erosion of a volcanic area are typified by volcanic necks, pipes or plugs, and wall-like dikes often rising above a generally flat countryside. These features consist of more resistant volcanic material that remain after softer lavas have been eroded away. However, narrow dikes and necks, no matter how great the durability of the rock, rarely produce any topographic effect, whereas thick dikes and necks are almost certain to do so. The Spanish Peaks district of Colorado is famous for its system of dikes revealed by erosion. Familiar examples of volcanic necks are Shiprock in New Mexico and the often-pictured Rocher Saint Michel at LePuy in France. Devil's Tower in Wyoming is believed to be a laccolith. Scores of volcanic necks are found through-out the high plateau country of Arizona, New Mexico, and Utah. Since these features are too numerous to show individually, a single symbol for that region has been shown on the world map.   Plates 5 and 6 (centerfold) were selected because they both show excellent remnant volcanic features. Plate 5 depicts the South Island of New Zealand where the Banks Peninsula, consisting of two eroded and breached volcanic cones, extends off the coast forming a large promontory. Plate 6 is a magnificent satellite photograph of the Jemez volcano in north-central New Mexico. While the cones of the Banks Peninsula of New Zealand appear to be extinct, the Jemez volcano still has heated magma at depth with numerous mineral springs in the vicinity.

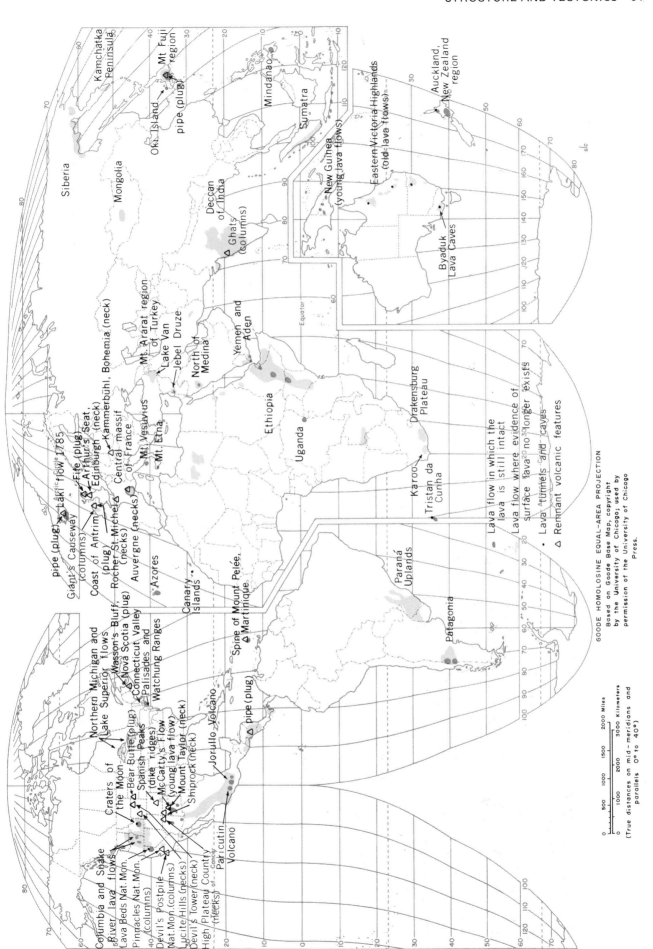

GOODE HOMOLOSINE EQUAL–AREA PROJECTION
Based on Goode Base Map, copyright
by the University of Chicago; used by
permission of the University of Chicago
Press.

MAP 2-14   LAVA FLOWS AND REMNANT VOLCANIC FEATURES OF THE WORLD

MAP 2-15   LAVA FLOWS AND REMNANT VOLCANIC FEATURES OF THE UNITED STATES

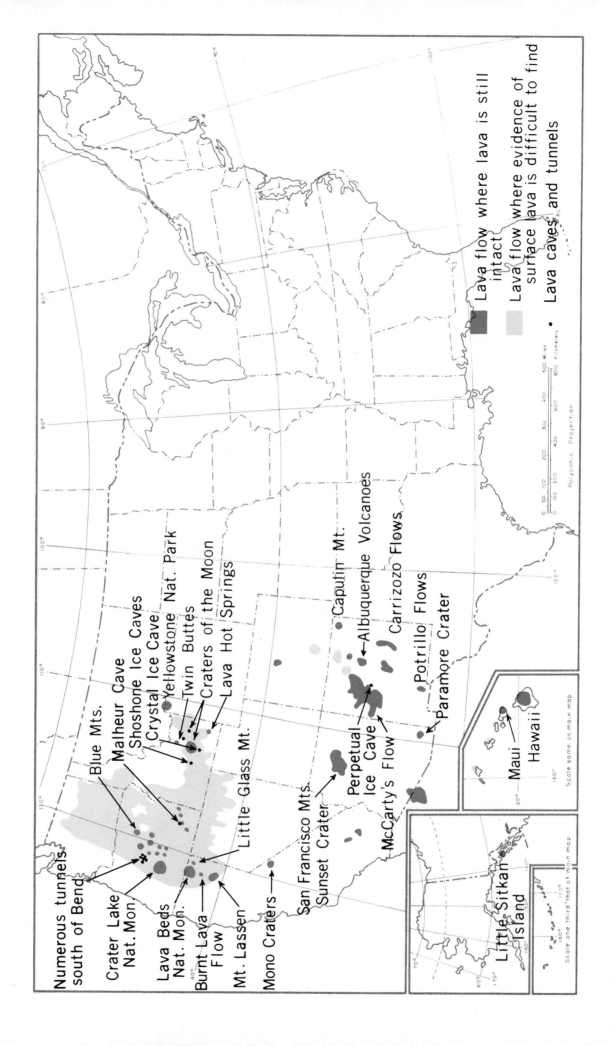

MAPS
2-16
2-17
2-18

THERMAL SPRINGS AND GEYSERS

A spring or well whose average temperature is noticeably above the mean annual temperature of the air at the same locality is called a thermal spring. Thermal springs are almost universal in their distribution. (It is amazing how widespread and dense their occurrence is! No continent is without them.) The definition of a hot spring differs from one area to another. European springs which have temperatures higher than 68°F (20°C) are classed as thermal, but in the United States springs are considered thermal if their temperature is, at least, 15°F (-9°C) above the mean annual temperature of the surrounding air. In areas where the mean annual air temperature is low, some springs that do not freeze in winter because of natural protective conditions are considered to be thermal. In tropical areas some springs that are only a few degrees warmer than the temperature of the surrounding air may be considered thermal. If an individual tracks down a spring from the above map he will not always find a steaming, boiling pool of water. In fact, most springs are difficult to locate because they issue small amounts of warm water along a rock crevice or small fault. The most notable feature of thermal spring distribution is close association with the main belts and areas of volcanoes of present or geologically recent activity. Compare the thermal spring map with the map of volcanoes. Thermal springs are common in extensive areas of lava flows of Tertiary and later geologic age, for example, in Yellowstone National Park, Wyoming, and in great lava covered areas of Idaho, eastern Oregon, and northern California. In the Auvergne region of France and in areas of volcanic rocks in Italy, there are thermal springs. They are common also in areas where rocks have been faulted and intensely folded in geologically recent times. The close relation of thermal springs to structure in intensely deformed mountain regions such as the Alps, Andes, Himalayas, and Pyrenees has been commented on by many writers. In regions of faulted block mountains in the western United States, thermal springs issue along or close to fault zones. Thermal springs can be found in unexpected regions. One such location is Hot Springs in central Arkansas.

These springs do not lie in a volcanic area, and they do not seem to be controlled by any major fault zone. The heat is thought to be derived from the mere movement of water through the small pores of sandstone, which is sufficient to raise its temperature to about 140°F (60°C). Some hot springs throw water out great distances as pressure builds up periodically underground (geysers). Three geyser areas deserve special mention: Iceland, Yellowstone National Park, and New Zealand. In Iceland, geysers and hot springs are found over an area of 5,000 square miles (12,950 square kilometers). Most of them are associated with volcanic activity due to the Mid-Atlantic Ridge running through the island. One area has more than a thousand hot springs in less than a square mile. Some of the hot springs are tapped and the water is led into Reykjavik where it is used for central heating — and for supplying swimming pools. The hot water is also used to heat greenhouses. The "Great Geysir," an Icelandic word meaning a "gusher" or "spouter," has been known for hundreds of years and is the source of our present word, which is applied to all similar phenomena. Most of the Icelandic geysers, like those in Yellowstone, stand on low mounds like inverted saucers, 70 feet (21 meters) or more in diameter, and there are many abandoned geyser mounds. Numerous hot springs, colored with algae, still contribute vast volumes of water to the adjacent streams. In Yellowstone Park hot springs and geysers represent the extreme eastern dying edge of the great Columbia lava fields. The geysers are largely concentrated in three distinct basins: Upper Geyser Basin, Lower Geyser Basin, and Norris Geyser Basin. These basins are structurally grabens, or down-faulted blocks, of the Yellowstone Plateau. The rimming walls of the basins are more or less abrupt fault scarps. The geysers and hot springs derive their water from the surrounding plateaus. The water seeps underground and rises again along the faults that have shattered the

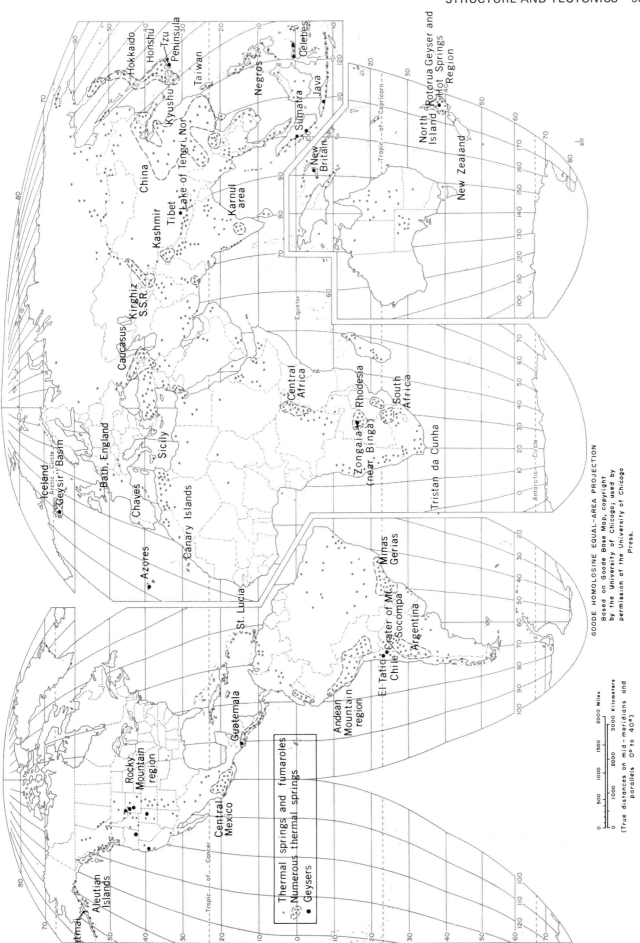

GOODE HOMOLOSINE EQUAL-AREA PROJECTION
Based on Goode Base Map, copyright
by the University of Chicago; used by
permission of the University of Chicago
Press.

MAP 2-16 THERMAL SPRINGS AND GEYSERS OF THE WORLD

floors of the basins. Many geysers and springs occur along other major lines of faulting that have broken the plateau into large blocks. In Yellowstone Park more than 200 springs have been listed and mapped, among them 72 geysers, of which 20 are known to spout to a height of not less than 50 feet (15 meters). Old Faithful rises 150 feet (46 meters), but there are several that throw their water still higher. There are two geysers in Oregon in Lake County, both man-made, and both resulting from well drilling in hot spring areas. "Old Perpetual," shoots a distance of 60 feet (18 meters) in the air at 20 second intervals. The one at Crump Ranch, Adel, spouts at about two hour intervals. The famous geyser area of New Zealand is situated along a zone of volcanoes in the central part of North Island. The largest geyser ever known was Waimangu, which during its greatest activity threw a column of water about 1,500 feet (456 meters) high, and is said to have once thrown a boulder weighing 150 pounds (68 kilograms) a quarter mile. However, it erupts no more after a short life of only two years, which began in 1901 following a terrific explosion. Smaller geyser regions include an area in Tibet at the lake of Tengri Nor, 15,744 feet (4,800 meters) above sea level, the geyser region of the Azores, the boiling lake of Dominica in the Lesser Antilles, and the water volcano of Guatemala. Isolated geysers also occur in other volcanic districts; for example, the Izu Peninsula near Atami in Japan, New Britain, Java, Sumatra, Celebes, in the crater of the 19,786 foot (6,080 meter) high volcano of Socompa, and in Northern Chile near the Argentina border.

Source: A major source consulted was Waring, G.A., "Thermal Springs of the United States and Other Countries of the World: A Summary." *Geological Survey Prof. Paper 492,* 1965.

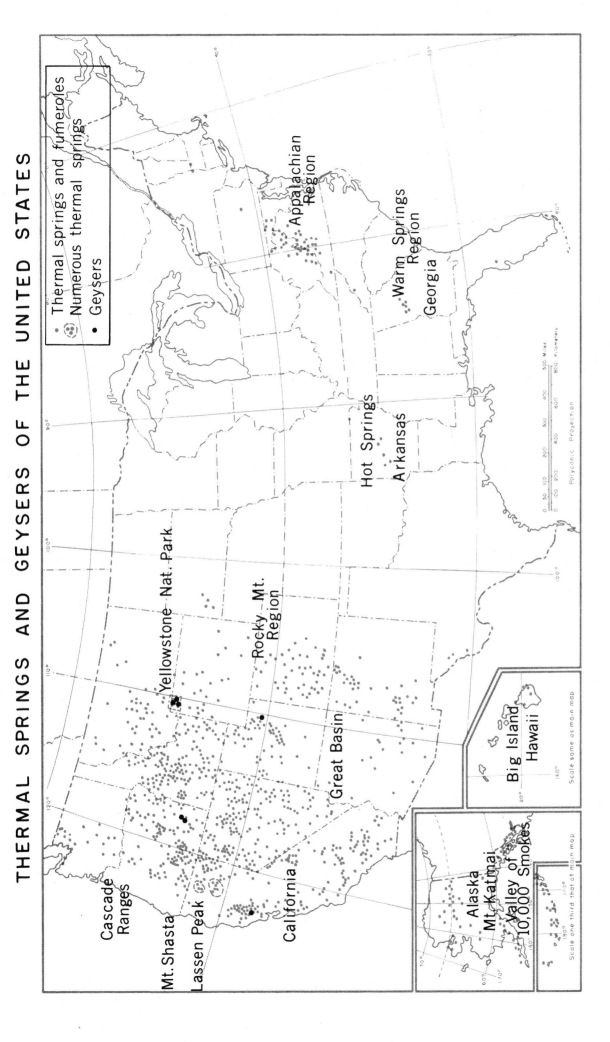

THERMAL SPRINGS AND GEYSERS OF THE UNITED STATES

Thermal springs and fumeroles
Numerous thermal springs
Geysers

Appalachian Region

Warm Springs Region

Georgia

Hot Springs

Arkansas

Yellowstone Nat. Park

Rocky Mt. Region

Great Basin

Cascade Ranges

Mt. Shasta

Lassen Peak

California

Big Island Hawaii

Scale same as main map

Alaska

Mt. Katmai

Valley of 10,000 Smokes

Scale one third that of main map

500 Miles

Kilometers

Polyconic Projection

MAP 2-17   THERMAL SPRINGS AND GEYSERS OF THE UNITED STATES

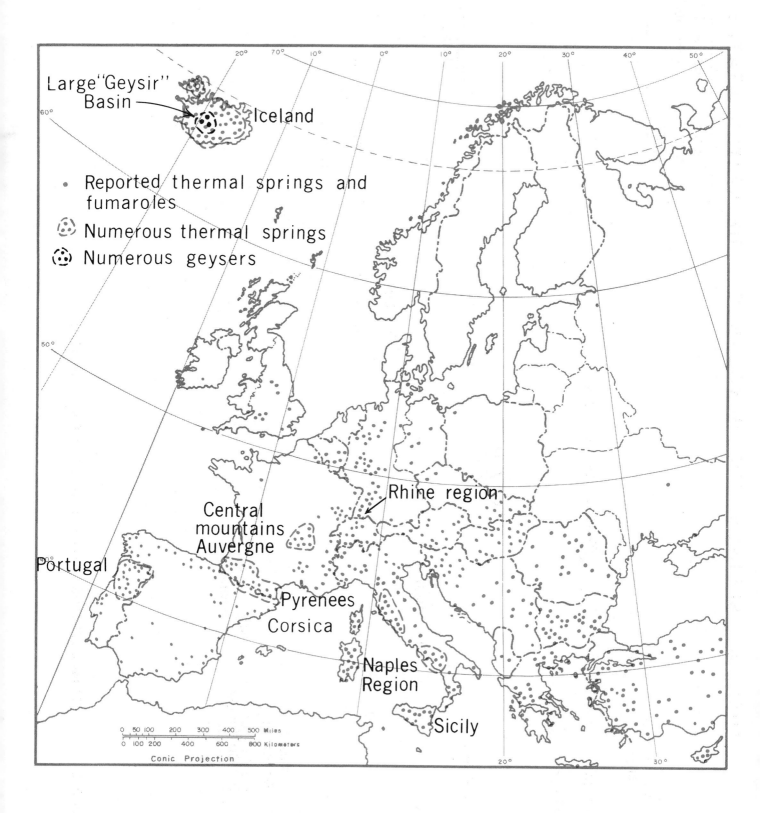

MAP 2-18   THERMAL SPRINGS AND GEYSERS OF EUROPE

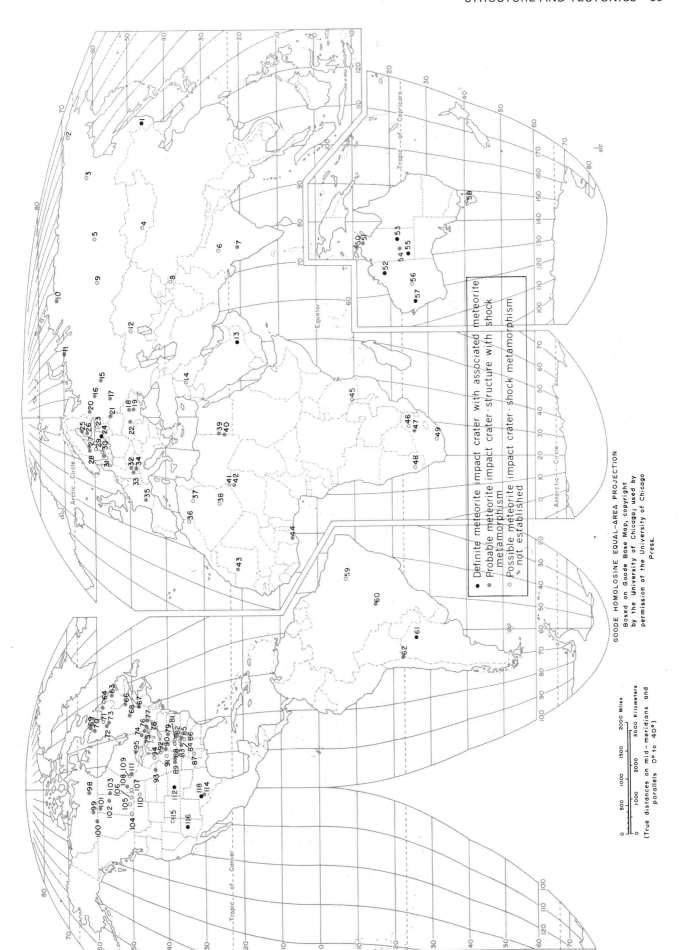

GOODE HOMOLOSINE EQUAL-AREA PROJECTION
Based on Goode Base Map, copyright
by the University of Chicago; used by
permission of the University of Chicago
Press.

- Definite meteorite impact crater with associated meteorite
- Probable meteorite impact crater-structure with shock metamorphism
- Possible meteorite impact crater-shock metamorphism not established

MAP 2-19   METEORITE IMPACT CRATERS

**MAP
2-19**

METEORITE IMPACT CRATERS

This map shows meteorite impact craters around the world. Such craters are produced by the impact and accompanying explosion of an object of extraterrestrial origin. For many years there was skepticism about this theory of origin, but it can no longer be questioned, especially after much research on lunar craters and the discovery of the mineral coesite. Coesite is produced by transformation of quartz, by shocks like the ones that meteorite impacts would produce. There are now about 50 authenticated impact structures, compared with less than 10 only 30 years ago. The great interest in the surface of the moon has led to detailed investigations on earth. For many of the known structures, impact is now favored over other suggested origins because of the presence of meteoritic materials and valid shock effects (Beals, 1958; Middlehurst and Kuiper, 1960). Recent publications suggest, after careful and systematic searching that about 250 impact structures should be found on the earth (Freeburg, 1969). There are 116 plotted on the world map. If impact occurred at frequencies comparable to the ones postulated for the moon, this number would be larger. Erosion, depositional cover, and ocean impacts have combined to keep the relative number expected at a level far below that of the moon. Probably the most famous of all the meteor craters is in Arizona, west of Winslow. Barringer Crater is a rimmed basin about 4,000 feet wide (1,219 meters) and 570 feet deep (174 meters). Its rim rises 130 to 160 feet (49 meters) above the surrounding desert. There are three small craters near Odessa, Texas, the large Ungava-Quebec (Chubb) crater in northern Canada, a group of 14 craters near Henbury, Australia, the Wolf Creek and Dalgaranga craters in western Australia, the Wabar craters in the barren Rub-al-Khali Desert of Saudi Arabia, and many depressions and small lakes at Camp del Cielo in the Chaco region of northern Argentina (Spencer, 1933). Circular lakes in Canada that could be or meteoritic impact origin are being studied. Plate 7 (centerfold) in east central Quebec, Canada depicts the large circular depression called Manicouagan Lake which is thought by geologists to be a possible meteorite crater scar. If this proves to be true it will have major implications because it is as large as many lunar craters. Intensive studies are being made at Deep Bay and West Hawk Lakes in central Canada (Short, 1970). It has even been suggested that Hudson Bay may be an impact crater, but no real evidence exists. At one time a series of lakes called Carolina Bays on the coastal plain of the southeastern United States were thought to be of meteorite impact origin. Recent evidence, however, points to their being the result of other factors (Cooke, 1934; Johnson, 1942; Thom, 1970). The great Tunguska Fall of 1908 is not included because it was a comet shower which left no impact crater.

Any presentation of meteoritic craters would be incomplete without the mention of cryptovolcanic structures, since it was in connection with these unusual landforms that the actual existence of fossil craters was first suggested (Bucher, 1963). The term "cryptovolcanic" was first used by two Germans, W. Branca and E. Fraas, in 1905, in describing the Steinheim Basin in south-central Germany. This basin, typical of other similar features of various sizes, is a ring-shaped depression one and one-half miles in diameter with a depth of 260 feet (79 meters) below the surrounding plain. Although no traces of volcanic materials have been reported, it has been generally assumed by geomorphologists that this and similar features discovered later are the result of concealed volcanic explosions. Hence the term "cryptovolcanic structure" (Dietz, 1963). However, there is a good deal of diversity of opinion concerning these land forms, since several depart from the true circular form. Recent evidence suggests they may have been formed by meteors or cometheads, because of their shape, and should therefore, be called "crypto-explosion structures" (Dietz, 1959). I have taken these suggestions and included these features with meteorite impact craters, but much must be learned before they can be properly classified. Cryptovolcanic features in the United States include Jeptha Knob in Shelby County, Kentucky, Kentland Dome in Newton County, Indiana, Wells Creek Dome in Houston and Stewart Counties, Tennessee, the Crooked Creek structure in Crawford County, Missouri (Bucher, 1936), Des Plains disturbance, Cook County, Illionois and Howell disturbance, Lincoln County, Tennessee.

METEORITE IMPACT CRATERS, CONTINUED

The following is the list of meteorite impact craters plotted on the world map. The numbers in brackets, such as (14), indicate the number of craters present. An asterisk indicates authenticated impact structures.

	LOCATION	LATITUDE	LONGITUDE	MILES	KILOMETERS
1.	Sikhote-Alin (22) * (1947) Primorye Terr. Siberia USSR	46°07'N	134°40'W	16	26.5
2.	El'gytkhyn, Chukotsk, USSR	67°30'N	172°00'E	7	12
3.	Labynkyr, Yakut, SSR, USSR	62°30'N	143°00'E	37	60
4.	Tabun-Khara-Obo, Mongolia	44°06'N	209°36'E	0.6	1.3
5.	Patomskii, Irkutsk Prov. USSR	59°00'N	116°25'E	0.06	0.09
6.	Ramgarh, India	25°20'N	076°37'30''E	2	3
7.	Lonar, India	19°58'N	076°31'E	1	1.83
8.	Murgab (2) * , Tadzhik, SSR, USSR	49°00'N	059°00'E	0.05	0.08
9.	Yenisei Ridge Crater, USSR	49°00'N	059°00'E	9	15
10.	Popigai, USSR	71°30'N	111°00'E	62	100
11.	Kara, USSR	approx. 69°30'N	064°00'E	31	50
12.	Zhamanshin, Aktyubinsk, USSR	49°00'N	059°00'E	9	15
13.	Nejed (Wabar) (2) * , Saudi Arabia	21°30'N	050°28'E	56	90
14.	Al Umchaimin, Iraq	32°41'N	039°35'E	2	3.2
15.	Puchezh-Katun, USSR	57°06'N	043°35'E	50	80
16.	Mishina, Gora, USSR	approx. 59°N	039°00'E	3	4.5
17.	Kaluga, USSR	approx. 54°30'N	036°15'E	9	15
18.	Rotmistrov, USSR	approx. 48°N	032°00'E	1	2
19.	Boltysh, Ukrainian, SSR, USSR	48°50'N	032°20'E	16	25
20.	Janisjarvi, USSR	61°58'N	303°55'E	6	10
21.	Kamen, USSR	approx. 55°N	029°00'E	16	25
22.	Ilinet, USSR	approx. 49°N	027°00'E	3	5
23.	Ilumetsa, (3) * , Estonian, SSR	57°58'N	025°25'E	0.05	0.08
24.	Kaalijarvi (7) * , Estonian, SSR	58°24'N	022°40'E	68	110
25.	Saaksjarvi, Finland	61°23'N	022°25'E	3	5
26.	Lappajarvi, Finland	63°10'N	023°40'E	6	10
27.	Dellen, Sweden	61°50'N	016°45'E	8	12
28.	Siljen, Sweden	61°05'N	015°00'E	28	45
29.	Tvaren Bay, Sweden	58°46'N	017°25'E	1	2

	LOCATION	LATITUDE	LONGITUDE	MILES	KILOMETERS
30.	Humeln, Sweden	57°22'N	016°15'W	0.6	1.2
31.	Mien, Sweden	56°25'N	014°55'E	3	5
32.	Riex, Germany	48°53'N	010°37'E	15	24
33.	Steinheim, Germany	48°02'N	010°04'E	2	3
34.	Kofels, Austria	47°13'N	010°58'E	3	5
35.	Rochechouart, France	45°49'N	000°46'E	11	18.5
36.	Michlifen, Morocco (2) *	32°00'N	003°00'E	1	1.9
37.	Talemzane, Algeria	33°18'N	004°06'E	1	1.75
38.	Amguid, Algeria	26°31'N	005°21'E	0.02	0.4
39.	"B.P." Structure, Libya	25°19'N	024°24'E	2	2.8
40.	"Oasis", Libya	24°35'N	024°24'E	6	11.5
41.	Temimichat, Mauritania	24°15'N	009°39'W	0.31	0.5
42.	Tenoumer, Mauritania	22°55'N	010°24'W	1	1.8
43.	Chinguetti; Aouelloul, Mauritania	20°15'N	012°41'W	0.15	0.25
44.	Bosumtwi, Ghana	06°32'N	001°23'W	6	10.5
45.	Nyika Plateau Crater, Zambia-Malawi	10°35'S	033°43'E	0.05	0.08
46.	Pretoria Salt Pan, S. Africa	25°30'S	028°00'E	0.6	1
47.	Vredefort, South Africa	27°00'S	027°30'E	62	100
48.	Roterkamm, Southwest Africa	27°45'S	016°17'E	1	2.3
49.	Kalkkop, South Africa	32°43'S	024°34'E	0.40	0.64
50.	Strangways, N.T., Australia	15°12'S	133°35'E	10	16
51.	Liverpool, N.T., Australia	12°24'S	134°03'E	1	1.6
52.	Wolf Creek, W.A., Australia (1) *	19°18'S	127°47'E	528	850
53.	Boxhole, N.T., Australia (1) *	22°37'S	135°12'E	109	175
54.	Gosses Bluff, N.T., Australia	23°48'S	132°18'E	14	22
55.	Henbury, N.T., Australia (14) *	24°34'S	133°10'E	93	150
56.	Lake Teague, W.A., Australia	25°50'S	120°55'E	17	28
57.	Dalgaranga, W.A., Australia (1) *	27°45'S	117°05'E	13	21
58.	Darwin Crater, Tasmania, Australia	42°15'S	145°36'E	0.6	1
59.	Serra da Canghala, Brazil	08°05'S	065°3''W	8	12

METEORITE IMPACT CRATERS CONTINUED

	LOCATION	LATITUDE	LONGITUDE	MILES	KILOMETERS
60.	Araguainha Dome, Brazil	16°46'S	052°59'W	25	40
61.	Campo del Cielo (Otumpa) (9) * Argentina	27°28'S	061°30'W	44	70
62.	Monturaqui, Chile	23°56'S	068°17'W	0.30	0.48
63.	Mistastin, Labrador, Canada	55°53'N	063°18'W	12	20
64.	Merewether, Labrador, Canada	58°02'N	064°02'W	0.12	0.2
65.	Lac La Moinerie, Quebec, Canada	57°26'N	066°36'W	5	8
66.	Manicouagan, Quebec, Canada	51°23'N	068°42'W	40	65
67.	Charlevoix, Quebec, Canada	47°32'N	070°18'W	22	35
68.	Ile Rouleau, Quebec, Canada	50°41'N	073°53'W	3	4
69.	New Quebec, New Quebec, Canada	61°17'N	073°40'W	2	3.2
70.	Lac Couture (Chubb Crater), Quebec, Canada	60°08'N	075°18'W	6	10
71.	Lac Kakiattukallak, Quebec, Canada	57°42'N	071°40'W	4	6
72.	Clearwater West, Quebec, Canada	56°13'N	074°30'W	19	30
73.	Clearwater East, Quebec, Canada	56°05'N	074°07'W	9	15
74.	Sudbury, Ontario, Canada	46°36'N	081°11'W	62	100
75.	Wanapitei, Ontario, Canada	46°44'N	080°44'W	5	8.5
76.	Brent, Ontario, Canada	46°05'N	078°29'E	3	4
77.	Holleford, Ontario, Canada	44°28'N	076°38'W	1	2
78.	Skeleton Lake, Ontario, Canada	45°16'N	079°27'W	2	3.5
79.	Serpent Mound, Ohio, USA	39°02'N	083°25'W	4	6.4
80.	Varsailles, Kentucky, USA	38°02'N	084°45'W	1	1.5
81.	Jeptha Knob, Kentucky, USA	38°06'N	085°06'W	2	3.2
82.	Middlesboro, Kentucky, USA	36°37'N	083°44'W	4	7
83.	Wells Creek, Tenn., USA	36°23'N	087°40'W	9	14
84.	Dycus, Tennessee, USA	36°22'N	085°45'W	......	------
85.	Flynn Creek, Tenn., USA	36°16'N	085°37'W	2	3.6
86.	Howell, Tennessee, USA	35°15'N	086°35'W	1	2.4
87.	Kilmichael, Mississippi, USA	33°30'N	089°33'W	8	13
88.	Crooked Creek, Miss., USA	37°50'N	091°23'W	3	5
89.	Decaturville, Miss., USA	37°54'N	092°43'W	4	6
90.	Kentland, Indiana, USA	40°45'N	087°24'W	4	6

	LOCATION	LATITUDE	LONGITUDE	MILES	KILOMETERS
91.	Glasford, Illinois, USA	40°22'N	089°48'W	3	5
92.	Des Plains, Illinois, USA	42°02'N	087°56'W	6	10
93.	Manson, Iowa, USA	45°35'N	094°31'W	19	30
94.	Glover Bluff, Wisconsin, USA	43°55'N	089°35'W	0.3	0.43
95.	Slate Island, Superior, Ontario, Canada	48°40'N	087°00'W	19	30
96.	Meen Lake, N.W.T., Canada	64°58'N	087°41'W	3	4
97.	Haughton Dome, N.W.T., Canada	75°22'N	089°40'W	19	30
98.	Nicholson Lake, N.W.T., Canada	62°40'N	102°41'W	8	12.5
99.	Pilot Lake, N.W.T., Canada	60°17'N	111°01'W	3	5
100.	Steen R., Alberta, Canada	59°31'N	117°38'W	16	25
101.	Carswell, Saskatchewan, Canada	56°27'N	109°30'W	19	30
102.	Gow Lake, Saskatchewan, Canada	56°27'N	104°29'W	3	5
103.	Deep Bay, Saskatchewan, Canada	56°24'N	102°59'W	6	9
104.	Eagle Butte, Alberta, Canada	49°42'N	110°30'W	6	10
105.	Elbow, Saskatchewan, Canada	50°58'N	106°45'W	5	8
106.	Viewfield, Saskatchewan, Canada	49°35'N	103°04'W	1.5	2.5
107.	Hartney, Manitoba, Canada	49°24'N	100°40'W	4	6
108.	St. Martin, Manitoba, Canada	51°47'N	098°33'E	15	24
109.	Poplar Bay, Manitoba, Canada	50°23'N	095°47'W	2	3
110.	Redwing Ck., N. Dakota, USA approx.	48°N	102°00'W	4	6
111.	West Hawk Lake, Manitoba, Canada	49°46'N	095°11'W	2	2.7
112.	Haviland, Kansas, USA (1) *	37°37'N	099°05'W	7	11
113.	Odessa, Texas, USA, (3) *	31°48'N	102°55'W	104	168
114.	Sierra Madera, Texas, USA	30°36'N	102°55'W	3	5
115.	Upheaval Dome, Utah, USA	38°26'N	109°54'W	3	5
116.	Canyon Diablo (Barringer or Meteor Crater) Arizona, USA, (1) *	35°02'N	111°01'W	746	1200

Sources: Numerous sources, but mainly compiled from a list by M.R. Dence, Earth Physics Branch, Gravity and Geodynamics Division, Energy, Mines and Resources, Ottawa, Canada and V.F. Buchwald, "World Map of Meteorites in 12 plates showing the sites of all recorded meteorites as of January 1, 1968." Center for Meteorite Studies, Arizona State University, Tempe, Arizona.

* Number of craters associated with the meteorite falls.

Section Three

Oceanographic and Hydrographic Features

MAP
3-1

SURFACE DRIFTS AND CURRENTS OF THE OCEANS

The circulation of the oceans is a matter of considerable import-ance to man because it has a great effect on climate, types of marine life, and the movement of ships. Through ocean currents warm water brings a moderating influence to coasts in Arctic latitudes, whereas cool currents greatly alleviate the heat of tropical deserts along narrow coastal belts. The map, drawn for the month of January, shows a number of circular movements, or gyres, centered about 25° to 30° (-3.9°C to -1.1°C) north and south of the equator. These currents move around subtropical high pressure cells. Toward the equator the currents become much straighter and run more parallel to it. A north equatorial current (N. Eq. C.) and south equatorial current (S. Eq. C.) are separated by an equatorial countercurrent (Eq. Cc.). The latter is well developed in the Pacific, Atlantic, and Indian oceans. Along the west sides of the oceans in low latitudes equatorial currents turn poleward, forming warm currents that parallel the coast. Examples are the gulf Stream (Florida current), the Kuroshio current (Japan current), and the Brazil current, all of which bring higher than average temperatures along these coasts. In middle latitudes along west coasts, onshore currents are poleward and equatorward. The equatorward flow becomes a cool current, augmented by upwelling of colder water from greater depths. This is best represented by the Peru current (Humboldt current) off the coast of Chile and Peru, by the California current off the west coast of the United States, by the Benguela current off the southwest African coast, and by the Canaries current off the Spanish and North African coast. In the northeastern Atlantic Ocean, the Gulf Stream is deflected poleward as a relatively warm current. This becomes the North Atlantic drift, which spreads around the British Isles, into the North Sea, and along the Norwegian coast. The ice-free port of Murmansk, several degrees north of the Arctic Circle, has year-round navigability by way of this coast, because of the North Atlantic drift.

Sources: Adapted from Strahler, A.N., and Alan H. Strahler, Figure 5.25, p. 78 in *Elements of Physical Geography,* John Wiley and Sons, Inc., New York, 1976; Sverdrup, H.U., M.W. Johnson, and R.H. Fleming, Chart VII in *The Oceans: Their Physics, Chemistry, and General Biology,* Prentice-Hall, Inc., New York, 1942; Van Riper, J.E., Figure 15.13 in *Man's Physical World,* McGraw-Hill Book Company, Inc., New York, 1962; and Fairbridge, R.W., *The Encyclopedia of Oceanography,* Encyclopedia of Earth Sciences Series, Vol. 1, Reinhold Publishing Corp., New York, 1966, Fig. 1, p. 591.

MAP 3-1   SURFACE DRIFTS AND CURRENTS OF THE OCEANS

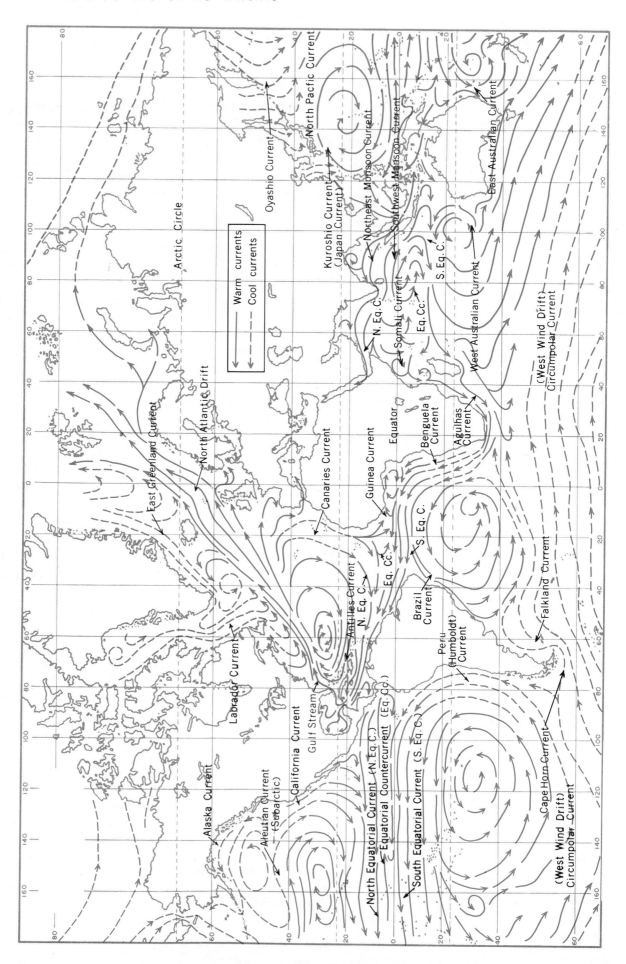

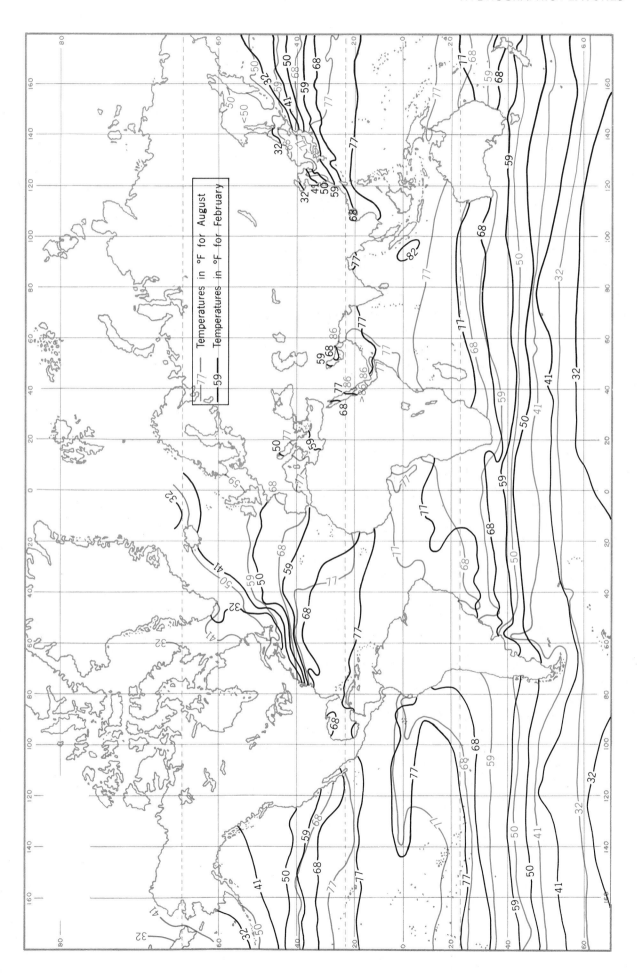

MAP 3-2 SEA TEMPERATURES

**MAP
3-2**

SEA TEMPERATURES

Because man is basically a land animal, he knows little about the range and distribution of sea surface temperatures. Sea temperatures range from about 28.4°F (-2°C) to 86°F (30°C). The lower limit is determined by processes of radiation and exchange of heat with the atmoshpere. In landlocked areas (for example, the Persian Gulf) the surface temperature may be as high as 96°F (35.6°C) but, in the open ocean, it rarely exceeds 86°F (30°C). In the open ocean the subsurface temperatures decrease rapidly with depth, at times to readings of 30.2°F (-1°C). The annual variation of surface temperature in any region depends on a number of factors; the most important ones are the annual variation of radiation income, the character of ocean currents, and prevailing winds. A comparison of Maps 8 and 9 shows that where there are cold oceans currents, the range of temperature between February and August remains small. The annual range of surface temperature is much greater in the North Atlantic and North Pacific oceans than in the southern oceans. In the southern oceans the temperature range is definitely related to the range in radiation income, whereas in the northern oceans no such definite relation appears to exist. Great ranges in northern oceans are associated with the character of the prevailing winds, and particularly with the fact that cold winds blow from the continents toward the ocean and greatly reduce winter temperatures. The highest open ocean temperatures occur in the region of the equator with sea water off the coast of Sumatra remaining over 82°F (27.8°C) annually.

Sources: *Fiziko-Feograticheskiy Atlas Mira* (Physical Geographic Atlas of the World), published in 1964 in Moscow by the Academy of Sciences USSR and the Main Administration of Geodesy and Cartography, State Geological Committee USSR. Translated in *Soviet Geography: Review and Translation,* American Geographical Society, New York, May-June 1965; Defant, A., Plate 3A and 3B in Physical Oceanography, Pergamon Press, New York, 1961; and Sverdrup, H.U., M.W. Johnson, and R.H. Fleming, The Oceans; Their Physics, Chemistry, and General Biology, Prentice-Hall, Inc., Englewood Cliffs, New Jersey, 1942.

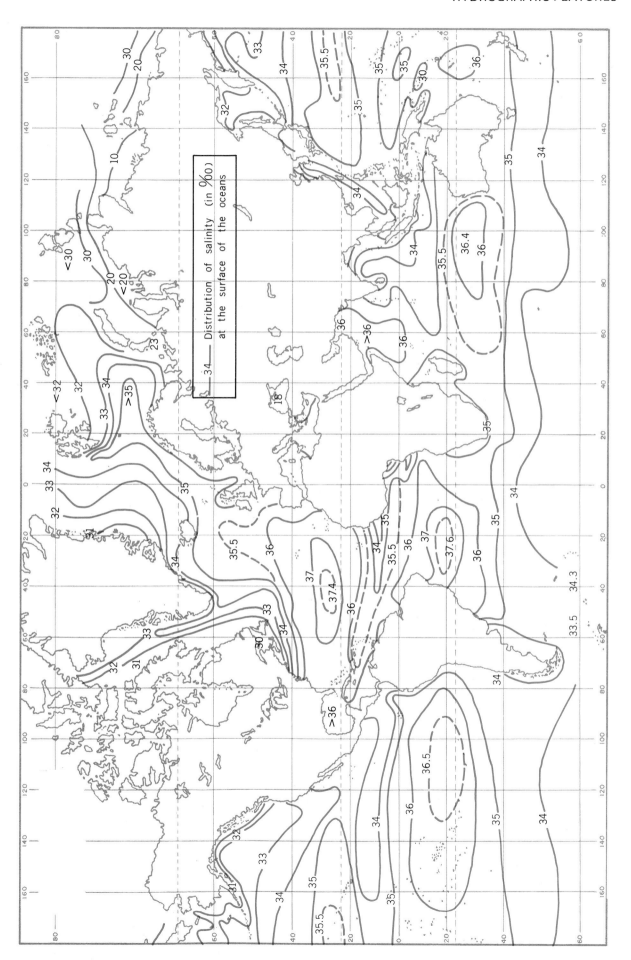

MAP 3-3 SURFACE SALINITY OF THE OCEANS

34 —— Distribution of salinity (in ‰)
at the surface of the oceans

**MAP
3-3**

SURFACE SALINITY OF THE OCEANS

A map that shows the degree of surface salinity for oceans and marginal seas is not often included in atlases. Surface salinity varies from 10 to 37.6 parts per thousand depending on the latitude of the locality and the nearness to shore. This map shows that surface salinity is considerably less in high latitudes, particularly the Arctic regions, in areas of high rainfall, or where there is a dilution by rivers. In certain semi-enclosed areas (for example, the Gulf of Bothnia) the salinity may approach zero. However, in isolated seas in intermediate latitudes, such as the Red Sea, where evaporation is excessive, salinities may reach 40 parts per thousand or more. As the range in the open oceans is rather small, it is sometimes convenient to use a salinity of 35 parts per thousand as an average for all oceans (Sverdrup, Johnson, Fleming 1942). Below 300 feet (91 meters), salinity in the ocean varies only slightly with depth, being in the range of 34 to 35 parts per thousand (Strakhov, 1967).

Source: Schott, W., "Zur Klimaschichtung der Tiefseesedlimente im aquatorialen Atlantischen Ozean," **Geol. Rdsch.,** Vol. 40, H. 1, 1952. Schott's map appears in Strakhov, N.M., Figure 17 in **Principles of Lithogeneses.** Vol. 1. Tomeieff, S.I., and Hemingway, J.E., editors, Edinburgh and London: Oliver and Boyd, Edinburgh, 1967.

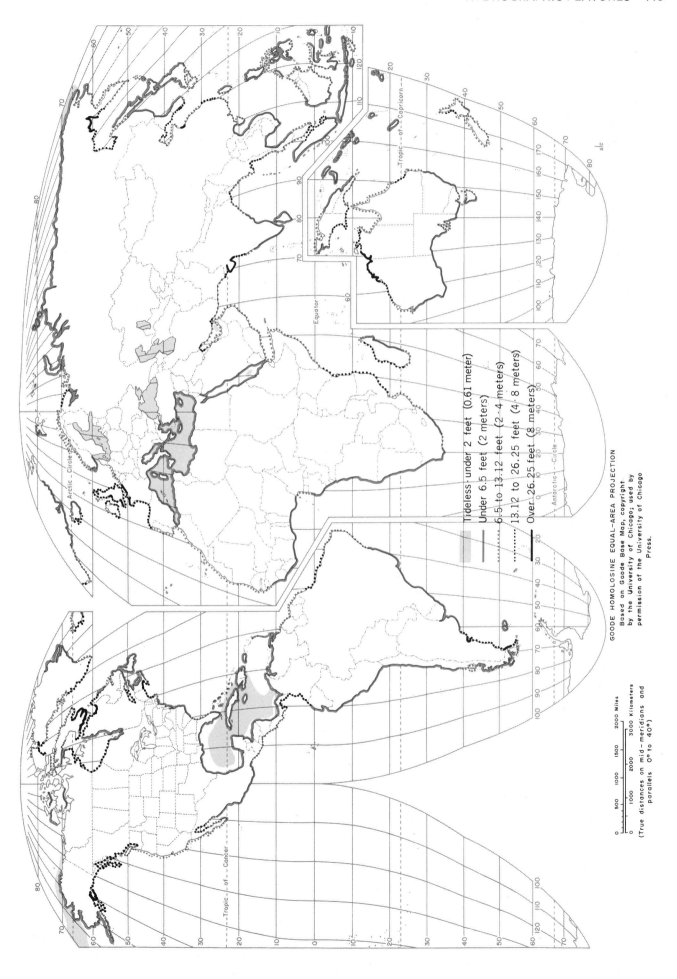

GOODE HOMOLOSINE EQUAL-AREA PROJECTION
Based on Goode Base Map, copyright
by the University of Chicago; used by
permission of the University of Chicago
Press.

Tideless: under 2 feet (0.61 meter)
Under 6.5 feet (2 meters)
6.5 to 13.12 feet (2-4 meters)
13.12 to 26.25 feet (4-8 meters)
Over 26.25 feet (8 meters)

0   500   1000   1500   2000 Miles
0   1000   2000   3000 Kilometers
(True distances on mid-meridians and
parallels 0° to 40°)

MAP 3-4   MAXIMUM TIDAL RANGES

**MAP
3-4**

MAXIMUM TIDAL RANGES

The tidal range between high and low water varies greatly around the world. This map shows in generalized form the mean maximum tidal ranges in feet and meters for coastal regions. The maximum tidal range takes into account the so-called "spring" tides or spring range when the moon is close to being full or new and the tides are at their extreme range. The neap range, on the other hand, is when the moon is in its first and third quarters and tides do not rise or fall as much as during spring tides. A mean spring tidal range is higher than the average or mean range, and a neap range is lower than the mean. In the Bay of Fundy, the mean or average range of the tide at Noel Bay is 44.2 feet (13 meters); at times of new and full moon, when spring tides occur, the range of the tide at the same site has a rise and fall of about 50.5 feet (15 meters) (Marmer, 1926). The higher figure is given on this map. The Bay of Fundy has the greatest known rise and fall of tide in the world. The second highest is at Puerto Ballegos, Argentina, with a spring tidal range of 45 feet (14 meters). Collier Bay in northwest Australia has a spring tidal range of almost 40 feet (12 meters). Heights of tides can change very rapidly along a coast. At Boston, the mean range of the tide is very nearly 10 feet (3 meters); on Nantucket island, less than 100 miles (161 kilometers) south, it is just a little more than one foot and about 400 miles (644 kilometers) northeast of Boston, it is over 40 feet (12 meters). Wherever a coastal area has an insignificant tidal range (less than 2 feet (.61 meters), the description "tideless" has been applied on the map. The Mediterranean and Baltic are the best known tideless seas. Their location behind restricting inlets limits tidal ranges. An entirely different cause underlies the virtual tidelessness of the oceanic centers. Palmyra Island and the Society Islands in the central Pacific are tideless (not shown). Unfortunately, only the accidental occurrence of islands furnishes a means of determining the tidal ranges within the central parts of the larger oceans. It is a surmise that these islands are located near the nodal lines of oscillating systems. Any open oceanic location near the node of an oscillation is thought to be associated with small tidal ranges.

Sources: Bauer, H.A., "A World Map of Tides," *The Geographical Review,* Vol. 23, pp. 259-270, 1933; Plate 32 in *Pergamon World Atlas,* Pergamon Press, New York, 1968; Sverdrup, H.U., M.W. Johnson, and R.H. Fleming, *The Oceans: Their Physics, Chemistry, and General Biology,* Prentice-Hall, Inc., Englewood Cliffs, N.J., 1942; and *Morsky Atlas* (Oceanic Atlas), edited by I.S. Isakov, Naval General Staff, Moscow, 1950.

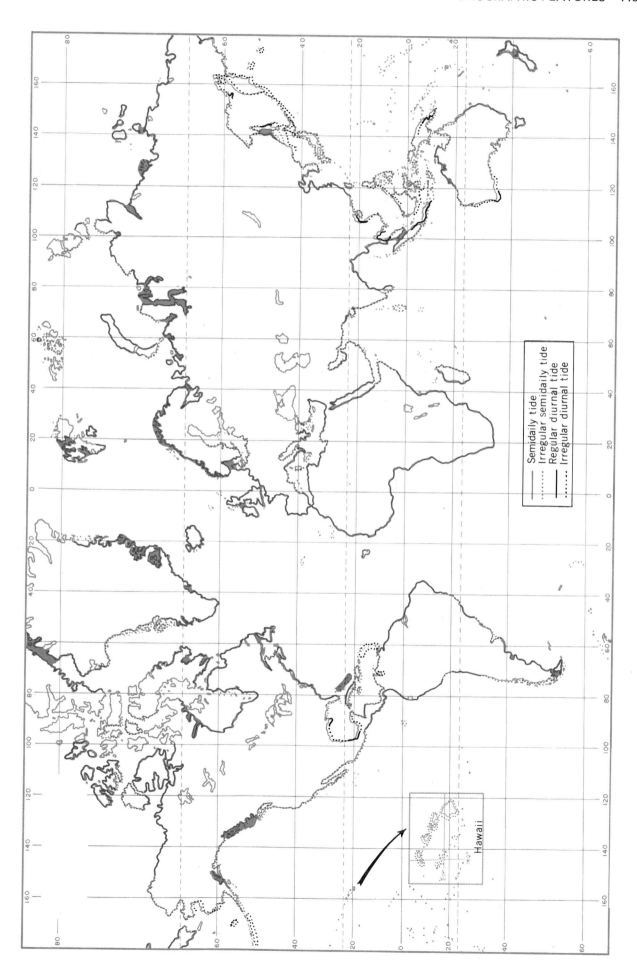

Semidaily tide
Irregular semidaily tide
Regular diurnal tide
Irregular diurnal tide

Hawaii

MAP 3-5   TYPES OF TIDES

MAP
3-5

TYPES OF TIDES

The tides of the earth are subject to the periodicity of the tide-producing forces of the moon and sun; hence, they undergo periodic changes in intensity while running through certain cycles conditioned by the motions and changing distances of the three bodies involved. The numerous existing kinds of tidal oscillation have been classified into four major types: semidaily, irregular semidaily, irregular diurnal, and regular diurnal (Zubov, 1947). When high tides occur approximately every 12 hours they are termed *semidaily (semidiurnal)* tides; those with high water every 24 hours are *diurnal (daily)* tides, but those characterized by a higher high water every 12 hours and a lower high water occurring between any two higher high waters, suggesting an interference of two oscillations or waves, are termed *irregular* or *mixed* tides. Thus, an *irregular semidaily* tide is one with a high tide every 12 hours but the highs are not even in their height from one 12 hour period to the next. An *irregular diurnal* tide has uneven highs every 24 hours. The irregular or mixed type of tide, reflecting the largest number of tide-producing components, is usually considered the most normal of all types of tide. This type is usually expected to exist in waters that offer the least obstruction to free oscillations of both a 12-hour and 24 hour period. Large sections of the Pacific Basin possess daily, semidaily, and irregular tides. Outside the Pacific basin there are irregular tides in the Arabian, Caribbean, Mediterranean, and Baltic seas. Semidaily (semidiurnal) tides with an oscillation of a 12-hour period are the most widespread types of tides, particularly around the Atlantic Ocean and in Arctic regions. It is interesting on a world scale the contrast between the semidaily tides of the Arctic and the diurnal and irregular tides of Antarctic shores. Regular diurnal tides must be called the most incomplete tides, since they reflect only the impulses given at intervals of about 24 hours. Their crippled character is also shown by the smallness of their ranges, generally 1 to 5 feet (.30 to 1.5 meters). Sections of the Gulf of Mexico, the East Indies, northern Japan and Kamchatka peninsula, and two portions of southwest Australia have this type of tide. All of Antarctica which is not shown on the world map, has a regular diurnal tide except one small portion of the Antarctic Peninsula which has an irregular diurnal tide.

Sources: Bauer, H.A., Plate III in "A World Map of Tides," *The Geographical Review,* Vol. 23, facing p. 266, 1933; Marmer, J.A., *The Tide,* D. Appleton and Company, New York, 1926; Zubov, N.N. *Dinamicheskaya Oceanology,* Hydrometerological Publishing House, Moscow, 1947, (Zubov's map appears in Doty, M.S., "Rocky Intertidal Surfaces," *Geological Society of America Memoir 67,* Vol. 1, Plate 1, 1957); Dietrich, G., *General Oceanography,* Wiley Interscience, New York, 1963; Davies, J. L., *Geographical Variation in Coastal Development,* Hafner Publishing Company, New York, 1973, pp. 49-52.

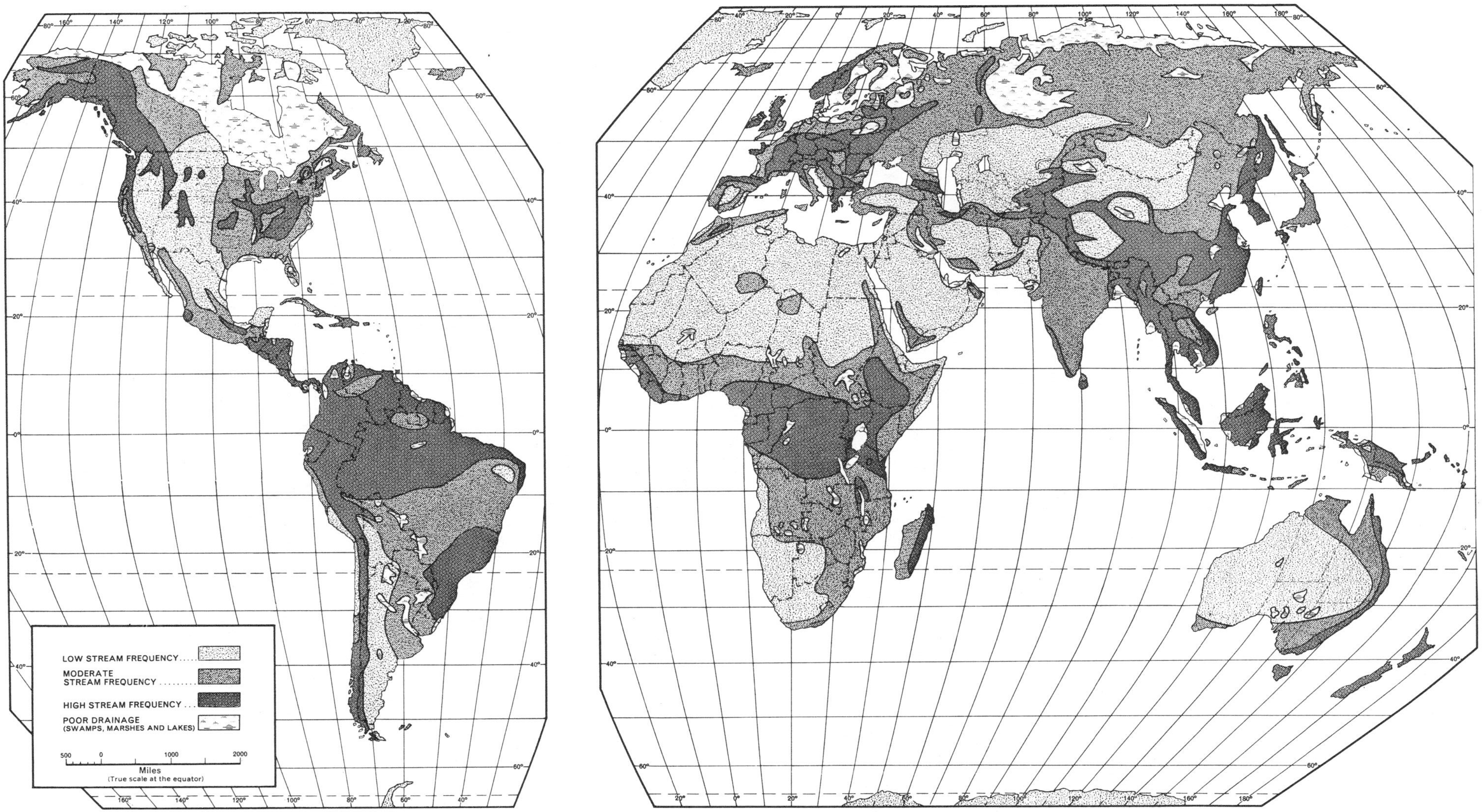

LOW STREAM FREQUENCY ....

MODERATE
STREAM FREQUENCY ....

HIGH STREAM FREQUENCY .....

POOR DRAINAGE
(SWAMPS, MARSHES AND LAKES)

Miles

500   0   1000   2000

(True scale at the equator)

MAP 3-6   WORLD MAP OF STREAM FLOW

**MAP
3-6**

WORLD MAP OF STREAM FLOW

A map showing stream flow is a useful map because it indicates the uneven distribution of surplus surface water around the world. For example, a comparison can be made between the pattern of high stream flow and the humid climatic regions of the world. The highest stream flow is found in areas of sloping terrain where there is considerable rainfall and where sufficient geologic time has elapsed for streams to develop elaborate tributary systems, for example, those found in mountainous regions and in humid hill lands. The hill lands of the eastern United States, eastern China, and southeastern Brazil offer excellent examples of close and intricate stream patterns. At the opposite end of the scale, areas of low stream flow include the arid and semiarid portions of the world. They have high evaporation rates, the rainfall is slight and irregular, and the water reservoirs of soils require constant replenishment. Stream channels are relatively infrequent in dry regions. However, areas of low stream flow are not wholly confined to dry regions, and small occurrences can be found in all parts of the world, regardless of climate. Sandy flats near coastlines and extremely porous limestone areas may lack any permanent surface streams, despite a humid climate. The Yucatan Peninsula of Mexico is included in the category of low stream flow because of its low elevation, pronounced dry period, and porous limestones. The author of this map, Joseph Van Riper uses the term stream frequency, but this is a difficult term to define. Therefore, the term stream flow is used. The division between low, moderate and high is also ambiguous and has no clear-cut boundary. In areas of poor drainage, lakes, swamps and marshes form a large part of the land surface. The largest areas of poor drainage are in the northern parts of North America and Eurasia — for example, the tundra plains that border the Arctic Ocean. Other large poorly drained areas show the effects of continental glaciation during the Pleistocene ice age. Most of the Canadian Shield is an area of poor drainage, with swamps and lakes found in nearly every depression between low hills. Swamps and marshes are also numerous along flat coastal plains, as in the eastern United States, northern China, the larger islands of Indonesia, and the Guianas of South America. Deltas of large rivers are usually poorly drained.

Source: Van Riper, J.E., Figure 5.7 in ***Man's Physical World,*** McGraw-Hill Book Company, Inc., 1962.

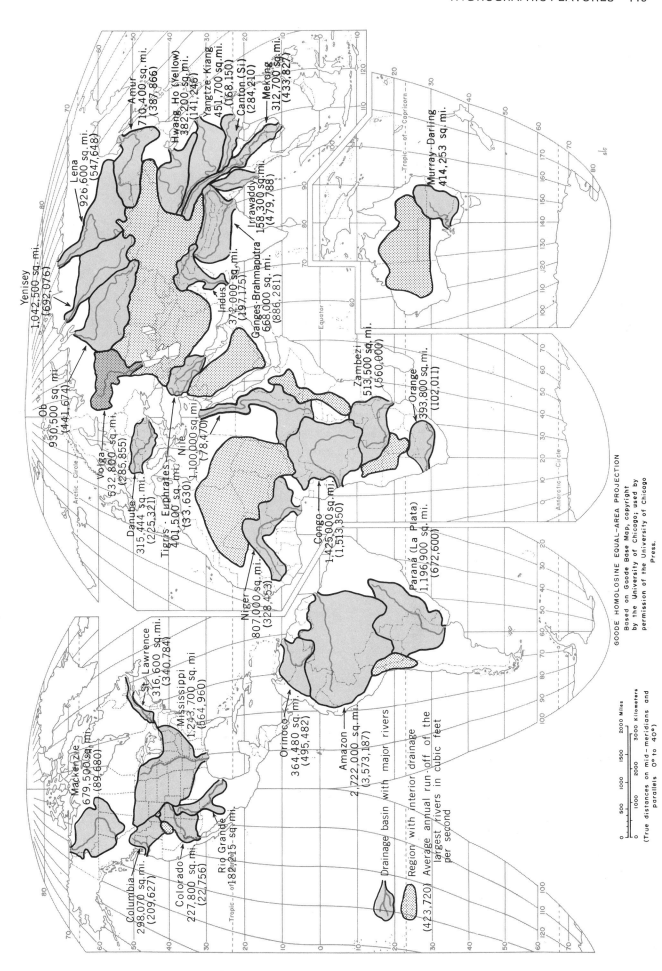

GOODE HOMOLOSINE EQUAL-AREA PROJECTION
Based on Goode Base Map, copyright
by the University of Chicago; used by
permission of the University of Chicago
Press.

Yenisey 1,042,500 sq. mi. (692,076)

Ob 930,500 sq. mi. (441,674)

Lena 926,600 sq. mi. (547,648)

Amur 710,400 sq. mi. (387,866)

Hwang Ho (Yellow) 382,200 sq. mi. (141,246)

Yangtze-Kiang 451,700 sq. mi. (568,150)

Canton (Si) (284,210)

Mekong 312,700 sq. mi. (433,827)

Irrawaddy 158,300 sq. mi. (479,788)

Ganges-Brahmaputra 668,000 sq. mi. (886,281)

Indus 372,000 sq. mi. (197,175)

Volga 532,800 sq. mi. (285,855)

Danube 315,444 sq. mi. (225,321)

Tigris-Euphrates 401,500 sq. mi. (33,630)

Nile 1,100,000 sq. mi. (78,470)

Niger 807,000 sq. mi. (328,453)

Congo 1,425,000 sq. mi. (1,513,350)

Zambezi 513,500 sq. mi. (560,000)

Orange 393,800 sq. mi. (102,011)

Murray-Darling 414,253 sq. mi.

Paraná (La Plata) 1,196,900 sq. mi. (672,600)

Amazon 2,722,000 sq. mi. (3,573,187)

Orinoco 364,480 sq. mi. (495,482)

Mackenzie 679,500 sq. mi. (89,680)

St. Lawrence 316,600 sq. mi. (340,784)

Mississippi 1,243,700 sq. mi. (564,960)

Columbia 298,070 sq. mi. (209,627)

Colorado 227,800 sq. mi. (22,756)

Rio Grande 182,215 sq. mi.

Drainage basin with major rivers

Region with interior drainage

Average annual run-off of the
largest rivers in cubic feet
per second

(423,720)

0 500 1000 1500 2000 Miles
0 1000 2000 3000 Kilometers
(True distances on mid-meridians and
parallels 0° to 40°)

MAP 3-7   DRAINAGE BASINS

Equator
Tropic of Cancer
Tropic of Capricorn
Arctic Circle
Antarctic Circle

THE FOLLOWING IS INFORMATION ON THE 63 LONGEST RIVERS OF THE WORLD:

NAME	LENGTH		DRAINAGE AREA		AVERAGE ANNUAL RUN-OFF	
	MILES	KILOMETERS	SQ. MILES	SQ. KILOMETERS	CUBIC FT./SEC	CU. METERS/SEC
1. Nile	4937	7950	1,100,000	2,849,000	78,470	2,222
2. Amazon	3900	6275	2,722,000	7,049,980	3,573,187	101,181
3. Mississippi-Missouri-Red Rock	3872	6230	1,247,300	3,230,500	664,960	18,829
4. Yangtze	3400	5471	451,700	1,169,900	168,150	4,761
5. Congo	2900	4666	1,425,000	3,690,750	1,513,350	42,853
6. Lena	2860	4602	926,600	2,400,000	547,648	15,508
7. Amur	2800	4505	710,400	1,840,000	387,866	10,983
8. Hwang Ho (Yellow)	2700	4344	382,200	990,000	141,246	39,996
9. Parana	2700	4347	1,196,900	3,100,000	672,600	19,046
10. Yenisey	2619	4214	1,042,500	2,700,100	692,076	19,597
11. Niger	2600	4187	807,000	2,090,100	328,453	9,301
12. Mackenzie	2517	4053	679,500	1,760,000	89,680	2,539
13. Mekong	2500	4022	312,700	809,900	433,827	12,285
14. Missouri	2466	3971	529,400	1,371,150	76,300	2,160
15. Mississippi	2348	3778	1,247,300	3,230,500	640,000	18,123
16. Volga	2300	3701	532,800	1,380,000	285,855	8,094
17. Ob	2270	3652	930,500	2,410,000	441.674	12,507
18. Euphrates	2235	3596	401,500	1,039,890 Tigris/Euphrates	33,630	952
19. Great Lakes-St. Lawrence	2150	3459	316,600	820,000	340,784	9,650
20. Indus	2000	3218	372,000	963,500	197,175	5,583
21. Purus	1900	3057				
22. Irtysh	1844	2967				
23. São Francisco	1802	2899				
24. Brahmaputra	1800	2896	668,000 Brahmaputra/Ganges	1,730,100	886,281	25,097
25. Danube	1800	2896	315,444	8,170,000	225,321	6,380
26. Rio Grande	1800	2896	182,215	472,000		
27. Yukon	1800	2896	327,600	842,480	240,000	6,796
28. Darling	1760	2832	414,253 Darling/Murray	1,075,000		
29. Salween	1750	2816				
30. Zambezi	1650	2655	513,500	1,329,960	560,000	15,857
31. Canton (Si)	1650	2655			284,210	8,048
32. Tocantins	1600	2574				

LONGEST RIVERS OF THE WORLD, CONTINUED

NAME	LENGTH		DRAINAGE AREA		AVERAGE ANNUAL RUN-OFF	
	MILES	KILOMETERS	SQ. MILES	SQ. KILOMETERS	CUBIC FT./SEC	CU. METERS/SEC
33. Kolyma	1550	2494				
34. Vilyuy	1512	2432				
35. Amu Darya	1500	2414				
36. Araguaia	1500	2414				
37. Ganges	1500	2414	668,000 Brahmaputra/Ganges	1,730,100	886,281	23,097
38. Negro	1500	2414				
39. Orinoco	1500	2414	364,480	944,000	495,482	14,030
40. Paraguay	1500	2414				
41. Ural	1490	2397				
42. Arkansas	1450	2333				
43. Colorado	1400	2252	227,800	590,000	22,756	644
44. Dneper	1400	2252				
45. Irrawaddy	1250	2011	158,300	410,000	479,788	13,586
46. Jurua	1250	2011				
47. Xingu	1238	1992				
48. Don	1220	1963				
49. Ucayali	1210	1947				
50. Columbia	1200	1931	298,070	772,000	209,627	5,936
51. Murray	1200	1931	414,253 Murray/Darling	1,073,000		
52. Syr Dorya	1200	1931				
53. Orange	1200	1931	393,800	1,020,000	102,001	2,889
54. Tigris	1180	1899	401,500 Tigris/Euphrates	1,039,890	33,630	952
55. Kama	1172	1886				
56. Aldan	1160	1866				
57. Angara	1150	1850				
58. Pechora	1133	1823				
59. Tobol	1042	1677				
60. Snake	1038	1670				
61. Red	1018	1638	93,244	241,500	62,300	1,764
62. Churchill	1000	1609	108,600	281,270	42,400	1,201
63. Uruguay	1000	1609				

MAPS
3-7
3-8

MAJOR DRAINAGE BASINS

This map shows the overall drainage basins, the average annual runoff in cubic feet per second at the months and the lengths of major rivers around the world. The Amazon is the largest river system (but not the longest river), draining 2,722,000 square miles (7,049,952 square kilometers); its average annual discharge at its mouth is, 3,573,187 cubic feet (101,181 cubic meters) of water per second. The Congo is second in drainage area, with 1,425,000 square miles (3,690,735 square kilometers); it annually discharges 1,513,350 cubic feet (42,853 cubic meters) of water per second into the sea. The Mississippi River is third in size, and the Parana (La Plata), with 1,196,900 square miles (3,099,959 square kilometers) of drainage basin is fourth; but the Parana discharges more water than the Mississippi. What is surprising is the size of the rivers in Siberia. The Yenisey River has a total drainage area of 1,042,500 square miles (2,700,064 square kilometers) and its mouth empties 692,076 cubic feet (19,597 cubic meters) of water per second. This is the largest of several rivers that flow into the Arctic Ocean. Both the Ob and the Lena are larger than the Mackenzie, which flows into the Arctic Ocean from Canada. The Nile is an interesting river because, although it is the longest river in the world (4,145 miles/6,671 kilometers), its size in drainage area is fifth and its annual discharge is only 78,470 cubic feet (2,222 cubic meters) per second at its mouth. This very small discharge compared to its area is the result of several factors. The Nile flows across the Sahara Desert losing a great deal of water through evaporation and, in turn, water for irrigation and from flooding spreads large amounts of the discharge over the land. Thus, the Nile is one of the most heavily used rivers by man.

Sources: National Geographic Society, Washington, D.C.; K.I. Iseri and W.B. Langbem, "Large Rivers of the United States," *Geological Survey Circular 686,* U.S. Government Printing Office, Washington, D.C., 1974, pp. 8-10.

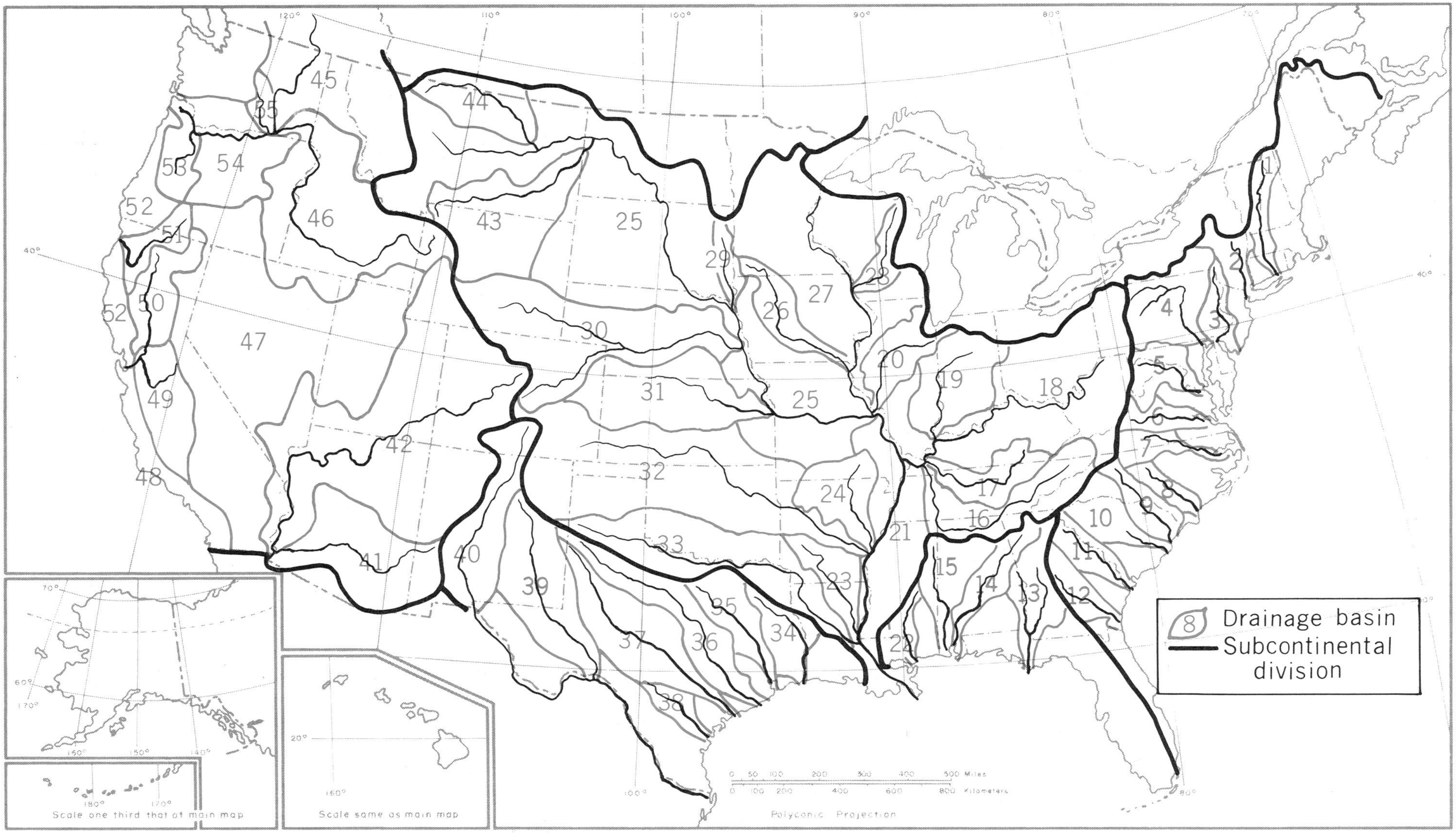

123.

MAP 3-8   PRINCIPAL DRAINAGE BASINS OF THE UNITED STATES

THE FOLLOWING IS A LIST OF THE PRINCIPAL DRAINAGE BASINS OF THE UNITED STATES:

	NAME	LENGTH		DRAINAGE AREA		DISCHARGE	
		MILES	KILOMETERS	SQ. MILES	SQ. KILOMETERS	CU. FT./SEC	CU.METERS/SEC
1.	Connecticut	407	655	11,085	28,710		
2.	Hudson	306	492	13,370	34,620	19,500	552
3.	Delaware	390	628	11,440	29,630	17,200	487
4.	Susquehana	444	714	27,570	71,410	37,190	1053
5.	Potomac	383	616	14,500	37,560		
6.	James	340	547	9,700	25,130		
7.	Roanoke	380	611	9,200	23,800		
8.	Cape Fear	202	325	4,550	11,800		
9.	Peedee	233	375	10,600	27,450		
10.	Santee	538	866	9,800	25,380		
11.	Savannah	314	505	11,100	28,750		
12.	Altamaha	392	631	14,100	36,520		
13.	Applachicola	524	843	19,500	50,500	24,700	699
14.	Alabama	735	1183	22,600	58,534	32,400	917
15.	Tombigbee	525	845	20,100	52,060	27,300	773
16.	Tennessee	652	1049	40,910	1,059,570	64,000	1812
17.	Cumberland	720	1158	18,080	46,827	26,900	762
18.	Ohio	981	1578	203,900	528,100	258,000	7306
19.	Wabash	529	851	33,150	85,860	30,400	861
20.	Illinois	420	676	27,900	72,260	22,800	646
21.	Lower Mississippi	2539	4085	1,247,300 Upper and Lower	3,230,510	640,000 Upper & Lower	18,123
22.	Pearl	411	661	4,400	11,400		
23.	Ouachita	605	973	10,700	27,700		
24.	White	720	1158	17,700	71,750	32,100	909
25.	Missouri	2315	3725	529,400	1,371,150	76,300	2,160
26.	Des Moines	327	526	11,700	38,070		
27.	Upper Mississippi	1171	1887	1,247,300 Upper & Lower	3,230,510	640,000 Upper & Lower	18,123

PRINCIPAL DRAINAGE BASINS OF THE UNITED STATES, CONTINUED

	NAME	LENGTH		DRAINAGE AREA		DISCHARGE	
		MILES	KILOMETERS	SQ. MILES	SQ. KILOMETERS	CU.FT./SEC	CU. METERS/SEC
28.	Wisconsin	430	692	6,000	15,540		
29.	Big Sioux	300	483	4,340	11,240		
30.	Platte	310	499	84,000	217,560	10,000	283
31.	Kansas	169	272	61,300	158,770	10,000	283
32.	Arkansas	1459	2348	160,000	415,950	45,100	1,277
33.	Red	1270	2043	93,244	241,500	62,300	1,764
34.	Sabine	380	611	12,000	31,080		
35.	Trinity	360	579	10,200	20,400		
36.	Brazos	870	1340	44,500	115,250	10,000	283
37.	Colorado (tr)	840	1352	41,500	107,480	10,000	283
38.	Guadalupe	300	483	6,000	15,540		
39.	Pecos	735	1183	38,300	99,100	10,000	283
40.	Rio Grande	1885	3033	182,000	471,380	10,000	283
41.	Gila	630	1014	58,100	150,480	210,000	1283
42.	Colorado	1450	2333	227,800	590,000		
43.	Yellowstone	671	1050	67,500	174,800		
44.	Milk	625	1006	28,000	72,520		
45.	Upper Columbia	890	1432	258,000 Upper & Lower	668,220	262,000 Upper & Lower	7419
46.	Snake	1038	1620	109,000	282,310	50,000	1416
47.	Great Basin			208,000	538,720		
48.	Coast Rivers South			23,000	59,570		
49.	San Joaquin	350	563	22,700	58,790		
50.	Sacramento	377	607	27,100	70,140		
51.	Klamath	250	402	10,000	25,900		
52.	Coast Rivers North			22,000	72,520		
53.	Willamette	270	434	11,200	29,010	35,660	1010
54.	Lower Columbia	1243	2000	258,000 Upper & Lower	668,220	263,000 Upper & Lower	7419
55.	Yakima	200	322	5,270	13,650		

MAP 3-9 SEDIMENT CONCENTRATION OF RIVERS FOR THE UNITED STATES

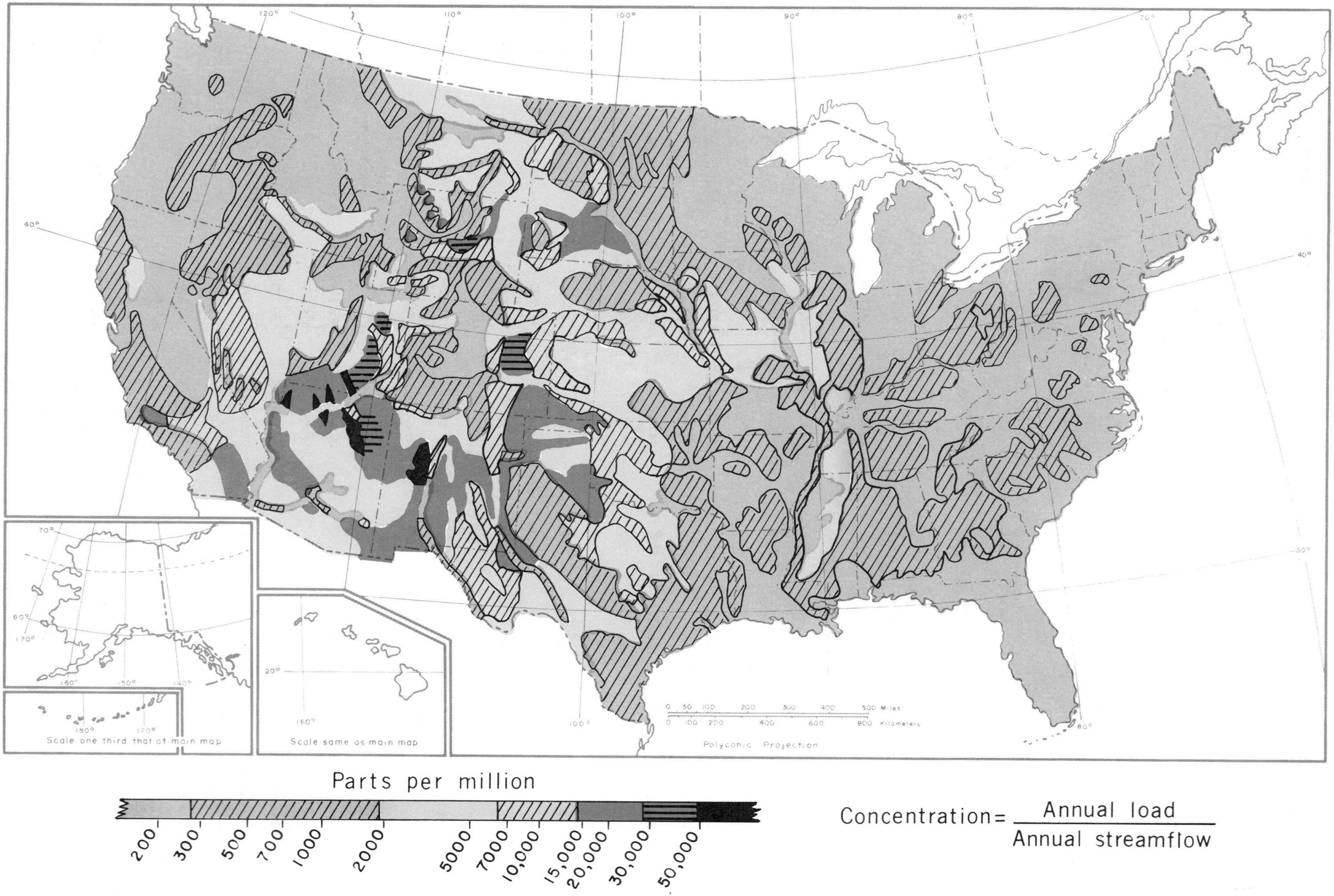

120°
110°
100°
90°
80°
70°
40°
30°
Scale one third that of main map
Scale same as main map
0 50 100 200 300 400 500 Miles
0 100 200 400 600 800 Kilometers
Polyconic Projection
Parts per million
200
300
500
700
1000
2000
5000
7000
10,000
15,000
20,000
30,000
50,000

$$\text{Concentration} = \frac{\text{Annual load}}{\text{Annual streamflow}}$$

SEDIMENT CONCENTRATION OF RIVERS FOR THE UNITED STATES

MAP
3-9

This detailed map has been redrawn from the United States Geological Survey map titled "Map of Conterminous United States Showing Sediment Concentrations of Rivers" published in 1962. This map is included in the atlas because it shows quite well how the more humid regions of the country, such as the eastern one third and the Pacific Northwest, have clearer running streams with 200 parts of sediment per million parts of water. In turn, several rivers in the Southwest have great concentrations of sediment amounting to 30,000 parts of sediment per million parts of water. Striking examples are the Rio Puerco in New Mexico, the Little Colorado River in Arizona, and Price River in Utah. The Mississippi River along much of its middle and lower course is carrying between 300 and 2,000 parts of sediment per million parts of water.

Source: Department of the Interior, United States Geological Survey, Map of Conterminous United States Showing Sediment Concentrations of Rivers, Hydrologic Investigations, Atlas , #A-61 Plate 3 of 3, 1962.

MAGNITUDE OF SOLID LOAD FROM DRAINAGE BASINS

MAP
3-10

This map shows the annual discharge of suspended solids in tons into oceans and lake basins (for 20 major drainage systems). These data complement those shown on the maps of drainage basins and river discharge. For example, the Mississippi River, with a drainage area of 1,243,700 square miles (3,221,170 sqaure kilometers) and a discharge of 564,960 cubic feet (15,998 cubic meters) per second discharges solids at the rate of 500 million tons annually. On the other hand, the Ganges-Brahmaputra drainage system, with an area of 668,000 square miles (1,730,113 square kilometers), discharges 886,281 cubic feet (25,096 cubic meters) of water per second and a solid load discharge of 1,800 million tons annually. The solid load discharge is more than three times as much as that of the Mississippi River. Certainly the much heavier rainfall over the Himalayas has an influence on this very great difference in erosion and mechanical denudation. Comparisons also can be made between large rivers in arctic, arid, and humid regions.

Source: Data from Stakhov, N.M., *Principles of Lithogenesis* Vol. 1, Tomkeieff, S.I. and Hemingway, J.E., editors, Edinburgh and London; Oliver and Boyd, Edinburgh, 1967 as it appears in Chorley, R.J., Figure 1.11.5 in *Water, Earth and Man,* Methuen and Company, Ltd., London, 1969.

MAP 3-10 MAGNITUDE OF SOLID LOAD FROM DRAINAGE BASINS

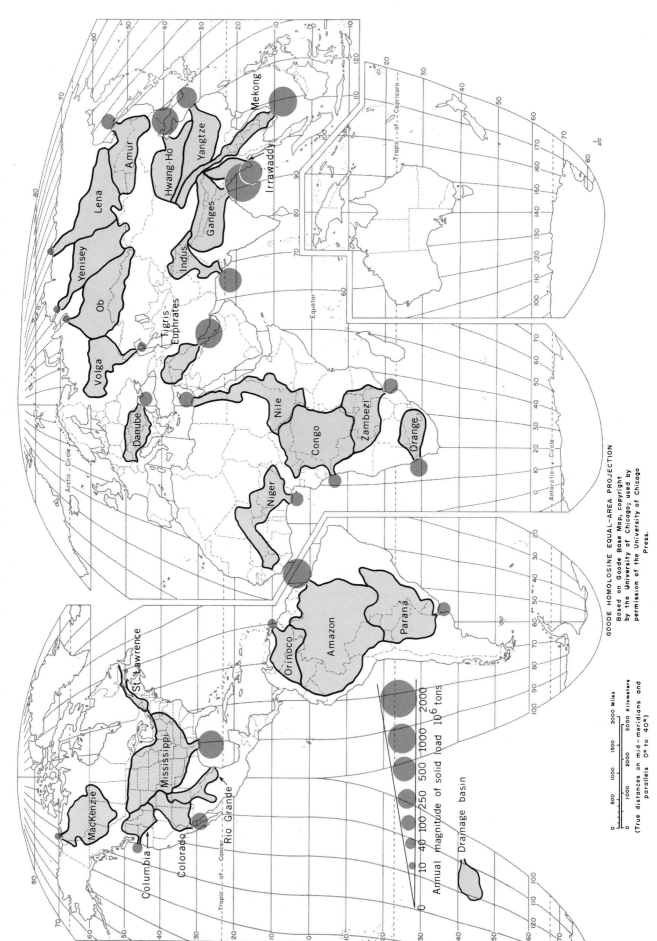

GOODE HOMOLOSINE EQUAL-AREA PROJECTION
Based on Goode Base Map, copyright
by the University of Chicago; used by
permission of the University of Chicago
Press.

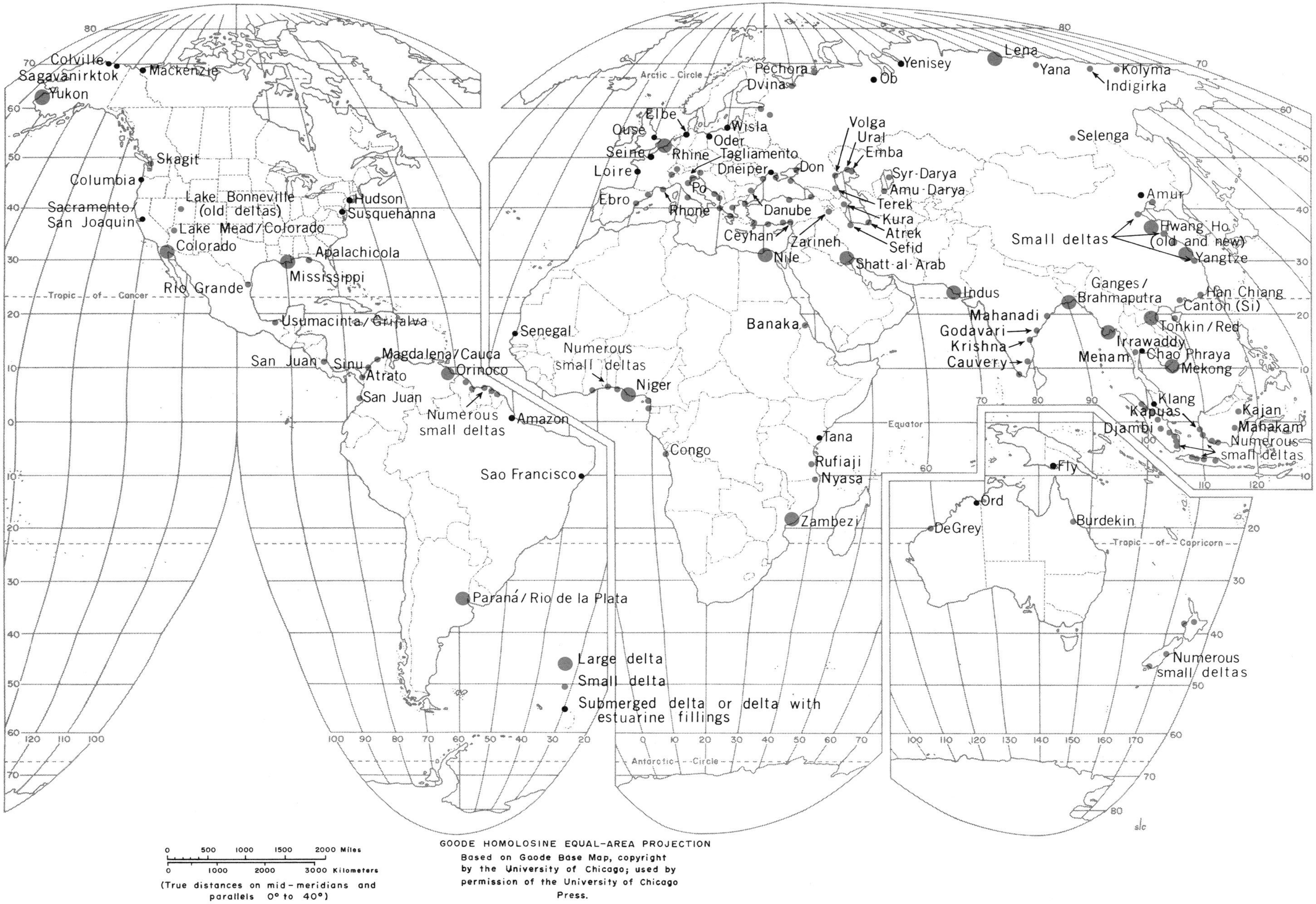

MAP 3-11 MAJOR DELTAS OF THE WORLD

GOODE HOMOLOSINE EQUAL-AREA PROJECTION
Based on Goode Base Map, copyright
by the University of Chicago; used by
permission of the University of Chicago
Press.

Large delta

Small delta

Submerged delta or delta with
estuarine fillings

0   500   1000   1500   2000 Miles

0   1000   2000   3000 Kilometers

(True distances on mid-meridians and
parallels 0° to 40°)

MAPS
3-11
3-12
3-13

MAJOR DELTAS

When a river enters the sea, it drops the material it has been carrying in suspension and forms a delta. The shape of the delta depends on the character of the river, its load, and the body of water in which the delta is built. The most common type of delta is arcuate in form, consisting of a convex-outward margin and intricate channels. The term *anastomosing* is applied to these streams that are braided into a network of small watercourses, the result of extensive deposition in the stream bed. Most of these small channels are shallow and change their positions frequently during high water. The most famous deltas of this type are the Nile, Rhine, Hwang Ho, Niger, Indus, Irrawaddy, Ganges, Mekong, Danube, Po, Rhone, Ebro, Volga, and Lena. Most deltas tend to be relatively simple fans or arcuate shaped. A delta with distributaries or passes radiating from a common point is termed *bird's foot.* The main channel of the river is controlled by resistant clays laid down by earlier deltas. The Mississippi River delta is a bird's foot delta. A number of submerged deltas represent estuarine fillings. The Mackenzie, the Elbe, the Wisla (Vistula), the Oder, the Susquehanna, the Seine, the Loire, the Ouse, the Ob, and the Hudson are examples of rivers depositing their loads in the form of long, narrow estuaries, which may constitute extensive flood plains or marshy land areas. Deltas built into lakes and inland seas are more perfect and less variable than the deltas built into the ocean with its strong currents. Some of the most perfect deltas known were those built into Lake Bonneville during glacial times. Small deltas are presently forming in Seneca Lake, a finger lake in central New York State, and in Lake Thun and Lake Geneva, Switzerland. There are far too many small deltas to plot them all on the world map. Numerous deltas occur along the eastern side of Sumatra, northern Java, southern Borneo, and the coast of the Gulf of Guinea. Parts of the Guiana coast of South America are deltaic in character, as well as the north coast of Iceland. Plate 8 (centerfold) is an excellent satellite photograph showing the delta of the Tigris and Euphrates Rivers as they join together and form one channel called the Shatt-al-Arab. Where the delta empties into the Persian Gulf three countries come together: Iraq, Iran, and Kuwait.

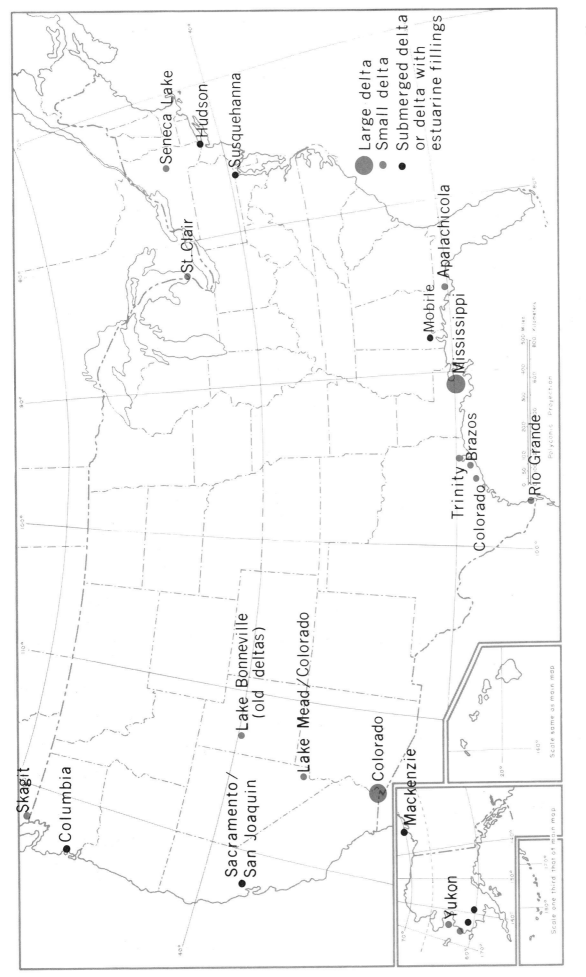

MAP   3-12   MAJOR DELTAS OF THE UNITED STATES

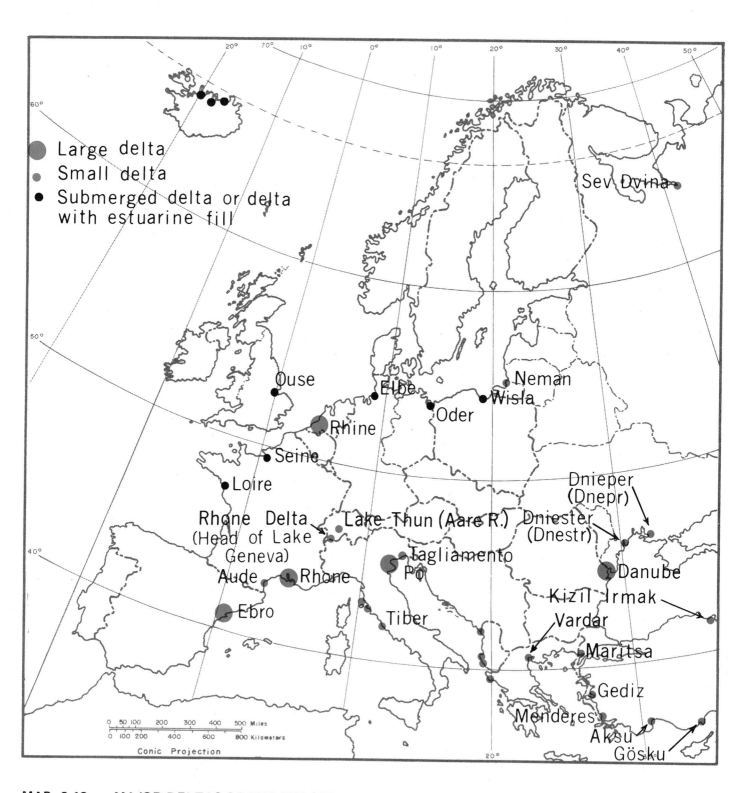

MAP 3-13   MAJOR DELTAS OF THE EUROPE

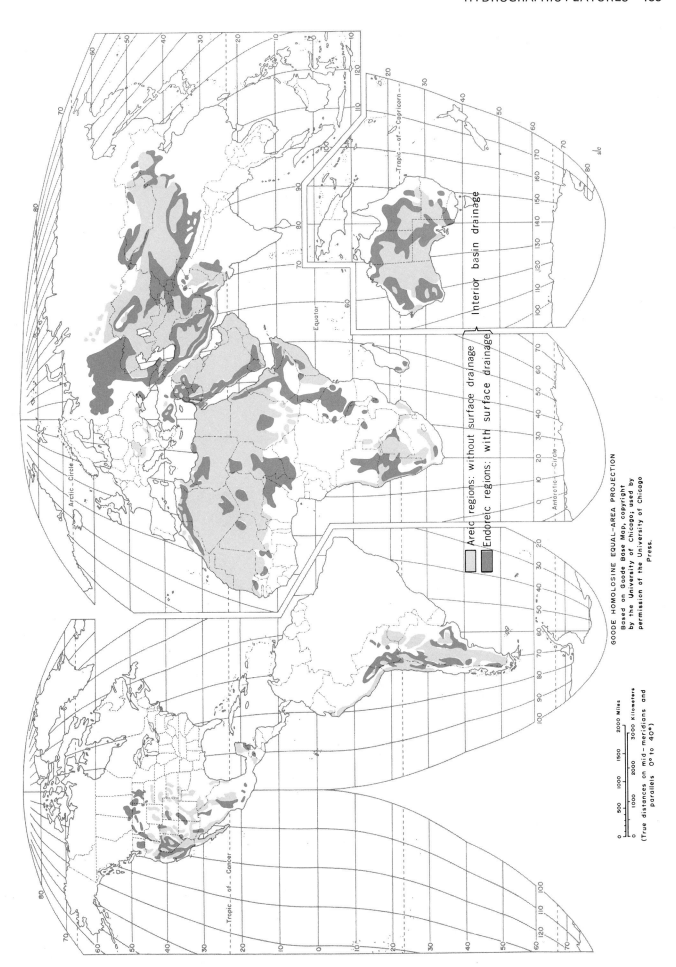

GOODE HOMOLOSINE EQUAL-AREA PROJECTION
Based on Goode Base Map, copyright
by the University of Chicago; used by
permission of the University of Chicago
Press.

Areic regions: without surface drainage

Endoreic regions: with surface drainage

Interior basin drainage

MAP 3-14   INTERIOR BASIN DRAINAGE

**MAP
3-14**

INTERIOR BASIN DRAINAGE

This map is presented in the atlas because it divides interior drainage into two main types: (1) areic regions without surface drainage, and (2) endoreic regions, where rivers flow at some time in every year on the surface, if only for a few days, but drain into interior basins. In principle, areic and endoreic regions are all areas that do not originate rivers that reach the sea, either directly or through flow into some other stream. A river crossing the region, as the Nile crosses Egypt, does not basically change the areic character of the region. The Nile is an alien feature of the Libyan Desert; it does not receive a single tributary stream once it enters the desert, but on the contrary loses a great part of its volume. The same is the case with the Sir Darya and Amu Darya rivers in Russian Turkestan. According to de Martonne, interior basin drainage covers more than 16 million square miles, more than 27 percent of the total land surface of the earth, and 33 percent if polar lands are excluded. These figures are quite surprising, because they indicate that a large percentage of surface water does not reach ocean basins. The map indicates that North America is the best drained continent, the endoreic and areic domain being about 10 percent. Surprisingly South America is not so well drained as North America, in spite of the enormous rainfall of the great Amazon basin. But South America is better drained than Europe. The anomaly is due to the northward extension of the Volga basin, which is a tributary of the Caspian Sea. Asia, Africa, and Australia are continents where drainage is less perfect. Interior basins resulting from faulting or warping are numerous and everywhere create systems of slopes converging toward closed basins that could drain toward the ocean only if the climate were very humid. Australia has the record for interior drainage with 64 percent; about 43 percent of this area is without surface drainage.

		ENDOREIC		AREIC	
Region	Area	Percent	Area		Percent
Europe	2,205	24	452		5
Asia	14,847	35	9,935		24
Africa	15,223	52	11,771		40
Australia	4,920	64	3,309		43
North America	2,136	10	1,070		5
South America	2,507	14	1,454		8
Northern Hemisphere	32,386	34	22,139		23
Southern Hemisphere	9,452	28	5,852		17
World	41,838	33	27,991		23

Plate 9 (centerfold) is a satellite photograph of the salars (salt lakes) of central Bolivia. These lakes, representing interior drainage, occur in a region called the Altiplano between high 5,200 meter (17,000 feet) peaks of the Andes Mountains.

Source: De Martonne, E., "Regions of Interior Basin Drainage," *The Geographical Review,* Vol. 17, pp. 397-414, 1927, and Aufrere, L., "L'orientation des dunes et la direction des Venes," *Ac. des Sci.,* C.R., Vol. 187, pp. 833-35, 1928.

**MAP
3-15**

MAJOR LAKES OF THE WORLD

There are numerous lakes throughout the world. Most are formed through the direct or indirect effects of glaciation, chemical action in limestone, dams built by man, damming by laval flows, and the formation of tectonic and structural basins. Because glacial, karst, and lava-dammed lakes are so numerous, this map can only show broad regions of occurence. Only large, well-known lakes can be shown. A seperate map for the world and the United States has been drawn showing large artificial lakes. By far the largest number of lakes occur in glaciated areas. Across Canada, the northern portion of the United States, and northern Europe there are literally thousands of lakes of many sizes and depths. Minnesota, for example, is called the "Land of 10,000 Lakes," and there may well be this many. Most glacial lakes are found in basins that have been deepened through the glacial erosion of bedrock. Some basins may have had an earlier structural or tectonic origin, altered through glaciation. When this has occurred, I have indicated the two origins by putting two symbols together. Many small lakes closely associated with kames and kettles, moraines, and eskers are the result of damming by glacial drift. A comparison of the lake regions of northern Europe and the lake regions of Ohio, Indiana, Wisconsin, and Minnesota can be made with moraines, kames and kettles (Map 5-15), and eskers (Map 5-17). Many glacial lakes result from crustal movements: when the large glaciers melted, an enormous weight was taken off the land, and the release of this weight caused large sections of the land to re-bound, separating basins and valleys from the sea and forming lakes. A number of lakes in Canada, New England, and Scandin-avia originated this way. In North America, the southern boundary of glaciated lakes roughly follows the edge of the last glaciation (see Map 5-2). This line extends from Nantucket Island west along Long Island, across southern New York State, and around the Great Lakes into northern Iowa and the Dak-otas. The Great Lakes are shaped by glaciers, as are the Finger Lakes in New York. In the western United States, rock basins, some occupied by lakes, occur in the high Sierra Nevadas of California and the Rocky Mountains, especially Glacier Nation-al Park and Rocky Mountain National Park. Smaller glacial lakes, many of them tarns in glacial cirques, can be found in the Cascades, Wasatch, and Grand Teton ranges. Large numbers of glacial lakes exist in Finland, Sweden, and Norway. In Finland, there are probably 250,000 lakes, but they are difficult to count because watersheds are extremely irregular and indefinite. Norway has about 200,000 glacial lakes, four of them the deepest in Europe; several are 1,700 feet (521 meters) deep. Another region of lakes in Europe extends from Denmark across the north European Plain. In Germany, there are two main lake belts associated with the two major moranic systems. The northern group of lakes is on the glacial drift of the Pomerania moraine, which stretches from northern Denmark into the Soviet Union. The southern group of lakes, south of Berlin, follows the Brandenburg moraine and extends into eastern Poland (see Map 5-16). There are numerous small cirque lakes high in the Alps, but the largest are the Piedmont Lakes at the southern fringe of the Alps. These are the beautiful finger lakes of Italy. Lakes are numerous in many high mountain areas where glaciers have scoured out rock basins. The Himalayas, Caucasus, Pyrenees, and the high mountains of central Asia contain scattered glacial lakes, as do the high Andean Ranges. Finger-lakes are rather unique; as they possess many of the features of fiords, but they formed inland by glaciers deepening long, narrow rock basins, which are then occupied by lakes. Many of them are preglacial valleys, and often they are on the opposite sides of mountains from fiords, formed by valley and piedmont glaciers moving down the eastern sides of high mountain ranges. For example, the great fiord region of the Alaskan and British Columbian coast has as its equivalent in the interior the finger-lake region of northern Canada and British Columbia. The fiords of Norway are balanced in the same way in the interior of Scandinavia, including the picturesque finger-lake district of Sweden. There are also large finger-lake regions in Chile and Argentina on the eastern slopes of the Andes: Lago Buenos Aires and Argentino are two of the largest. Other finger-lakes are Lake Geneva, Lac De Neuchatel, and the Zurich Sea in Switzerland, as well as the lake districts of England, Scotland, and Wales. The lochs of Scotland are, in some instances, true fiords; but in other cases they represent fresh water lakes, barely separated from tidal waters by a low neck of land.

Areas of limestone rock often have many lakes. Lakes abound in the Lakeland-Orlando-Gainsville Florida region, in basins in soluable limestone. Because Florida is not very high above sea level, groundwater easily accumulates in the rounded depressions, forming lakes. Unlike Florida, Yunnan Province of China is high, and limestone rock and soil cause the water to sink to a submerged water table, so that many of the depressions, called dolines, are large dry open basins. When the water table rises, lakes are formed. Dolines are also found in large sections of the Yucatan Peninsula.

In volcanic regions, where irregular distribution of volcanoes tend to form intermontane basins, lakes are found. In the central Mexican highlands they are numerous. The Valley of Mexico was once covered by several bodies of water, and the capital city Tenochtitlan, now Mexico City, was built on islands in the lakes. The largest lakes of this region existing today are the Lake Chapala, Lake Patzcuaro, and Lake Cuitzeo. In Central America the large lakes in the Nicaraguan and Managuan lowlands have been formed by faulting and lava flows. The Armenian region of Turkey around the cone of Mount Ararat has volcano-formed lakes. Lake Van and Lake Urmia do not have outlets and consequently are salt lakes. Lake Sevan is another lake to the northeast of Mount Ararat in the Soviet Union.

Lakes sometimes form in large joints or cracks in massive rocks. The most common pattern is where two or more joint systems intersect at more or less right angles, forming a rectangular basin. Structurally-formed lakes can be found in the Adirondack Mountains: Lake Placid is such a lake. Large structurally formed lakes occur in long trench-like depressions known as rift valleys or grabens (see Maps 2-8 and 2-9). The Dead Sea in Israel and Jordan, Lakes Tanganyika and Nyasa in Africa, Lake Baikal (Baykal) in central Siberia, and Lake Tahoe between California and Nevada are good examples of lakes occupying rift valleys.

Large lakes found in enclosed desert basins are mostly remnants of former, much larger, Pleistocene lakes (Maps 5-19; 5-20 and 5-21).

LAKES

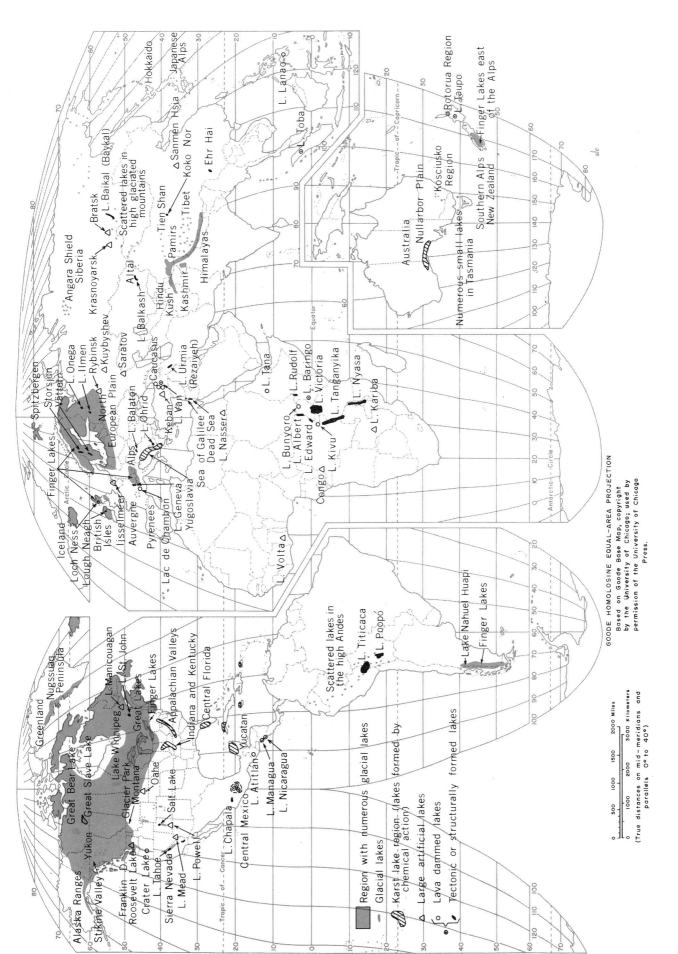

GOODE HOMOLOSINE EQUAL-AREA PROJECTION
Based on Goode Base Map, copyright
by the University of Chicago; used by
permission of the University of Chicago
Press.

0   500   1000   1500   2000 Miles

0   1000   2000   3000 Kilometers

(True distances on mid-meridians and
parallels 0° to 40°)

Region with numerous glacial lakes

Glacial lakes

Karst lake region (lakes formed by chemical action)

△ Large artificial lakes

○ Lava dammed lakes

● Tectonic or structurally formed lakes

MAP 3-15   MAJOR LAKES OF THE WORLD

LARGEST MAN-MADE LAKES AND RESERVOIRS

Large artificial lakes are becoming numerous everywhere as storage basins and recreation centers. Thousands of man-made lakes can be seen when one flies across the United States and the spread of large dams and man-made reservoirs in other countries of the world is phenomenal. On the world map, 30 large lakes and reservoirs are shown with a storage reservoir capacity of more than 25,000,000 acre feet (10,115,000 hectares). The Soviet Union and China have built or are building presently, extremely impressive bodies of water. Bratsk dam on the Angara River has a reservoir capacity of 137,214,000 acre feet (55,516,784 hectares). Large dams and reservoirs are being constructed in numerous African countries. The Cabara Basa dam and reservoir on the Zambezi River in Mozambique will have a storage capacity of 129,389,000 acre feet (52,350,789 hectares) when completed. Canada, is constructing a number of large reservoirs. The Daniel Johnson dam and reservoir on the Manicougan — St. Lawrence, completed in 1968, has a storage capacity of 115,000,000 acre feet (46,529,000 hectares).

The United States has by far the largest number of artifical lakes. Nearly every major river has a dam with a man-made lake behind it. A total of 81 major ones with a storage capacity of more than 1,000,000 acre feet (404,600 hectares) of water in the reservoirs are shown. Many more could be plotted if space on the map allowed. Large bodies of water now exist in remote semi-arid and arid regions. New Mexico, which lacked lakes except in the high mountain regions, now has at least five large lakes, two are shown on the map, and Texas is dotted with both small and large bodies of water. Ten of the largest reservoirs are shown in Texas alone. California, with numerous dams and lakes along the western slopes of the Sierra Nevada, has nine with a storage capacity over 1,000,000 acre feet (404,600 hectares).

Sources: *Reclamation Project Data (Supplement),* United States Department of the Interior, Bureau of Reclamation, U.S. Government Printing Office, Washington, D.C., 1966, pp. 258-270, (Revised May, 1975); *The World Almanac and Book of Facts, 1976,* Newspaper Enterprise Association, Inc., New York, 1976, pp. 590-591.

THE FOLLOWING IS A LIST OF 30 OF THE LARGEST MAN MADE LAKES AND RESERVOIRS

	NAME OF DAM	NAME OF LAKE OR RESERVOIR	YEAR	RIVER & BASIN	COUNTRY	RES. CAPACITY (1000 ACRE FEET)	RES. CAPACITY (1000 HECTARES)
1.	Tankiangkow		1962	Tan and Han	China	41,833	16,901
2.	Sanmen Hsia		1962	Hwang Ho-Yellow	China	52,700	21,322
3.	Zeya		UC	Zeya	USSR	55,080	22,285
4.	Irkutsk	Irkutsk	1956	Angara	USSR	37,290	15,088
5.	Bratsk	Bratskoye	1964	Angara	USSR	137,214	55,517
6.	Ust-Llim	Ust-Llimsk	UC	Angara	USSR	48,100	19,461
7.	Krasnoyarsk		UC	Yenisei	USSR	59,425	24,043
8.	Kapchagay		1970	Lli	USSR	22,813	9,230
9.	Sayansk		UC	Yenisei	USSR	25,353	10,258
10.	Rybinsk	Rybinskoye	1941	Volga-Caspian Sea	USSR	20,590	8,331
11.	Volga-22nd Congress USSR	Kuybyshevskoye	1958	Volga-Caspian Sea	USSR	27,160	10,989
12.	Volga V.I. Lenin	Volgogradskoye	1955	Volga-Caspian Sea	USSR	47,020	19,024
13.	Keban		UC	First (Euphrates)	Turkey	25,110	10,160
14.	High Aswan (Saad-El-Aali)	Lake Nasser	1971	Nile	UAR	133,000	53,812
15.	Owen Falls *	Owen Falls	1954	Lake Victoria-Nile	Uganda	166,000	67,164
16.	Kariba	Kariba Lake	1959	Zambesi	Rhodesia-Zambia	130,000	52,598
17.	Cabora Basa		UC	Zambesi	Mozambique	129,389	52,351
18.	Akosombo-Main	Lake Volta	1965	Volta	Ghana	120,000	48,552
19.	Kossou		1973	Bandama	Ivory Coast	24,000	9,710
20.	Itaipu		UC	Parana	Brazil-Paraguay	23,511	9,513
21.	Daniel Johnson		1968	Manicougan-St. Lawrence	Canada	115,000	46,529
22.	Churchill Falls	Churchill Falls	1971	Churchill	Canada	26,200	10,601
23.	Mica	Mica Creek Res.	UC	Columbia	Canada	20,000	8,092
24.	W.A.C. Bennett	Peace River Res.	1967	Peace-Mackenzie	Canada	57,006	23,065
25.	Iroquois	Iroquois	1958	St. Lawrence	Canada	24,288	11,041
26.	Oahe	Oahe Res.	1963	Missouri	USA	23,600	9,549
27.	Garrison	Garrison	1956	Missouri	USA	24,500	9,913
28.	Fort Peck	Fort Peck	1940	Missouri	USA	19,400	7,849
29.	Glen Canyon	Lake Powell	1964	Colorado	USA	27,000	10,924
30.	Hoover	Lake Meade	1936	Colorado	USA	29,755	12,039

UC   —   Under Construction
*   —   Represents increase in major natural lakes

MAP 3-16 MAJOR MAN–MADE LAKES AND RESERVOIRS OF THE WORLD

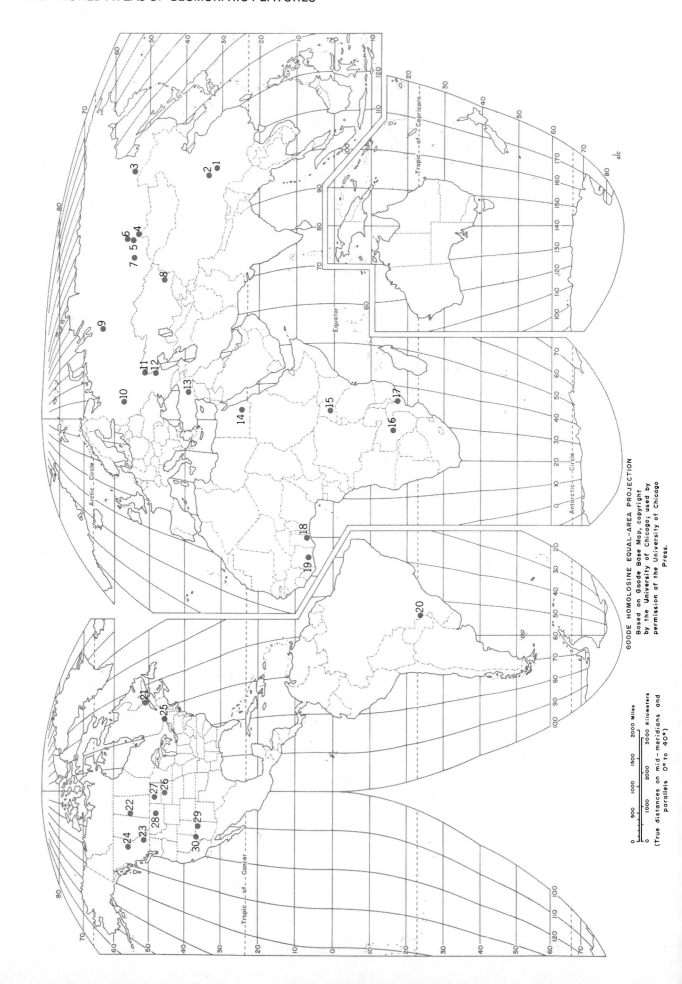

GOODE HOMOLOSINE EQUAL-AREA PROJECTION
Based on Goode Base Map, copyright
by the University of Chicago; used by
permission of the University of Chicago
Press.

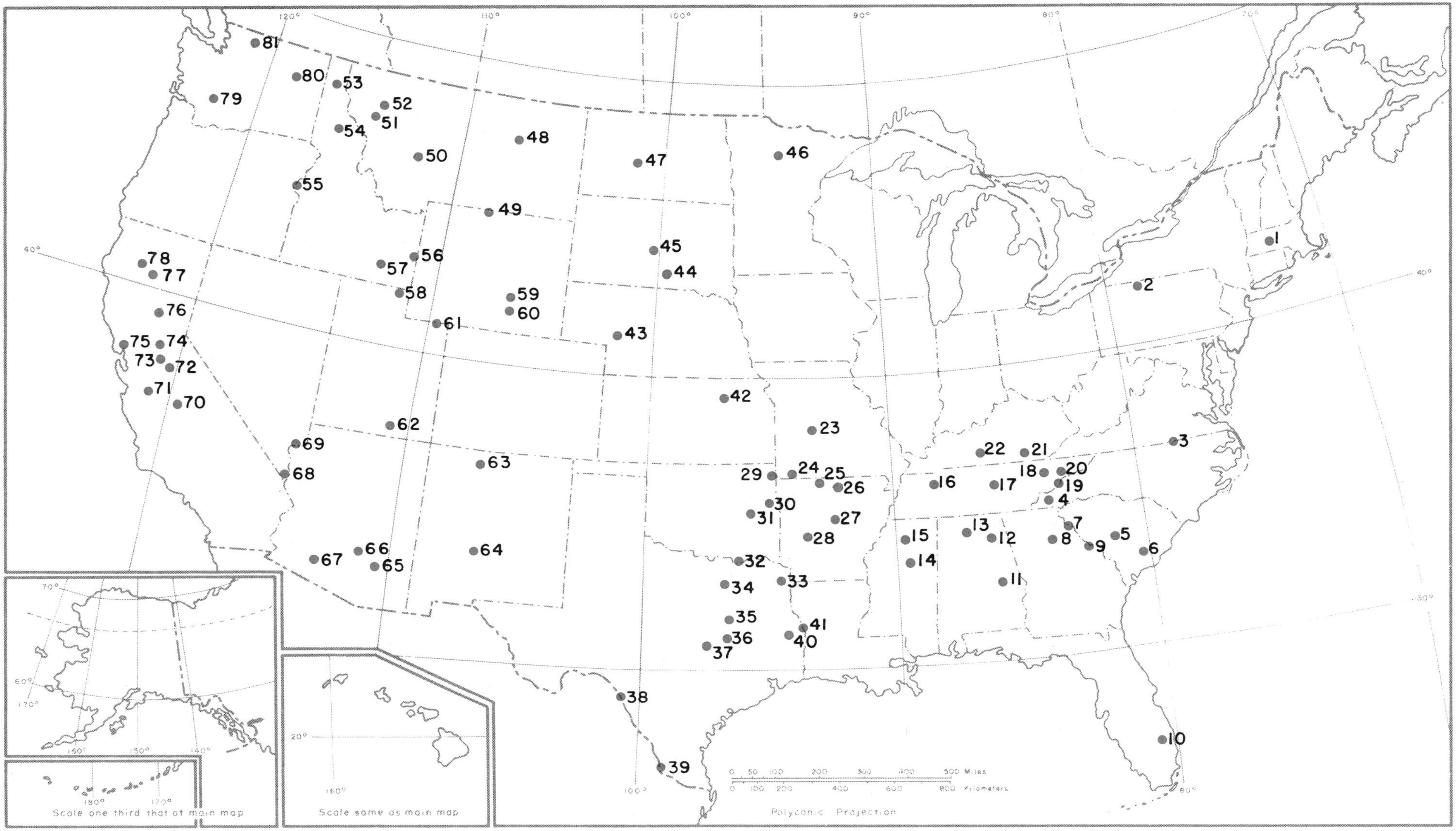

MAP 3-17 · MAJOR MAN—MADE LAKES AND RESERVOIRS OF THE UNITED STATES

MAJOR MAN-MADE LAKES AND RESERVOIRS IN THE UNITED STATES

	NAME OF RESERVOIR	NAME OF DAM	STATE	RIVER	ACRE/FEET	HECTARES
1.	Quabbin		Massachusetts	Swift R.	1,236,000	500,086
2.	Allegheny		Pennsylvania	Allegheny R.		
3.	John H. Kerr		Virginia	Staunton R.	2,110,500	853,908
4.	Fontana		North Carolina	Little Tenn. R.	1,157,300	468,244
5.	Lake Murray		South Carolina	Saluda R.	1,614,000	653,024
6.	Lake Marion		South Carolina	Santee R.	1,099,900	445,020
7.	Hartwell		Georgia	Savannah R.	1,708,000	691,057
8.	Lake Sidney Lanier	Buford Dam	Georgia	Chattahoochee R.	1,686,400	682,317
9.	Clark Hill		Georgia	Savannah R.	1,730,000	699,958
10.	Lake Okeechobee		Florida	Lake	1,300,000	525,980
11.	Lake Martin	Cherokee Bluffs Dam	Alabama	Tallapoosa R.	1,375,000	556,325
12.	Lake Guntersville		Alabama	Tennessee R.		
13.	Lake Wheeler		Alabama	Tennessee R.		
14.	Lake Grenada		Mississippi	Yalohusha R.	1,251,700	506,438
15.	Lake Sardis		Mississippi	Tallahatchie R.	1,478,000	597,999
16.	Kentucky Lake		Tenn./Kentucky	Tennessee R.	4,010,800	1,622,770
17.	Center Hill		Tennessee	Carrey Fork R.	1,254,000	507,368
18.	Norris Lake		Tennessee	Powell R.	2,281,000	922,893
19.	Douglas Lake		Tennessee	French Broad R.	1,419,700	574,411
20.	Cherokee Lake		Tennessee	Holslon R.	1,473,100	596,016
21.	Lake Cumberland	Wolf Creek Dam	Kentucky	New R.	4,236,000	1,713,886
22.	Barren		Kentucky	Barren R.		
23.	Lake of the Ozarks	Bagnell Dam	Missouri	Osage R.	1,246,000	504,132
24.	Table Rock		Missouri	James/King R.	3,462,000	1,400,725
25.	Bull Shoals		Arkansas	Beaver R.	5,399,000	2,184,435
26.	Northfork		Arkansas	Northfork R.	1,980,500	801,310
27.	Greer Ferry		Arkansas	Little Red R.	2,843,860	1,150,626
28.	Lake Ouachita	Blakeley Mtn. Dam	Arkansas	Lake	2,602,000	1,052,769
29.	Lake of the Cherokees	Pensacola Dam	Oklahoma	Neosho R.	2,017,000	816,078
30.	Tenkiller		Oklahoma	Illinois R.	1,230,500	497,860
31.	Eufaula		Oklahoma	Canadian R.	2,951,000	1,193,975
32.	Lake Texoma	Denison Dam	Oklahoma/Texas	Red/Wildhorse R.	4,424,000	1,716,313
33.	Lake Texarkana		Texas	Sulfur R.	2,654,700	1,074,092
34.	Garza-Little Elm		Texas	Elm Fork-Trenitz	1,002,700	405,692
35.	Lake Whitney		Texas	Brazos R.	2,012,400	814,217
36.	Lake Belton		Texas	Leon R.	1,097,700	444,129
37.	Lake Travis	Marshall Ford Dam	Texas	Colorado R.	1,922,000	777,641
38.	Amistad		Texas	Rio Grande R.		
39.	Falcon		Texas	Rio Grande R.	3,277,880	1,326,230
40.	Sam Rayburn		Texas	Attoyae-Angelina R.	4,478,800	1,182,122
41.	Toledo Bend		Texas/Louisiana	Sabine R.		

MAJOR MAN-MADE LAKES AND RESERVOIRS IN THE UNITED STATES, CONTINUED

NAME OF RESERVOIR	NAME OF DAM	STATE	RIVER	ACRE/FEET	HECTARES
42. Tuttle Creek		Kansas	Little Blue-Big Blue	2,367,000	957,688
43. Lake McConaughty		Nebraska	North Platte R.	1,948,000	788,161
44. Fort Randell		South Dakota	Missouri R.	6,093,000	2,465,228
45. Oahe		South Dakota	Missouri R.	23,628,000	9,559,889
46. Red Lake		Minnesota	Red Lake R.	1,905,000	770,763
47. Garrison		North Dakota	Missouri	24,500,000	9,912,700
48. Fort Peck		Montana	Little Dry-Missouri and Musselshell R.	19,400,000	7,849,240
49. Hungry Horse		Montana	N. Fork Flathead R.	1,219,000	493,207
50. Flathead	Kerr Dam	Montana	Flathead R.	3,468,000	1,403,153
51. Canyon Ferry Lake		Montana	Missouri R.	2,043,000	826,598
52. Yellow Trail		Montana/Wyoming	Bighorn-Missouri		
53. Pend Oreille Lake		Idaho	Pend Oreille R.	1,155,000	467,313
54. Dworshak		Idaho	N. Fork Clearwater	3,453,000	1,397,084
55. Brownlee		Idaho	Snake R.	1,000,000	404,600
56. Palisades		Idaho	Snake R.	1,202,000	486,329
57. American Falls		Idaho	Snake R.	1,700,000	687,820
58. Bear Lake		Idaho	Bear R.	1,420,000	574,532
59. Path Finder		Wyoming	North Platte R.	1,016,000	411,074
60. Seminoe		Wyoming	North Platte R.	1,011,600	409,293
61. Flaming Gorge		Utah/Wyoming	Green R.	3,789,000	1,533,029
62. Lake Powell		Utah/Arizona	Green/Colorado R.	27,000,000	10,924,200
63. Navajo		New Mexico	San Juan R.	1,036,000	419,166
64. Elephant Butte		New Mexico	Rio Grande R.	2,195,000	888,097
65. San Carlos Lake	Coolidge Dam	Arizona	Gila R.	1,205,500	487,745
66. Roosevelt Lake		Arizona	Salt R.	1,382,000	559,157
67. Painted Rock		Arizona	Gila R.	2,492,000	1,008,263
68. Lake Mohave	Davis Dam	Arizona	Colorado R.	1,809,800	732,245
69. Lake Mead	Hoover	Arizona/Nevada	Colorado/Virgin R.	29,755,000	12,038,873
70. Pine Flat		California	Kings R.	1,013,400	410,022
71. San Luis		California		2,110,000	853,706
72. Don Pedro		California		2,030,000	821,338
73. New Melones		California	N. Fork-Stanislaus	2,400,000	971,040
74. Folsom Lake		California	Americana R.	1,010,300	408,767
75. Lake Berryessa		California	Putah Creek	1,592,000	644,123
76. Oroville		California	Feather R.	3,498,000	141,529
77. Shasta Lake		California	Pit-Squaw-McCloud	4,500,000	1,820,700
78. Trinity Lake		California	Trinity R.	2,437,000	986,010
79. Mossyrock		Washington	Columbia R.	1,300,000	525,980
80. FDR Lake	Grand Coulee	Washington	Kettle-Columbia R.	9,724,000	3,934,330
81. Ross Lake		Washington	Skagit R.	1,404,100	568,099

**MAPS
3-18
3-19**

WATERFALLS

There are tens of thousands of waterfalls scattered over the earth, hundreds of them of considerable magnitude. It would be impossible to plot and name every waterfall. The problem is to decide which waterfalls should be included on the world map. The first criterion is that all waterfalls shown are major features easily identified. Height alone does not indicate the importance. Other significant factors are volume of flow, steadiness or variability of flow, width of crest, whether the water drops vertically or flows over a sloping surface, and whether it plunges in a single leap or in a succession of leaps. Relatively low falls in succession over a considerable length of stream bed are called cascades. Perhaps the most spectacular of all is Angel Falls in southeastern Venezuela, the highest known waterfall. Located on a tributary of the Caroni River, these falls drop more than 3,212 feet (979 meters) over a face of flat-lying quartzite. The descent is made in two falls, the upper one measuring 2,648 feet (807 meters). Victoria Falls, located on the Zambezi River in southern Africa, is more than a mile wide and plunges 355 feet (93 meters) through flat lava layers into a narrow chasm. Grand Falls, on the Hamilton River in Labrador, cascades a large flow 245 feet (75 meters) through resistant granitic rock into a narrow gorge. The Iguazu River, on the border of Argentina and Brazil, falls about 210 feet (64 meters) with great volume. The greatest in terms of both volume of water and height is Guaira Falls, on the upper Parana River between Brazil and easternmost Paraguay, where a series of rapids and falls descend nearly 130 feet (40 meters) with a discharge of more than 470,000 cubic feet (13,308 cubic meters) a second, more than twice that of the Niagara River. Less well known high waterfalls are Gavarnie Falls in the Pyrenees of southwestern France, which drop 1,385 feet (422 meters) into a glacial amphitheatre, and Takakkaw Falls in Yoho National Park, southeastern British Columbia, which have several falls, the highest being 1,200 feet (366 meters). There are many areas of small waterfalls. In the fiord regions of Norway, British Columbia, Chile, and New Zealand, they are abundant. The central highlands of Ceylon and the "Ghats" of India have waterfalls of considerable volume when the monsoon rains occur, but during the dry months most of these falls almost disappear.

By far the best-known waterfall in the United States is Niagara. Goat Island divides Niagara into Horseshoe Falls (Canadian) and the American Falls. Horseshoe Falls receives nine-tenths of the Niagara River's Flow. It is about 2,500 feet (762 meters) wide and 186 feet high (57 meters). Niagara Falls has receded several miles in the 10,000 years since glacial ice melted from the region. Some of the highest waterfalls issue from hanging tributary valleys above main valleys deepened by glaciers. In Yosemite National Park, California there are several spectacular falls of this type: Ribbon Falls drops 1,612 feet (491 meters); Yosemite Falls has an upper fall of 1,430 feet (436 meters) and a lower fall of 320 feet (97 meters); and Bridalveil is 620 feet (189 meters) high. Yellowstone Falls, in Yellowstone National Park, Wyoming has upper and lower falls of 109 and 308 feet (33 and 94 meters) controlled by vertical dikes of lava intruded into less resistant volcanic sediments; these falls are retreated very slowly. The Finger Lake region of central New York State has a number of small, picturesque waterfalls, even more spectacular in winter when they are covered by ice. Some falls are not high but have great volume. Niagara Falls, with a discharge of more than 230,000 cubic feet (6,513 cubic meters) per second is such a waterfall. Another is 75-foot (23 meters) high Great Falls on the Missouri River in Montana.

Sources: Numerous references consulted, mainly drawn from *The World Almanac and Book of Facts, 1976,* Newspaper Enterprises, Assoc., Inc., New York, 1976, p. 583; G. Crossette, "Waterfalls Nature's Extravaganza," *Explorers Journal,* Vol. 53, No. 4, 1975, pp. 146-153; and *Encyclopedia Americana,* America Corporation, New York, 1968, pp. 462-465.

THE FOLLOWING IS A LIST OF THE MAJOR WATERFALLS OF THE WORLD:

C = Cascade R = River L = Lake

	NAME	LOCATION	HEIGHT IN FEET	HEIGHT IN METERS
1.	Kegon	L. Chuzenji, Japan	330	101
2.	Nachi	Nachi R., Japan	430	131
3.	Khone	Mekong R., Laos	70	21
4.	Gersoppa (Jog)	Sharavati R., India	830	253
5.	Cauvery	Mysore District, India	320	98
6.	Tully	Queensland, Australia	450	137
7.	Coomera	Queensland, Australia	210	64
8.	Wollomombi	New South Wales, Australia	1100	335
9.	Wentworth	Murray R., New South Wales, Victoria, Australia	518	158
10.	Bowen (from Glaciers)	South Island, New Zealand	540	165
11.	Sutherland	Arthur R., South Island, New Zealand	1904	580
12.	Browne	South Island, New Zealand	2746	837
13.	Tugela (5 Falls)	Natal, South Africa	2800	853
14.	Howick (Karkloof),	Umgeni R., Natal, South Africa	311	95
	and Albert	Umgeni R., Natal, South Africa	350	107
15.	Maletsunyane	Lesotho	630	192
16.	Aughrabies (King George's)	Orange R., South Africa/Southwest Africa	450	137
17.	Victoria Falls	Zambezi R., Rhodesia	355	108
18.	Rua Cana	Southwest Africa/Angola	406	124
19.	Duque de Braganca	Lucala R., Angola	344	105
20.	Kalambo	Tanzania-Zambia	726	221
21.	Stanley	Zaire (Congo)	200	61
22.	Murchison (Kabalega)	Victoria Nile, Uganda	140	43
23.	Baratieri Dal Verme	Ganale Dorya R., Ethiopia	459	140
	Dal Verme	Ganale Doyra R., Ethiopia	98	30
24.	Tisisat (Tesissat)	Blue Nile, Ethiopia	140	43
25.	Gavarnie (C)	Pyrenees, France	1385	422
26.	Toce (C)	Italy	470	143
27.	Terni	Verlino R., Italy	525	160
28.	Gietroz (Glacier) (C)	Switzerland	1640	500
	Giessbach	Switzerland	1312	400
	Trummelbach	Switzerland	1312	400
	Staubbach	Switzerland	984	300
	Reichenbach	Switzerland	656	200
	Stauber	Brunnibach R., Switzerland	590	180
	Simmen	Simme R., Switzerland	459	140
	Iffigen	Switzerland	394	120
	Diesbach	Switzerland	394	120
	Pissevache	La Salnfe R., Switzerland	213	65
	Handegg	Aare R., Switzerland	151	46
	Rhine	Switzerland	65	20
29.	Gastein	Ache R., Austria (Upper-207, Lower-279)	373	114
	Golling	Schwarzbach R., Austria	200	61
	Krimmi (Glacier)	Austria	1250	381

MAJOR WATERFALLS OF THE WORLD CONTINUED

	NAME	LOCATION	HEIGHT IN FEET	HEIGHT IN METERS
30.	Trollhattan	Sweden	108	33
31.	Tannforsen	Are R., Sweden	120	37
32.	Stora Sjofallet	Lule R., Sweden	130	40
33.	Harspranget (Harsprang)	Sweden	246	75
34.	Handol	Handol Creek, Sweden	345	105
35.	Kjelfossen	Norway	2600	792
	Kile	Norway	1840	561
	Eastern Mardal	L. Eikesdal, Norway	1696	517
	Western Mardal	L. Eikesdal, Norway	1535	468
	Vetti (Vettis)	Morkedola R., Norway	1214	370
	Rjukan	Norway	983	300
	Skykkje	Skykkjua R., Norway	820	250
	Voring	Bjoreia R., Norway	597	182
	Skjeggedal	Norway	525	160
	Skjaeggedalsfos	Tysso R., Norway	525	160
	Silvlefossenfalls	(near Stalheim) Norway	525	160
36.	Pistyll Rhaiadr	Wales, United Kingdom	240	73
37.	Pistyll Cain	Wales, United Kingdom	150	46
38.	Glomach	Scotland	370	113
39.	Detti	Jokul R., Iceland	144	44
	Gull (Gullfoss)	Hvita R., Iceland	101	31
40.	Herval	Brazil	400	122
41.	Iguazu (Cataratas del Iguazu)	Rio Iguazu, Brazil/Argentina	210	64
42.	Guaira (Salto das) Sete Quedas)	Alto Parana R., Brazil/Paraguay	130	40
43.	Urbupunga	Alto Parana R., Brazil	40	12
44.	Patos-Maribondo	Rio Grande, Brazil	115	35
45.	Paula Afonso Falls	Rio Sao Francisco, Brazil	270	82
46.	Kaieteur	Potaro R., Guiana	741	226
47.	King George III & IV	Utshi R., Guiana	1600	488
48.	King Edward VIII	Semang R., Guiana	840	256
49.	Kukenaan (Cuquenan)	Guiana	500	152
50.	Marina	Ipobe R., (tributary of Kuribrongi R., and Potaro R.) Guiana	2000	610
51.	Angel	Carrao R., Venezuela	3212	979
52.	Catarata de Candelas	Cusiana R., Colombia	984	300
53.	Tequendama	Bogota R., Colombia	427	130
54.	Agoyan	Pastaza R. (Agoyan), Ecuador	200	61
55.	Cacalotenango	Mexico		
56.	Juanacatian	Santiago R., Mexico	72	22
57.	El Salto	Mexico	218	66
58.	Great Falls (C)	Potomac R., Maryland, USA	90	27
59.	Taughannock	Finger Lake Region, N.Y., USA	215	66
60.	Niagara Horseshoe	Niagara R., New York, USA (US Falls) Canadian Falls	186	57
61.	Montmorency	Montmorency R., Quebec, Canada	251	77

MAJOR WATERFALLS OF THE WORLD CONTINUED

NAME	LOCATION	HEIGHT IN FEET	HEIGHT IN METERS
62. Grand Falls	Labrador, Canada		
63. Yellowstone Park Tower	Yellowstone Lower, Yellowstone R., Wyoming, USA	132	40
64. Great (Missouri)	Missouri R., Montana, USA	75	23
65. Takakkaw	Yoho National Park, British Colombia	1650	503
	Highest Fall	1200	366
66. Virginia	S. Nahanni R., Mackenzie District Northwest Territories, Canada		
67. Della	Strathcona Provincial Park, British Colombia, Canada	1443	440
68. Shoshone	Snake R., Idaho, USA	212	65
69. Multnomah	Colombia R., Oregon, USA	620	189
	Highest Fall	542	165
70. Yosemite Nat'l. Park	California, USA		
Yosemite		2425	739
Upper		1430	436
Lower		320	98
Ribbon		1612	491
Silver Strand		1170	357
Bridalveil		590	180
Nevada		590	180
Illilouette		370	113
Vernal		317	97

**Regions Where Numerous
Small Waterfalls Are Found:**

1. Central Honshu, Japan

2. The Grass Mountain region of Tawain

3. Central Luzon, Phillipines

4. Eastern Thailand

5. Southern Alps and Auckland districts
    of New Zealand

6. Eastern Australia

7. Central New Guinea

8. Central hills of Ceylon

9. The western Ghats and Cauvery River Valley
    of India

10. Himalayan and Karakorum Mountain Ranges
    of India, Pakistan, Soviet Union and Tibet

11. Drakensburg Ranges of South Africa

12. Rift Valleys of Kenya, Uganda, and Ethiopia

13. Caucasus Ranges of the USSR

14. Alps of Switzerland, Austria, Italy and France

15. Scandinavian Alps of Norway and Sweden

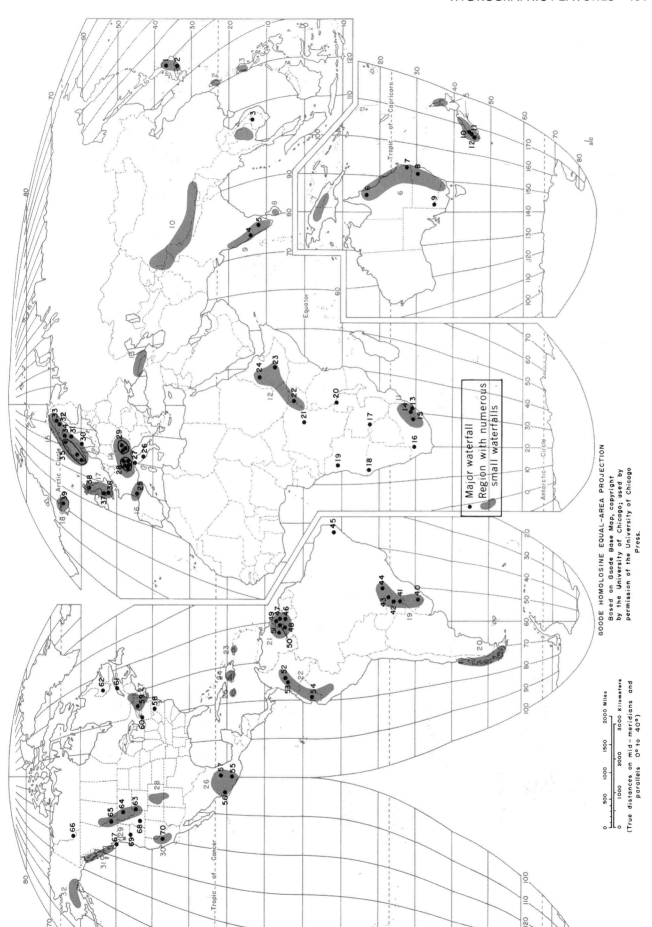

GOODE HOMOLOSINE EQUAL–AREA PROJECTION
Based on Goode Base Map, copyright
by the University of Chicago; used by
permission of University of Chicago
Press.

MAP   3-18   MAJOR WATERFALLS OF THE WORLD

MAP 3-19   MAJOR WATERFALLS OF THE UNITED STATES

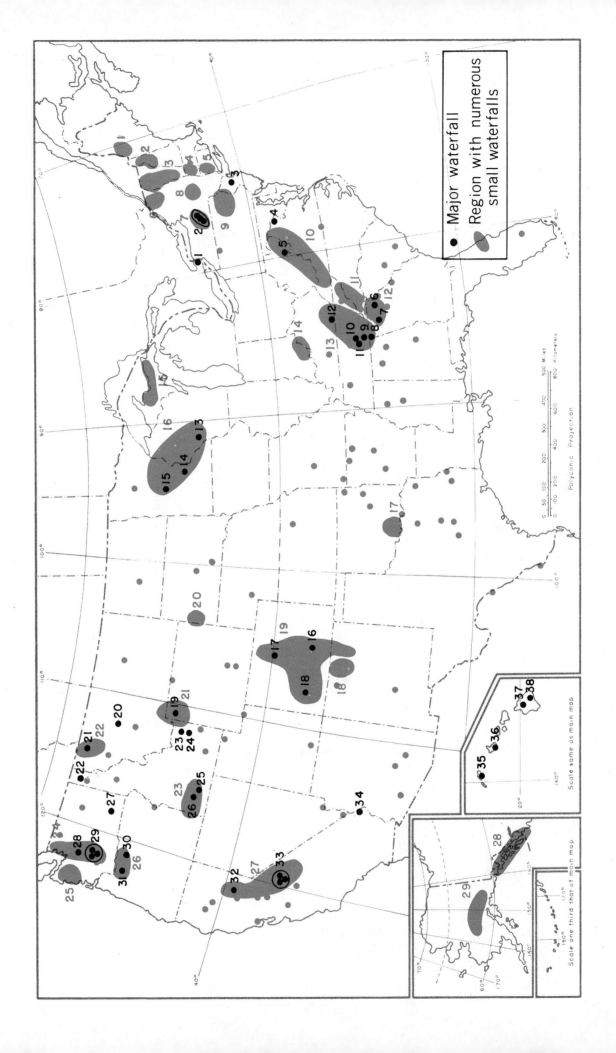

Legend:
● Major waterfall
Region with numerous small waterfalls

Polyconic Projection

Scale same as main map

Scale one third that of main map

THE FOLLOWING IS A LIST OF MAJOR WATERFALLS OF THE UNITED STATES.

C = Cascade R = River L = Lake

	NAME	LOCATION	HEIGHT IN FEET	HEIGHT IN METERS
1.	Niagara	Niagara R., New York		
	U.S. Falls		193	59
	Canadian Falls (Horseshoe)		186	57
2.	Finger Lake Region	New York	215	66
	Taughannock	Near L. Cagua	156	48
	Chequaga	Catherine Creek		
	Buttermilk Falls			
3.	Passaic	Passaic R., New Jersey	70	21
4.	Great (C)	Potomac R., Maryland/Virginia	90	27
5.	Black Water Falls	Blackwater R., West Virginia	57	17
6.	Tallulah	Tallulah R., Georgia	251	77
7.	Amicola	Amicola R., Georgia	729	222
8.	Ruby Falls	Lookout Mountain Caves, Chattanooga, Tennessee	145	44
9.	Cane Creek	Cane Creek, Tennessee	256	78
10.	Fall Creek	Fall Creek, Tennessee	256	78
11.	Rock House Creek	Rock House Creek, Tennessee	125	38
12.	Cumberland	Cumberland R., Kentucky	68	21
13.	Manitou	Black R., Wisconsin	165	50
14.	Anthony	Mississippi R.,	50	15
15.	Minnehaha	Minnehaha R., Minneapolis, Minnesota	54	16
16.	Seven	South Cheyenne Creek, Colorado	266	81
17.	Chasm	Fall River (Rocky Mountain Nat'l. Park) Colorado	25	8
18.	Bear Creek	Bear Creek, Colorado	227	69
19.	Yellowstone National Park	Wyoming		
	Yellowstone Park Tower		132	40
	Yellowstone Upper		109	33
	Yellowstone Lower		308	94
20.	Great Missouri	Missouri R., Montana	75	23
21.	Trick	Glacier Nat'l. Park, Montana	98	30
22.	Moyie	Kootenai R., Idaho	57	17
23.	Upper Mesa	Henry's Fork, Idaho	114	35
24.	Lower Mesa	Henry's Fork, Idaho	65	20
25.	Shoshone	Snake R., Idaho	210	64
26.	Twin	Snake R., Idaho	125	38
27.	Palouse	Palouse R., Washington	198	60
28.	Snoqualmie	Snoqualmie R., Washington	270	82

MAJOR WATERFALLS OF THE UNITED STATES, CONTINUED

	NAME	LOCATION	HEIGHT IN FEET	HEIGHT IN METERS
29.	Mount Ranier National Park	Washington		
	Fairy		700	213
	Comet	Van Trump Creek	320	98
	Sluiskin	Paradise R.	270	82
	Silver	Ohanapecosh R.		
	Chenuis	Chenuis Creek		
30.	Multnomah	Columbia R. Tributary, Oregon	620	189
31.	Latourelle	Merced R. Tributary, Oregon	224	68
32.	Feather	Fall R., California	640	195
33.	Yosemite National Park		2425	740
	Yosemite			
	Upper		1430	451
	Middle		675	206
	Lower		320	98
	Ribbon		1612	491
	Silver Strand		1170	357
	Bridalveil		620	189
	Nevada		594	181
	Illilouette		370	113
	Vernal		317	97
	Tuolumme	Tuolumme R.		
34.	Mooney	Havasu Creek, Arizona	220	67
35.	Wailua	Kauai, Hawaii	80	24
36.	Sacred	Hawaii	87	27
37.	Akaka	Hawaii, Hawaii	420	128
38.	Rainbow	Hawaii, Hawaii	120	37

Regions Where Numerous Small Waterfalls Are Found

1. Mt. Kataden, Main
2. White Mountains of New Hampshire
3. Green Mountains of Vermont
4. Berkshires of western Massachusetts
5. Western Connecticut
6. Adirondacks of New York
7. Finger Lake Region of New York
8. Catskills of New York
9. Poconos of Pennsylvania
10. Appalachian Mountains of Virginia and West Virginia
11. Smoky Mountains of North Carolina and Tennessee
12. Northeast Georgia
13. Eastern Tennessee
14. Southern Indiana
15. Upper Peninsula of Michigan
16. Northwest Wisconsin and eastern Minnesota
17. Arbuckle Mountains of Oklahoma
18. Jemez Region of New Mexico
19. Southern Rocky Mountains, Colorado and New Mexico
20. Black Hills of South Dakota
21. Yellowstone Region, Wyoming
22. Glacier National Park, Montana
23. Snake River Region, Idaho
24. Northern Cascades, Washington
25. Olympic Peninsula, Washington
26. Columbia River Valley, Oregon
27. Sierra Nevada Ranges, California
28. Fiord Region of Alaska
29. Alaska Ranges, Alaska

Plate 1—This Landsat space photograph illustrates very well the landforms shown on Map 1-8 High Mountain Peaks of the World, and also Map 2-7 Folded Mountain Ranges. Here rises the great Himalayan Ranges from the Ganges Plains to the crest of Mount Everest at 29,028 feet (8,848 meters). This famous peak lies just at the edge of the image (See Map 1-8). This region is rightly given the name "top of the world." Going north from the Ganges Plains one can pick out the Siwalik Hills at the base of the Himalayas at an elevation of from 3,300 feet to 4,300 feet (1,000 to 1,300 meters). The relief becomes much more rugged in the Lesser Himalayas which rise to 10,000 feet (3,000 meters). Farther north is the main crest ranges containing many peaks 20,000 to 29,000 feet (6,000 to 8,800 meters). The highest mountains are listed in the text. The Himalayans are one of the most active mountain building regions of the world as the Indian subcontinent continues to slam into Asia and then plunges under it along a great thrust fault (See Map 2-8). Because this is a tectonically active region numerous earthquakes occur (See Map 2-10). Latitude 27° 21'N; Longitude 86° 01'E. (Satellite—color infrared—NASA Landsat.)

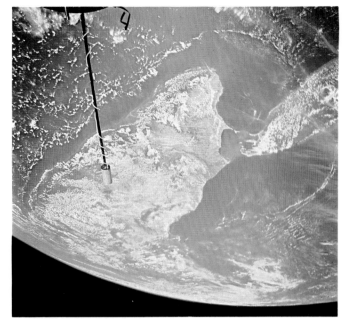

Plate 3—Ideas on continental drift have been forthcoming for the last twenty years. Now there is considerable evidence that continents are moving. This excellent space photograph of India and Ceylon was taken during the Gemini XI mission. The location and shape of the peninsula of India and Ceylon stand out sharply. The shield of south India is slowly moving north forcing up the Great Himalayan Ranges which are faintly visible far to the north at the horizon, 2,300 miles (3,680 kilometers) away. The island of Ceylon, now the country of Sri Lanka, has separated from India but its central mountain core is similar to the Nilgiri and Cardamon Hills of south India (See Map 2-2). Latitude 12° 88'N; Longitude 77° 36'E. (Satellite—color—NASA Gemini XI.)

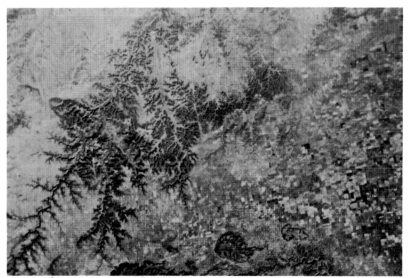

Plate 4—Southern California represents an excellent location to study landforms made by faulting. One of the most obvious features in this image is the intersection of the linear patterns produced by the San Andreas and the Garlock fault zones, shown where the darker tones of the mountains meet the lighter tones of the Mohave desert. At least a dozen major fault systems can be seen in this image. The reader has only to look for obvious linearity and then refer to a geologic map for confirmation. The city of Los Angeles and its many surrounding communities appear in the lower-right corner of the image. They are located on the largest lowland of this coastal region. Topographically the basins and valleys of southern California are ringed by mountains of the Transverse and Peninsula Ranges. Isolated hill areas outcrop throughout the lowlands, often aligned with fault zones and petroleum-producing areas. The Whitter fault extends through the northern part of Los Angeles. Because of the numerous faults, this is a very high earthquake zone (See maps 2-10, 2-11, 2-12). Latitude 34° 34'N; Longitude 118° 37'W. (Satellite—color infrared—NASA Landsat.)

Plate 2—Many examples could have been used for Map 1-15 Distribution of Erosion but the High Plains of northeastern New Mexico present a good example of a region undergoing heavy erosion by the Cimarron River and its tributaries. This was the region of the Dust Bowl in the 1930's and much of the surface is still too dissected or too dry for farming. The eastern part of this blown-up space photograph extends into Texas where more well-water is used and farming becomes much more intense than in New Mexico. The headward cutting of the Cimarron River and its tributaries is clearly depicted. The very fine branching of the small streams and gullies is called a pinnate drainage pattern. Latitude 36° 27'N; Longitude 104° 0.6'W. (Satellite—color infrared—NASA Landsat.)

Plate 5—About midway along the coast of the South Island of New Zealand there is a large promontory which jets into the sea. This is the Banks Peninsula where two large, now extinct, volcanoes have been shaped by erosion and glaciation. Over time the rims of the old volcanoes were breached and with the rise of sea level at the end of the Pleistocene the sea invaded their craters. These water-gaps now form Lyttleton and Akaroa Harbors (Map 2-14). Lyttleton is the major harbor for the city of Christchurch which lies just to the north of the volcanic cones. Some of the most spectacular scenery and geology in the western Pacific can be found in New Zealand. South Island is dominated by the Southern Alps, shown to the west and covered with snow. Much of this region has been heavily glaciated (Map 5-6). The browns and dull reds represent different vegetation patterns on this color infrared photograph. Latitude 43° 05'S: Longitude 172° 29'E. (Satellite—color infrared—NASA Landsat.)

Plate 7—In east central Quebec, Canada, there is a large circular depression called Manicouagan Lake. Some believe this to be a volcanic-tectonic structure but the probability that it is a meteorite crater scar has recently gained wide acceptance (Map 2-19). The theory that the Manicouagan structure is the result of a great meteoritic impact will have major implications if verified, for it is as big, some 41 miles (66 kilometers) wide, as many of the large lunar craters. Precambrian crystalline rocks constitute most of the higher central core but these are overlapped by volcanic rocks with the depression. The circular depression serves as a water reservoir, here covered with ice, for the hydro-electric power station located at the dam in the lower portion of the photograph. Latitude 51° 13'N; 69° 05'W. (Satellite—color infrared—NASA Landsat.)

Plate 6—This space photograph is one of the most spectacular images taken by a satellite. It is a color infrared meaning that the vegetation, such as the forests and agricultural lands, appear as bright red. The image is a blowup of the Jemez Mountains in north central New Mexico. This is a large volcanic complex which is dormant at the present time (Map 2-15). In the center of the Jemez Mountains, composed mostly of late Tertiary volcanic rocks, is the large 13 to 15 mile (21 to 24 kilometer) diameter, Valles Grande, a craterlike (or caldera) volcanic feature produced in Pleistocene times by catastrophic blowouts and subsidence and then subjected to resurgence by upwelling of more lavas to form a ring of domes and cones. To the west of the crater are the Nacimiento Mountains bounded on the west by the Nacimiento fault. Los Alamos, the nuclear research center, lies just to the east of the Jemez crater. Latitude 36° 02'N; Longitude 106° 26'W. (Satellite—color infrared—NASA Landsat.)

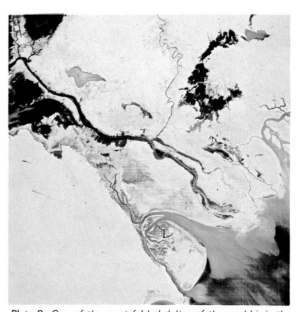

Plate 8—One of the most fabled deltas of the world is in the Middle East where the Tigris and Euphrates Rivers meet just north of the swamplands around Hawr al Hammr and form a single channel known as the Shatt-al-Arab (Map 3-11). This channel then flows past Al Basrah in Iraq and Abadan in Iran until it empties across a great delta into the Persian Gulf. This satellite image presents striking color contrasts between the white desert sands, the infrared irrigated farmlands along the channel of the river, and the black and blue swamps, ponds, and deltas at the northern end of the Persian Gulf. Part of the sheikdom of Kuwait, reputedly underlain by one of the richest concentrations of oil in the world, is shown in the southern (lower) portion of the photograph. Latitude 30° 20'N; Longitude 48° 08'E. (Satellite color infrared—NASA Landsat.)

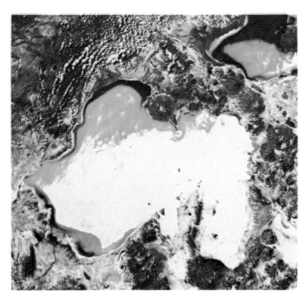

Plate 9—An area of interior basin drainage occurs between the high Andes Mountains of Bolivia. This region is called the Altiplano and represents the largest basin of interior drainage in South America (Map 3-14). This satellite photograph shows vast salt flats such as the Salar de Coipasa in the northwest (upper left) and the large Salar de Uyuni, in the center of the image. Salar de Uyuni represents the largest playa-type ephemeral lake in the world. These salars (salt lakes) represent the remnants of a drainage and lake network dating from the more pluvial glacial periods. Today, Coipasa occasionally receives overflow from the Lake Titicaca drainage system to the north during periods of flooding. The dark circular features in the western and central part of the photograph are volcanic peaks, many of which rise to 17,000 feet (5,200) meters). Latitude 20° 03'S; Longitude 67° 40'W. (Satellite color infrared—NASA Landsat.)

Plate 11—The easternmost land in Africa is shown in this Gemini VI space photograph which looks east over Somalia and the Indian Ocean. Of interest is the very large tombolo called Ras Hafun which extends 30 miles (48 kilometers) seaward to connect an island with the mainland. Tombolos are familiar coastal features but this one is notable for its size. The coastline is of geomorphic interest for another reason. One can depict on the space photograph a number of marine terraces formed when uplift of the land occurred along this section of coast. The drainage pattern is complex being a composite of dendritic and parallel patterns probably complicated by wind erosion (see Map 4-11). Latitude 10° 39'N; Longitude 51° 08'E. (Satellite—color infrared—NASA Gemini VI.)

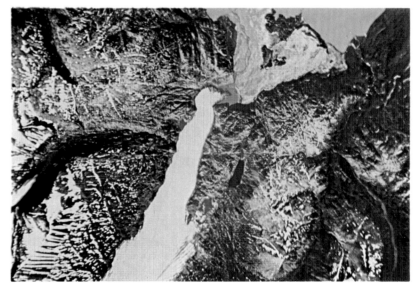

Plate 10—This space photograph illustrates excellent examples of depositional coastal landforms. Long Island, New York and coastal New Jersey represent barrier island fringed shorelines along the east coast of the United States (Map 4-11). Ocean currents have reworked sands and gravels derived from the glacier outwash plains and short rivers along the south coast of Long Island and the north coast of New Jersey to form several long offshore bars (barrier islands). Montauk Point, at the east end of Long Island (off the image) and the Navesink Highlands of New Jersey are also major contributors to these islands. While westerly moving currents have moved material along the south shore of Long Island a north flowing current has built out Sandy Hook at the entrance of New York harbor. The heavily built up area of New York City can be seen on Mantattan island and on western Long Island. Once can also discern the New Jersey portion of the port of New York. Latitude 40° 20'N; Longitude 73° 41'W. (Satellite color infrared—NASA Landsat.)

Plate 12—There are numerous satellite photographs over high mountainous regions depicting glaciation but this plate is one of the most spectacular enlargements of a glacier ever produced. This image shows Tustamena glacial tongue in Alaska with an outwash delta emptying into Tustamena Lake. The glacier is coming down the western slopes of the Kenai Mountains. At one time the ice flowed directly into the lake but recently it has been retreating. This entire region is a glacially dominated landscape with U-shaped valleys and scouring characteristics typical of recent glaciation. Many mountain glacial landforms could be pointed out. For example, one can note the shape of the glacier and its change in size with distance down valley, the morphology of the ice surface showing flow structures, the dissected upland terrain, numerous bedrock structures, and the braided outwash streams (See Map 5-6). Latitude 60° 12'N; Longitude 150° 50'W. (Satellite—color infrared—Technology Application Center—NASA.)

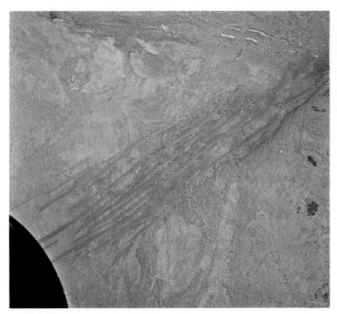

Plate 13—Is a satellite photograph showing long linear sand dunes called seifs or longitudinal dunes (Map 6-1). This image is over western Mauritania near the Atlantic Ocean. This is a huge erg area (erg is a term signifying a sand-covered portion of the North African Desert). Several of the seifs extend for approximately 75 miles (120 kilometers) without a break. Such large dunes are formed by the persistent southwest to northeast winds which rework and transport alluvial sands. The tan areas on the photograph are desert playas and sand flats and the dark regions are old crystalline rocks making up the Precambrian shield of north Africa. The orange color of these hugh dunes tells us that the sands are quite old having been oxidized by wind and water. Latitude 17° 02'N; Longitude 05° 31'W. (Satellite—color infrared— NASA Gemini VI.)

Plate 15—This is an enlargement of a satellite photograph over northern Arizona. This blown-up section gives a closer view of the world famous Colorado River canyon and gorge that has been cut over millions of years by a single river and its tributaries (See Map 7-1). In the center of this image the Colorado River flows in a canyon almost 1 mile (1.6 kilometers) deep and roughly 12 miles (19 kilometers) wide from rim to rim. Although there are other canyons in the world deeper and wider, this is certainly one of the most spectacular and most widely known gorges. Note the asymmetry of the main canyon as the Colorado River has been cutting its south side more than its north side. This results because of a gentle southward slope of the Colorado Plateau whose north rim is more than 1000 feet (305 meters) higher than the south rim. The canyon is incised in essentially horizontal Paleozoic sedimentary rocks exposing a relatively undisturbed stratigraphic cross-section unequaled elsewhere in the world. The well-defined canyon of the Little Colorado joins the Colorado River west of Cameron, Arizona, as the Colorado emerges from its Marble Canyon section. Several fault lines, mostly with a northeast-southwest trend can be followed across the image. Tributary rivers, particularly in the center of the photograph are controlled by these faults. Snow covers the high 9000 foot (2700 meter) Kaibab Plateau. Latitude 36°08'N; Longitude 112°26'W. (Satellite—color infrared—NASA Landsat.)

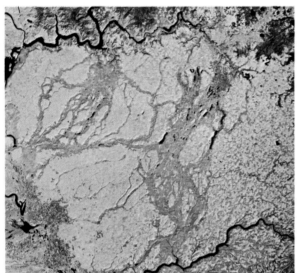

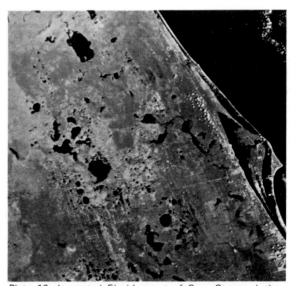

Plate 14—Is over the Palouse area of eastern Washington State which is covered by fine wind-blown silts and clays, called loess (See Map 6-4). There are also some deposits of light-colored volcanic ash. These deposits are largely derived from old glacial outwash plains and disintegrated volcanic lavas. Miocene and Pliocene basaltic volcanic flows underlie much of the Columbia Plateau. From the lava flows, dark soils have been derived forming a rich winter wheat farming region. This is part of the channelled scablands, a strongly dissected terrain consisting of canyons and outliers in basalts overlain by loess. These large channels cross the central part of the image. In the eastern part of the photograph, the loess and ash are dissected and gashlike markings can be faintly detected. Because of the columnar nature of the loess deposits, many of the gullies and river valleys have very steep sides. These are not visible on the space photograph. Latitude 47° 21'N; Longitude 118° 10'W. (Satellite—color infrared—NASA Landsat.)

Plate 16—In central Florida, west of Cape Canaveral, there are hundreds of lakes, a few of which are shown on this satellite infrared photograph. These lakes occupy solution-weathered sinkholes, now filled with water, creating a pock-marked surface known as karst topography (See Map 7-3). Three of the larger lakes are Lake George in the north, Lake Apopka in the center, and Lake Kissimmee in the south. Much of the limestone underlying this portion of central Florida is honey-combed with water-filled sub-terranean passageways, some of which are associated with artesian springs. The linear features trending northwest to southeast, parallel to the coast, are emerged beaches produced as the sea retreated from Florida, which it covered until fairly recent time. Denser vegetation along water courses such as the St. Johns River are shown as darker red tones on this color infrared image. Cumulus clouds occur in several places over the mainland as well as out over the open water. (Latitude 28° 45'N; Longitude 81° 21'W. (Satellite—color infrared NASA Landsat.)

Section Four

Coastal Landforms

MAP
4-1

A GEOGRAPHIC CLASSIFICATION OF COASTS
IN TERMS OF PLATE TECTONICS

This map is in close association with the maps of Continental Drift (Map 2-2) The Plates of the Earth's Lithosphere (Map 2-1) and the map on Major Types of Coastal Instability (Map 4-2). The old distinction between Pacific type coasts, where the structural grain is parallel to the coast, and Atlantic type coasts, where it is discordant, has taken on a new significance since the development and wide acceptance of the concepts of plate tectonics. These concepts envisage lateral movement of enormous crustal plates away from zones of spreading towards zones of convergence, a process which has fundamentally affected the evolution of world coasts on a grand scale (see Maps 2-1 and 2-2). Inman and Nordstrom (1971) have discussed first order crustal evolution in relation to these ideas of plate tectonics and have proposed a broad classification shown on the world map. Collision coasts are formed where two plates converge while marginal sea coasts occur where a plate-imbedded coast faces an island arc. Trailing edge coasts are found where a plate-imbedded coast faces a spreading zone. The Neo-trailing edge coasts form where a new zone of spreading is separating a land mass; while an Afro-trailing edge coast is where the opposite continental coast is also trailing; and an Amero-trailing edge coast is a collision coast. The Red Sea and Gulf of California are Neo-trailing edge coasts where spreading of a land mass is taking place. Almost the entire coast of Africa except in the north and along the Red Sea is a trailing edge coast, while the United States coast is a trailing edge coast on the east and a collision coast on the west. Certainly the wide continental shelf of the east coast of the United States is very different from the narrow continental shelf and tectonically active coast on the west.

The coasts around the Pacific Ocean are largely collision coasts with island arc collision coasts dominating from New Zealand north through Japan and the Aleutians. Collision coasts elsewhere include those of the East and West Indies and those of the Mediterranean and Baluchistan coasts of Pakistan and Iran. Marginal sea coasts are the most diverse in character of the categories presented. They are frequently modified by fluvial plains and deltas; their hinterland may vary considerably in relief; and their adjoining shelves often vary greatly in width. The Australian coast presents a problem because the authors did not know how to classify it. Since the western and southern coast of Australia is fairly stable, it has been called an afro-trailing edge coast, while the eastern northern portions are called the marginal sea coast lying behind the Tonga-Kermadec-New Zealand convergence line. Active trenches are included on this map instead of Map 4-2 — Major Types of Coastal Instability, because they are close to island arc collision coasts. The Puerto Rican trench reaches depths of 26,000 feet (7,920 meters) while the Marianas trench in the Pacific Ocean reaches depth of 35,000 feet (10,670 meters).

Sources: J. L. Davies, *Geographic Variation in Coastal Development,* Hafner Publishing Company, New York, 1973, Fig. 2, p. 9 and Fig. 4, p. 11; and D.L. Inman and W.R. Nordstrom, "On the Tectonic and Morphologic Classification of Coasts," *Journal of Geology,* Vol. 79, 1971, pp. 1-21.

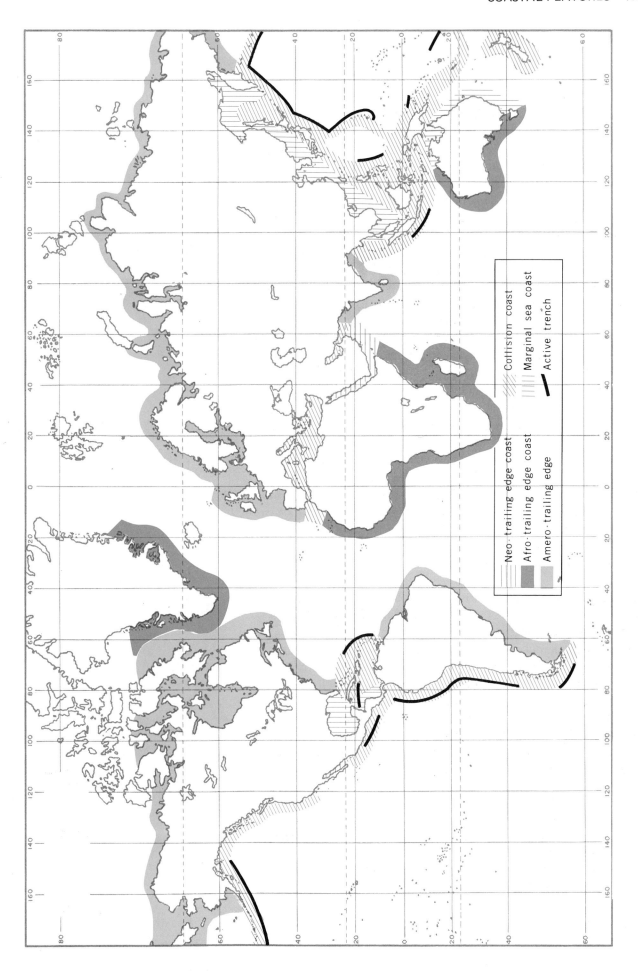

MAP 4-1   A GEOGRAPHIC CLASSIFICATION OF COASTS IN TERMS OF PLATE TECTONICS

**MAP
4-2**

MAJOR TYPES OF COASTAL INSTABILITY

This map depicts what appears to be happening along coasts today. It is a difficult map to compile because there are so many variables which must be taken into account. One of the major problems is, sea level has not been static and during the last 12,000 years + it appears to have come up as much as 330 feet (91 meters) (Russell, 1964; Shepard, 1964 and 1968; Fairbridge, 1968). This map is drawn to indicate where permanent change in shore-line configuration is an explicit indication of coastal instability caused by a shift in one or more of such variables as changes in tide levels, tectonic activity or crustal guiescence, or oceanographic or atmospheric factors. Broader coastal instability deals mainly with the effects of endogenic crustal and sub-crustal processes affecting the surface of the lithosphere as it relates to shoreline configuration. Methods for measuring instability over a time interval include leveling (Geodetic) surveys, tide gauges, historical data, pre-literate folklore and proto-literate sagas, and archaeological evidence. This map can only present a general picture of what is taking place along coasts adjacent to the sea. Local crustal up warping and down warping can affect a coast within just a few miles. A number of early Eskimo sites in northern Canada associated with raising marine strandlines are now found up to 98 feet (30 meters) above sea level and up to several miles inland from the coast. Another Eskimo site on the southwestern tip of Banks Island in the Canadian Archipelago is now a wash at high tide.

The map clearly points to most areas covered with glacial ice being on the rebound, (rising coasts) (see Map 5-1) central and northern Canada, northern Europe and the coast of Siberia. Subsiding coasts include the east coast of the United States south of Massachusetts, the Gulf Coast and the southern Californian coast. Other noteable subsiding coasts include the north European plain, west coast of Central America from Guatemala to Panama and west coast of Japan. There are really very few stable coasts. The Australian coast is complicated with only three stage sections. Sections of the Brazil coast and around Africa appear to be the most stable.

Sources: Largely drawn from R.W. Fairbridge, *The Encyclopedia of Geomorphology,* Reinhold Book Corp., New York, 1968, pp. 150-156; H. Valentin, "Die Kusten der Erde," *Peter. Geogr. Mitt.,* 246, 1952, p 118; P.H. Kuenen, "Sea Level and Crustal Warping," *Geol. Am. Spec. Papers,* 62 (Crust of Earth), 1955, pp. 193-204.

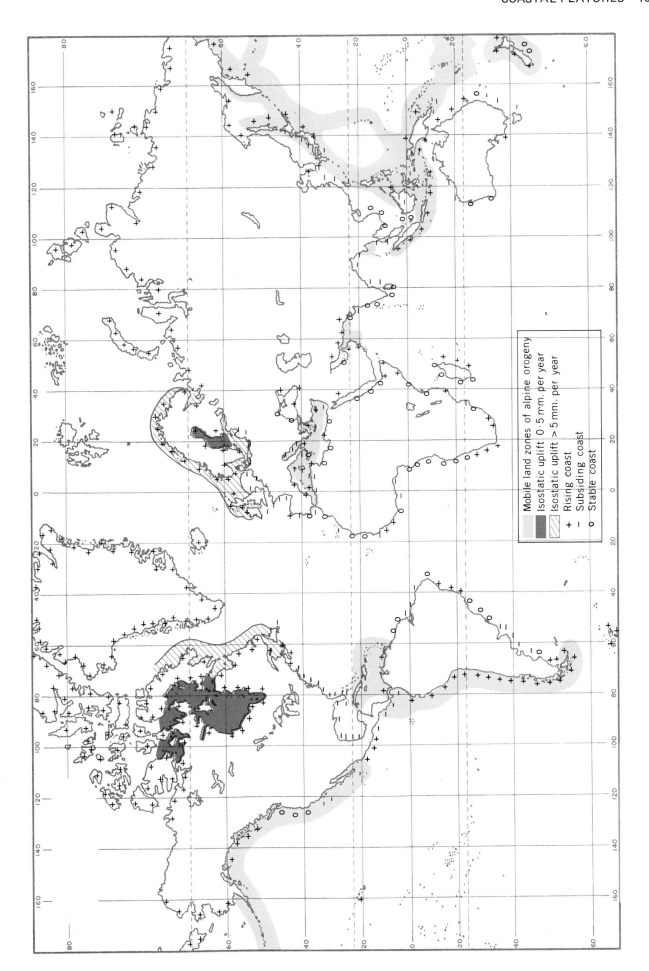

MAP 4-2 MAJOR TYPES OF COASTAL INSTABILITY

**MAP
4-3**

MAJOR WORLD WAVE ENVIRONMENTS

This map is drawn from a sketch in J.L. Davies volume titled *Geographical Variation in Coastal Development,* (1973), which attempts to define very genetically certain broad wave environments. Although it is a simple model, still it gives a starting point for further refinement as more information becomes available. Dr. Davies states that the dominant quadrants of wave approach have been checked against The British Monthly Meteorological Charts of the Oceans, the Sea and Swell Charts produced by the United States Hydragraphic Office, and Becker (1936), Schubart and Mockel (1949), Bruns (1953), and Helle (1958). Although micro-studies reveal wave environments much more varied, in practice, the alignment of the coasts themselves normally filters out all but a few possible directions of wave approach and this often simplifies complication. Because of the map projection only a small portion of Antarctica is shown. Except for the Antarctic Peninsula the coast of Antarctica is a protected sea environment. similar to large sections of the Arctic coast.

A significant proportion of waves in storm wave environments are generated by local winds having gale force, these being short, high energy waves of varying direction. Swell waves form a background persistant in the Southern Hemisphere, but they tend to disappear from the Northern Hemisphere during summer months.

In west coast swell environments gale force winds are rare and, within the tropics, local winds tend to blow onshore. In tropical South America and Africa these are the doldrum coasts. Mean

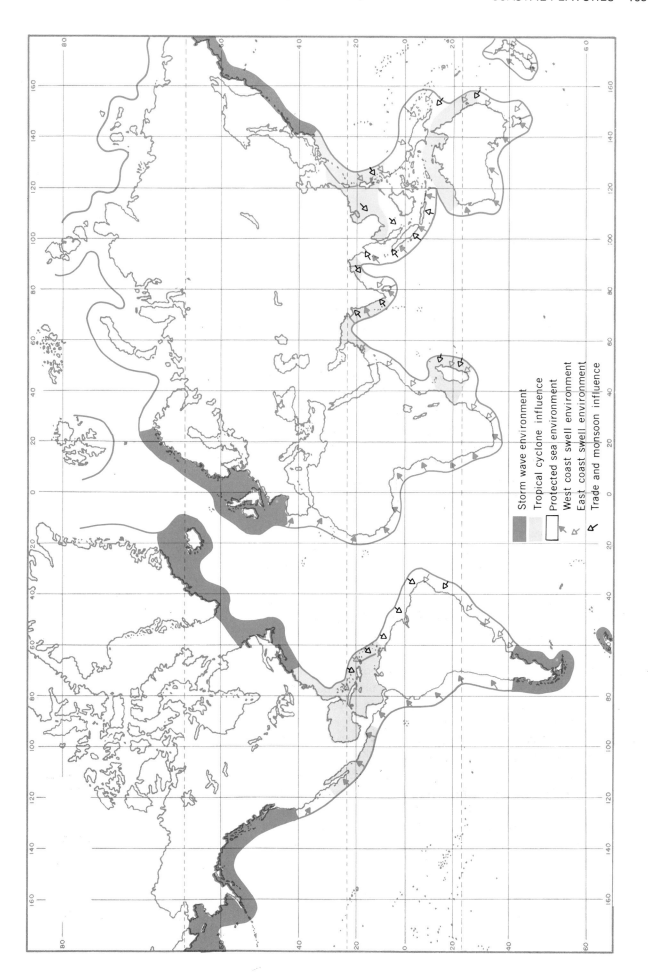

MAP 4-3   MAJOR WORLD WAVE ENVIRONMENTS

Storm wave environment

Tropical cyclone influence

Protected sea environment

West coast swell environment

East coast swell environment

Trade and monsoon influence

wave energy varies from high levels in the higher latitudes to moderate levels in the tropics and low levels around the Gulf of Panama and between Dakar and Freetown in West Africa. In spite of differences in mean energy values, this type of coast may be the most homogeneous in regard to wave environment, but part of the Mexican coast must be distinguished as being subject to tropical cyclones and, on the Indian coasts, a swell is reinforced seasonally by monsoon generated waves.

For east coast swell environments gale force winds are rare and swell from temperate storm belts appears to be weaker and less regular in occurrence. However, it remains relatively consistent in direction. These are the onshore winds of the tropics with an extremely regular inflow. However, segments of this environment are subject to very high energy input from tropical cyclones for short durations at infrequent and irregular intervals.

Protected sea environments are those coasts of seas where little oceanic swell penetrates and which are either outside the temperate storm belts or are protected by ice cover. Generally they are protected by the seas being more or less enclosed, but the coast of Antarctica is a major exception because it is sheltered from swell by the way in which waves tend to be deflected equatorward along great circle courses, and by the damping effects of sea ice. Generally, these are low energy environments, but, some, such as the east coast of Malaya, are subject to seasonal monsoon generated waves of higher energy and others, such as those of the Gulf of Mexico, which are liable to occasional very high energy waves from tropical cyclones.

MAP
4-4

SEDIMENT TRANSPORT LONG TERM NET MOVEMENT

This map is presented in the Atlas because it reveals some interesting information as to where rock and fixed coastline plus depositional coasts can be found. Although extremely generalized, the map presents overall regions where silt, sand, and shingle coasts are most prominent. It is impossible to plot all the minor coastal features such as pocket beaches but along many sections of rock coast these can be found between rocky headlands. From northern Massachussets to Honduras there is a nearly continuous stretch of silt, sand, and shingle coast. This is the longest non-rocky coastline in the world. On the other hand, nearly all of Scandinavia is bordered by rock or fixed coastal landforms. The coast of Australia has great diversity with alternating sections of bedrock and sand, silt, and shingle beaches.

The arrows indicate the direction of predominant longshore drift due to persistent swells and waves. The movement of this drift controls the net sediment movement around the coastlines of the world. Because of the scale of the map, drastic simplifications have been necessary. It needs to be stressed that the arrows indicate the direction of longshore drift should sediment be available and no rate of transport is implied. The net movement that is indicated does not preclude reversals of drift. The plus symbol is used for those areas where drift is substantially retarded to zero but cannot be used to indicate accretion in the area. For example, near the delta of the Mississippi in the Gulf of Mexico, drift is largely retarded but accretion at times does take place from sediment deposited by the Mississippi. On the other hand, the plus symbol in the Bay of Plenty off the North Island of New Zealand indicates hardly any drift but there is also almost no accretion.

MAP 4-4   SEDIMENT TRANSPORT—LONG TERM NET MOVEMENT

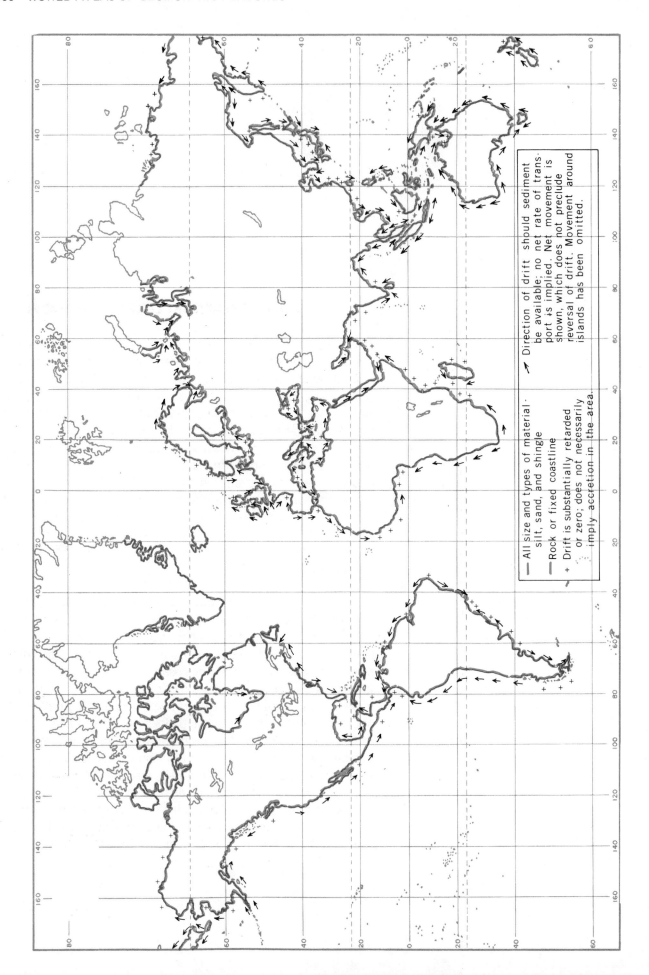

All size and types of material - silt, sand, and shingle

Rock or fixed coastline

+ Drift is substantially retarded or zero; does not necessarily imply accretion in the area.

Direction of drift should sediment be available; no net rate of transport is implied. Net movement is shown, which does not preclude reversal of drift. Movement around islands has been omitted.

MAPS
4-5
4-6
4-7

EROSIONAL COASTAL FEATURES

The next five topics present features that are related to coastal regions. The first map is concerned with marine erosion and the landforms that result from the energy of waves and currents. The author has divided the erosional coastal features into two main types. The first category shows where high-cliffed and/or terraced coasts occur. These features represent steep coasts with sheer cliffs and uplifted terraces that have considerable relief. A coast of this kind usually has deep water immediately offshore because the waves are moving directly against sheer walls of rock. The second category depicts the location of erosional coastal platforms, or benches, with low cliffs or terraces. Often these have dunes and vegetation immediately inland. This is also often a coast with pocket beaches. Energy of the waves is spent crossing wide, low abrasion surfaces. In addition, the map shows the location of sea caves, stacks, and arches; smaller landforms often associated with cliffed coasts. Sea cliffs and wave-cut terraces are two of the most common features created by marine erosion. Sea cliffs vary greatly, depending upon the kind, structure, and attitude of the rocks in which they are being cut. A cliff will look quite different depending whether rocks dip seaward, landward, or are nearly horizontal. Cliffs cut in granite differ in appearance from those cut in basalt. Cliffs in easily erodible material, such as glacial till or nonindurated clays and sands, will be bold and marked by much slumping and landsliding. Some very resistant rocks that face deep water where abrasive materials are absent stand up for long periods of time against the most powerful storm waves. High, exposed cliffs near Cape St. Vincent, at the southwestern corner of Portugal, remain practically unnotched (Russell, 1963). Examination of the hard rocks of the Cornish coast indicates that they probably have changed little over the past ten thousand years (Bascom, 1964). Abrasion platforms and benches, often with low terraces, or dunes immediately inland, are distinguished from high cliffs and uplifted terraces, because this coastal type is frequently found where erosion over a long period of time has straightened and smoothed a rocky coast. Good examples are extensive sections of east Africa, northern Australia and New Zealand.

All types and shapes of sea caves, sea stacks, and sea arches can be found around the world. Only the larger, well-known examples are shown on these maps. Sea caves often originate at the base of sea cliffs where wave action has enlarged natural lines of weakness, forming first a notch, then a hole, and finally a cave.. Along the south coast of Portugal bright red sandstones and limestones have been cut by waves into intricate mazes of interconnected caves and tunnels. In some areas, the caves are connected to openings in the ceiling, out of which water spouts when there is a large surge of high water, or where there is at least a detectable inhalation and expulsion of air associated with wave surges. At several locations around the islands of Hawaii there are beautiful spouting horns with hissing air where wave surges have cut deep into the recent volcanic lava flows. As erosion continues cutting a sea cave, both ends may become open to the sea, producing one type of arch or natural bridge. Large sea arches are often formed by waves cutting all the way through the cliff. In time, as a sea arch grows larger, its roof may collapse, leaving the seaward side as a sea stack. Thus, sea stacks are tall isolated columns of rocks that have been separated from a sea cliff by wave erosion. They represent erosional remnants that have not been fully destroyed by wave and, therefore, stand as tall islands off the shore. In most cases, sea stacks are made of the same type material as the nearby cliff. In the United States sea stacks are most common along the rugged coast of Oregon. In Europe, large sections of the coast of the British Isles, as well as the coasts of Normandy and Brittancy in France, have excellent examples of sea caves, arches, and stacks.

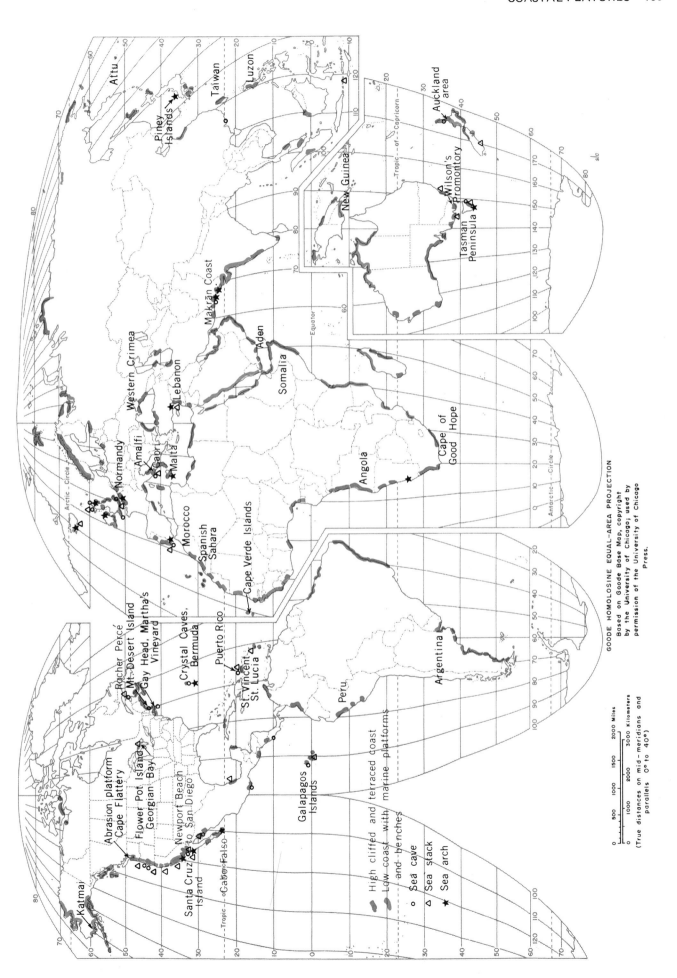

GOODE HOMOLOSINE EQUAL-AREA PROJECTION
Based on Goode Base Map, copyright
by the University of Chicago; used by
permission of the University of Chicago
Press.

MAP 4-5   EROSIONAL COASTAL FEATURES OF THE WORLD

MAP 4-6   EROSIONAL COASTAL FEATURES OF THE UNITED STATES

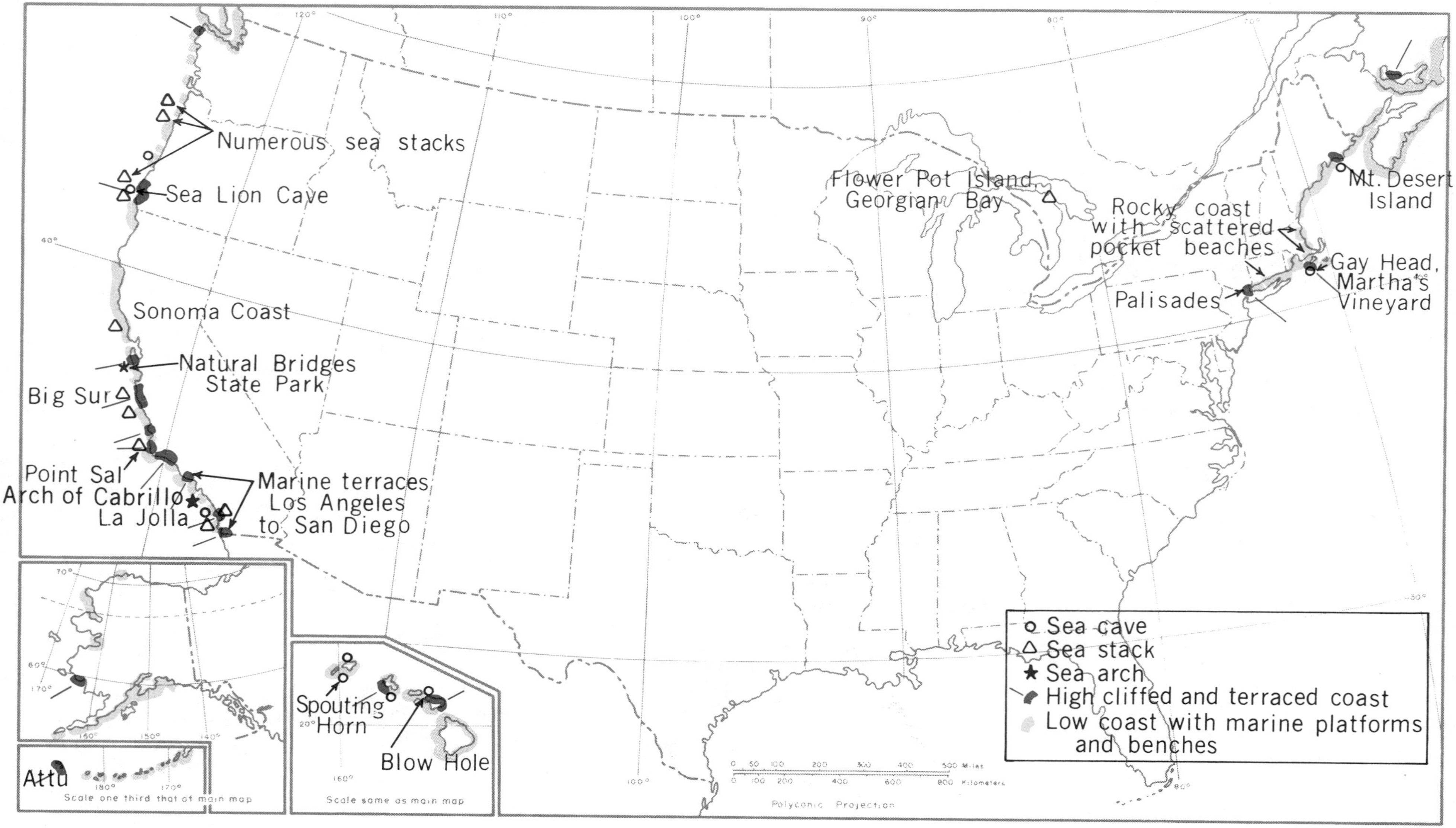

Mt. Desert
Island

Gay Head,
Martha's
Vineyard

Rocky coast
with scattered
pocket beaches

Palisades

Flower Pot Island
Georgian Bay

Numerous sea stacks

Sea Lion Cave

Sonoma Coast

Natural Bridges
State Park

Marine terraces
Los Angeles
to San Diego

Big Sur

Point Sal
Arch of Cabrillo
La Jolla

Sea cave
Sea stack
Sea arch
High cliffed and terraced coast
Low coast with marine platforms
and benches

0   50 100   200   300   400   500 Miles
0 100 200   400   600   800 Kilometers

Polyconic Projection

Blow Hole

Spouting Horn

Scale same as main map

Attu

Scale one third that of main map

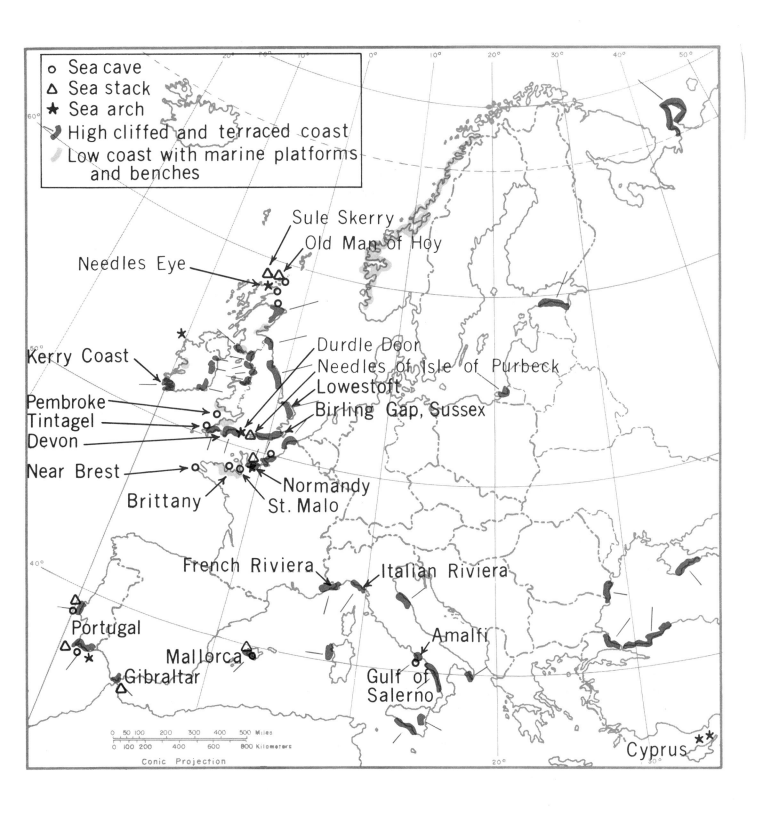

Sea cave
Sea stack
Sea arch
High cliffed and terraced coast
Low coast with marine platforms and benches

Sule Skerry
Old Man of Hoy
Needles Eye
Durdle Door
Needles of Isle of Purbeck
Lowestoft
Birling Gap, Sussex
Kerry Coast
Pembroke
Tintagel
Devon
Near Brest
Normandy
Brittany
St. Malo
French Riviera
Italian Riviera
Portugal
Amalfi
Mallorca
Gibraltar
Gulf of Salerno
Cyprus

0 50 100 200 300 400 500 Miles
0 100 200 400 600 800 Kilometers
Conic Projection

MAP 4-7 EROSIONAL COASTAL FEATURES OF EUROPE

**MAPS
4-8
4-9
4-10**

DEPOSITIONAL COASTAL FEATURES—WIDE SANDY BEACHES, BEACH RIDGES, AND SAND DUNES

Along the boundary between land and sea, the solid underlying materials are covered with a depositional layer of rock fragments. These fragments range in size from fine sand to large cobbles, in thickness from a few inches to hundreds of feet, in color from white to opaque black. They are beach materials deposited by waves and currents (Bascom, 1964). The open-sea beaches that border much of the United States from Cape Cod south along the east coast to Florida, and along the California coast south of Point Conception, are for the most part composed of coarse, light-colored sand, produced by the weathering of granitic rocks into two main constituents: quartz and feldspar. Other beaches are very different (Johnson, 1919). Hundreds of miles of beach along the Oregon-Washington coast are composed of fine-grained and dark gray-green colored sand. This material is weathered basalt; it forms beaches that are as firm as a racetrack. Much of the Florida coast is equally hard and fine-grained, drived from the disintegration of coral. On the other hand, the beach at Cannes in southern France is largely composed of coarse pebbles, and much of the English coast is lined with small flat stones called shingle (Steers, 1962). On Tahiti, the windward side of the island has black volcanic sand but on the leeward side, where a wide coral reef furnishes the beach material, the sand is blindingly white. In fact, beaches are made up of nearly any material that is present in quantity including shell fragments.

Beaches of sand, gravel, and shingle are widespread throughout the world but wide sandy beaches are not as numerous and are usually found where there is a good source of material. The word sandy needs to be defined because it is used here to include not only loose granular quartz materials, resulting from the disintegration of rocks, but also fragmented volcanic materials such as the black sandy beaches on the large island of Hawaii and the sands formed from weathered coral reefs. Coral and shell sands are mainly composed of organic particles that

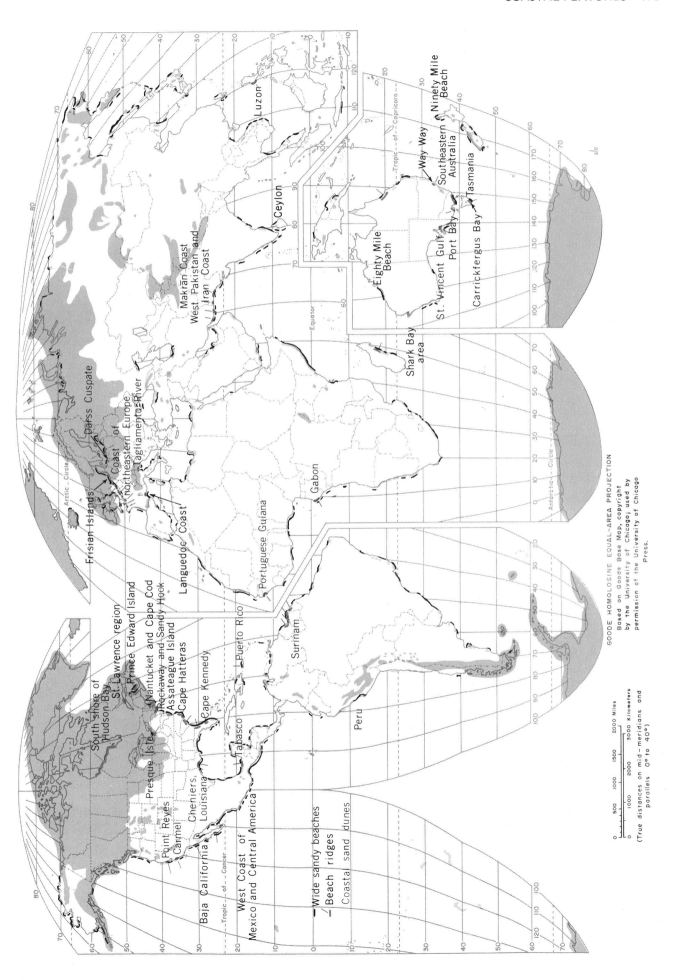

GOODE HOMOLOSINE EQUAL-AREA PROJECTION
Based on Goode Base Map, copyright
by the University of Chicago; used by
permission of the University of Chicago
Press.

MAP 4-8 DEPOSITIONAL COASTAL FEATURES OF THE WORLD

MAP 4-9   DEPOSITIONAL COASTAL FEATURES OF THE UNITED STATES

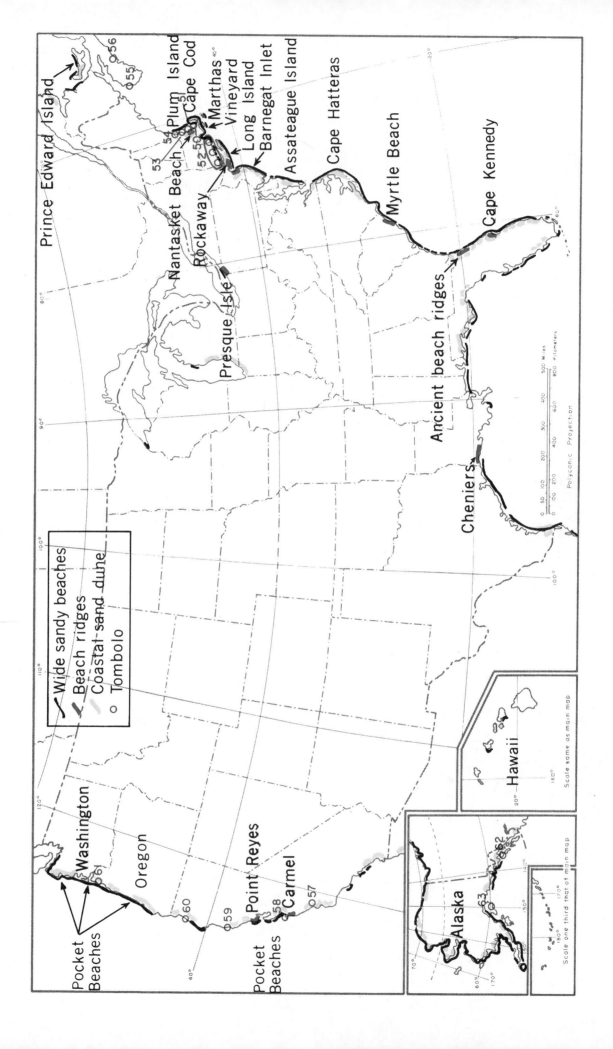

Prince Edward Island

○56

○55

Nantasket Beach

Plum Island
Cape Cod
Marthas Vineyard
Long Island
Barnegat Inlet
Assateague Island

Rockaway

Presque Isle

Cape Hatteras

Myrtle Beach

Cape Kennedy

Ancient beach ridges

Cheniers

Wide sandy beaches
Beach ridges
Coastal sand dune
Tombolo

Washington

Oregon

Pocket Beaches

○61

○60

○59

Point Reyes

Pocket Beaches

○58
Carmel

○57

Hawaii

Scale same as main map

Alaska

○62

○63

Scale one third that of main map

Polyconic Projection

0   100   200   300   400   500 Miles
0   100   200   300   400   500   600   700   800 Kilometers

have been worn into angular grains or fine powder. Beaches composed of coarse gravel and shingle are not included in this definition. The word 'wide' needs also to be defined because this is the term which restricts the number of beaches shown on the map. A wide beach is one generally over 300 feet (91 meters) in width from the low water line to the beginning of vegetation. Most of these beaches have a gentle slope that is not more than one to three degrees. There are hundreds of beaches which fit this definition but most cannot be shown on a map of this scale. Only those wide sandy beaches which extend for considerable distances along the coast are indicated.

On a prograding (seaward advance) beach there often occur a series of ridges composed of coarse sand or shingle. These linear beach ridges may occur singly or in series, the youngest being shoreward. Their heights are generally less than 20 feet (6 meters), and they are often covered by dune grasses and shrubs. Where the climate is warm, they are often cemented by calcium from shells. Conflicting theories are given as to their origin, but high states of the sea, uprush of waves beyond ordinary tidal limits, and creation of higher level berms (nearly horizontal portion of the beach of backshore formed by the deposit of material by wave action) are involved in their formation. In some places, beach ridges are called ridge and swale topography. The swales represent the low depressions between the ridges. Cheniers are similar to and often compared with beach ridges, but their origin is the result of deposition of fine sediment, such as delta deposits, on the seaward side of a ridge. These features have formed along the southern coast of Louisiana west of the Mississippi River. When little sediment moves west from the Mississippi delta, a beach ridge (chenier) forms. When the supply of sediment is large, a marsh builds out in front of the beach isolating it from the shore (Russell, 1967).

Beach ridges and runnels (the depressions behind the ridges), as they are called in Europe, are fairly widely distributed. They are generally found where the tidal range is considerable, for example, along the coasts of the Irish and North seas, and along parts of the Firth of Clyde. In the North Sea they are found at Druridge Bay in Northumberland, the Lincolnshire coast, and parts of East Anglia. In Europe they are found from north Holland to Cherbourg, being especially well developed near Le Touquet in France. The Dungeness area of southeast England is an excellent example of beach ridges (King, 1961).

The location of coastal sand dunes is shown on this map, but they are also plotted on the map showing active and stable sand regions (Maps 6-1 and 6-2). Coastal sand dunes are widespread. Small dunes can be found behind most sandy beaches. The above map portrays only the larger coastal dune areas.

In North America, sand dunes are found in scattered areas along much of the Atlantic coast of the United States and around the Gulf of Mexico into Mexico. Although low dunes are quite common, large sand deposits can only be found at a few locations: Plum Island, Crane's Beach, the tip of Cape Cod, Horseneck Beach in Massachusetts, Island Beach State Park in New Jersey, Assateague Island in Virginia, and Cape Henry, Virginia. Many dunes are stabilized by vegetation. The highest dunes along the east coast of the United States are at Nags Head, North Carolina where several large dune fields reach elevations of 200 to 300 feet (61 to 91 meters) (Case, 1921). Very white quartz dunes can be found scattered along the coast of Florida, especially along the northwest Gulf coast. A large sand dune area is around the eastern and southern sides of Lake Michigan, where crescentric dunes can be seen in Indiana Dunes State Park. Along sections of the Oregon and Washington coast, especially in the vicinity of Reedsport, Oregon, large transverse dunes are common (Cooper, 1958). Sand dunes can be found along sections of Baja, California, and in scattered locations around the Caribbean. In South America, large dunes are found along long sections of the Peruvian and Chilean coasts, as well as along the northern sections of the Patagonian coast and in southeastern Brazil. In Africa, large dune complexes are found on both the Atlantic and Red Sea coasts of the Sahara Desert, particularly along the

MAP 4-10   DEPOSITIONAL COASTAL FEATURES OF EUROPE

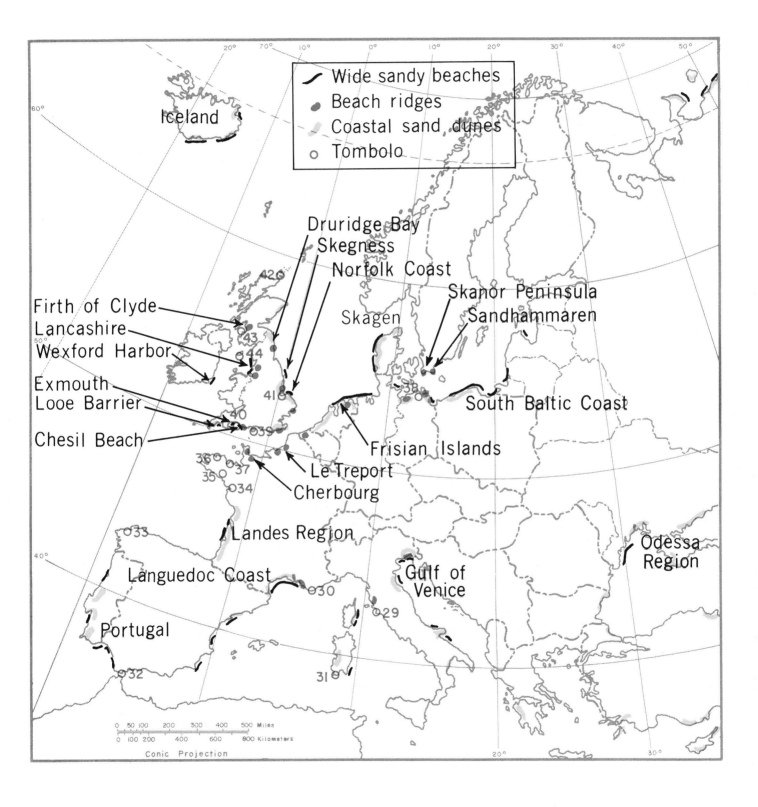

coast of Morocco. Low dunes are found along the coast of Libya and the Sinai Peninsula. The sand dunes on the Egyptian coast are derived from calcareous deposits found in the neighborhood. The southern coast of Somalia has numerous dunes, and the Namib coast of Southwest Africa has large dune fields. In Australia, at least nine long coastal sections have large coastal dunes (Cotton, 1941; Bird, 1964). Excellent parabolic dunes are found on King Island, Tasmania, and dunes are found on the northern end of the North Island of New Zealand (Cotton, 1942). Few large areas of coastal dunes exist in Asia. However, the Red Sea coast has extensive dune fields as well as the south and Persian Gulf coast of the Arabian Peninsula. Scattered dune fields, in places reaching 100 to 200 feet (30 to 61 meters) high, are found along the coast of Iran and West Pakistan. Large dunes are found in the Rann of Kutch area of India. Sand dunes are found on the north side of the Caspian Sea near the mouth of the Volga River, on the southeast side of the Aral Sea, and around sections of Lake Balkhash. In Europe a number of small dune areas are found. There are small dunes along sections of the coast of Ireland, the Raabjerg Mile area near Skagen, Denmark, along the north coast of Germany, the coast of the Netherlands, Belgium, and northern and central France. A large dune area with parallel rows of dunes up to 300 feet (91 meters), now partially stabilized by pine forests, is along the Bay of Biscay south of Bordeaux in the Landes Region of southern France. Stabilized Pleistocene sand dunes, many of them thousands of years old, are found at a number of scattered locations around the world. The coastal plain between Brisbane and Rokhampton in Australia has stabilized Pleistocene dunes approximately 1,000 feet high (304 meters) (Bird, 1964). These are probably the largest dunes in the world. Small areas of Pleistocene dunes can also be found along the coast of Pakistan and Iran, in Puerto Rico, and on Bermuda.

Tombolos are shown on Map 4-11.

MAP
4-11

DEPOSITIONAL COASTAL FEATURES—
BARRIER BEACHES, BARRIER ISLANDS,
BARRIER SPITS, BAY BARRIERS, AND TOMBOLOS

Since many beaches separate lagoons, bays, marshes, and swamps from the mainland barrier islands are shown on this map. In turn, tombolos are a type of depositional barrier which connect an island with the mainland or other islands so these too are included on this map. Coastal barriers represent an offshore ridge or bar, usually, of sand, which is essentially parallel to the shore, with a crest exposed during high water. They may be unattached at one or both ends and they are frequently broken by inlets. The author uses the terms: barrier beaches, barrier islands, barrier bars, barrier spits, and bay barriers, to include the closely related unconsolidated deposits that are formed by waves and currents.

These features occur along one third of the coasts of the world. Their shape and size varies greatly. Many are straight and no more than a few hundred feet wide, but others are several miles in width. Where the sands on the barrier are still moving, only dune grasses and scattered shrubs are able to grow, but on older barrier bars the vegetation may consist of clumps of trees or low, thick shrubs. Many of the beaches along the east coast of the United States are on barrier islands that are separated from the mainland by tidal marshes and lagoons (U.S. Geological Survey, 1970). Plate 10 (centerfold) is a space photograph of Long Island and northeastern New Jersey. Depositional coastal landforms in the form of long narrow barrier islands fringe nearly the entire outer coastal zone shown on this photograph. Most of the sand barrier islands are ephemeral meaning that storm waves and currents are continually changing their shape by eroding away their beaches or breaching them with new tidal inlets. Barrier island also occur for long stretches along the Gulf of Mexico, along sections of the Mediterranean Sea, the coast of Ceylon, and along the coast of Australia (McGill, 1958). Beach ridges, coastal dunes, and mangroves are closely associated with barrier islands. This is particularly true of mangroves, because in tropical regions they grow in lagoons behind the barriers.

A spit or bar may be constructed in protected waters between islands, or between islands and the mainland. The connecting unconsolidated bar, is called a tombolo. (The word "tombolo" is Italian, from the fine examples found along the Italian coast.) These features may be single or double, or even triple or V-shaped, when the island is attached to the mainland by two or more bars. When several are united with each other and with the mainland by a series of bars, the whole series or cluster is called a complex or composite tombolo. Although most tombolos are composed of sand some may be formed by coarse gravel or rocky shingle. There are numerous tombolos around the world. This map names and shows the distribution of only the largest, well-known features. Maps 4-9 and 4-10 show the location of tombolos in the United States and Europe. The northeastern United States has several small tombolos, largely in Massachusetts. At Marblehead, a sand bar—a single tombolo—connects a rocky island with the mainland. Big and Little Nahant are a complex tombolo consisting of three small bars connecting two small islands with the mainland (Johnson, 1965). Nantasket Beach consists of a series of sand bars that connect several drumlins in Boston Harbor (Guilcher, 1958). Spits and bars between drumlins form excellent examples of complex tombolos. Small tombolos occur along other sections of the northeast coast, for example, along the south coast of Rhode Island and Connecticut. Europe has a number of well-known tombolos. Probably the most famous in England is the coarse shingle barrier at Chesil Beach which connects the Bill of Portland with the mainland (King, 1961). The Isle of Wight has a tombolo, and there is one at the southeastern corner of the Isle of Man (Gresswell, 1962). A triple tombolo is found at Monte Argentario, Italy, north of Rome, where three bars join a hill to the mainland (Loebeck, 1958). In France, double tombolos are found at Bourg-de-Batz (Loire Inferieure), at Giens near Toulon, and at Guerande. Large tombolos occur along the Makran coast of Pakistan and Iran. The tombolos of Ormara and Gwadar are at least ten miles (16 kilometers) long and several miles wide, as is Ras Tank tombolo along the coast of Iran. Plate 11 (centerfold) is an excellent view of the tombolo called Ras Hafun on the north coast of Somalia. This is reported to be the largest tombolo in the world, being about 30 miles long.

The following is a list of the major tombolos found around the world. Large tombolos are shown with a large dot, small features are shown with a small dot.

1. San Fernando, Philippines

2. Da Nang, South Vietnam

3. Tq. Meblu, Bali, Indonesia

4. Lucinda, Queensland, Australia

5. North Bowen, Queensland, Australia

6. Near Proserpine, Queensland, Australia

7. North of Yeppoon, Queensland, Australia

8. Palm Beach, Australia

9. Near Wilson's Promontory, Australia

10. Yanakie Isthmus, Australia

11. Karikari Peninsula, North Island, New Zealand

12. Mount Maunganui, North Island, New Zealand

13. Mahior, North Island, New Zealand

14. Miramar, North Island, New Zealand

15. Pepin Island, near Nelson, South Island, New Zealand

16. Kaikoura Peninsula, South Island, New Zealand

17. Banks Peninsula, South Island, New Zealand

18. Karitane Peninsula, South Island, New Zealand

19. Otago Peninsula, South Island, New Zealand

20. Bluff Peninsula, South Island, New Zealand

21. Old Neck, Stewart Island, New Zealand

22. Ras Ormara, Pakistan

23. Gwadar, Pakistan

24. Ras Tank, Iran

25. Ras Hafun, Somali

26. Near Capetown, South Africa

27. Punta da Marca, Angola

28. Cape Verde, near Dakar, Senegal

29. Monte Argentario, Italy

30. Bourg-de-Beatz (Loire Inferieure) Geins
    near Toulon, France

31. Sante Antioco, Sardinia

32. Near Gibralter, Spain

33. Molene Archipelago, Finisterre, Spain

34. G'verande', Loire, France

35. Quiberon, Brittany, France

36. Lle Molene, Brittany, France

37. Paimpal, Brittany, France

38. Swienemunde, East Germany

39. Isle of Wight, United Kingdom

40. Bill of Portland, United Kingdom

41. Flamborough Head, United Kingdom

42. The Long Ayre, Takerness, Orkney Islands, Scotland,
    United Kingdom

43. S.E. Coast of Isle of Man, United Kingdom

44. Howth, N.E. of Dublin, Ireland

45. Hualpen National Park, Chile

46. Mario de Puerto Santo, Venezuela

47. Cabo Rojo, Puerto Rico

48. Magdalena Bay, Mexico

49. Punta Sargento, Mexico

50. Duxbury and Saquish Neck near Plymouth,
    Massachusetts, U.S.A.

51. Nantasket Beach, Boston Harbor, Massachusetts, U.S.A.

52. Small tombolos along Connecticut
    and Rhode Island Coast, U.S.A.

53. Big and Little Nahant, Massachusetts, U.S.A.

54. Marblehead, Massachusetts, U.S.A.

55. Yarmouth, Nova Scotia, Canada

56. Near Halifax, Nova Scotia, Canada

57. Morro Bay, California, U.S.A.

58. Point Sur, California, U.S.A.

59. Badiga Head, California, U.S.A.

60. Trinidad Head, California, U.S.A.

61. Haystack Rack, near Cannon, Oregon, U.S.A.

62. Chichogof Island, Alaska, U.S.A.

63. Rernard Island, Alaska, U.S.A.

MAP 4-11   DEPOSITIONAL COASTAL FEATURES: BARRIER BEACHES AND TOMBOLOS

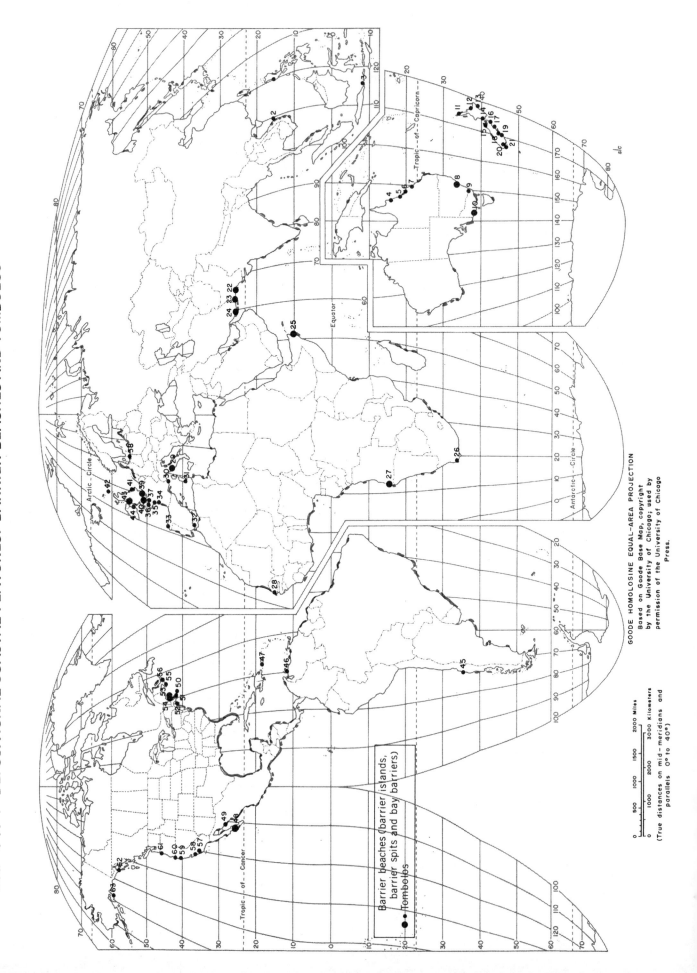

GOODE HOMOLOSINE EQUAL-AREA PROJECTION
Based on Goode Base Map, copyright
by the University of Chicago; used by
permission of the University of Chicago
Press.

Barrier beaches (barrier islands,
barrier spits and bay barriers)

Tombolos

MAPS
4-12
4-13

MANGROVE VEGETATION

Mangrove trees grow in the tidal zone of most low alluvial coasts in the humid tropics. The peculiar morphological adaption of these plants to a saline environment, as well as the characteristic hydrology and landforms of the swamps they grow in, makes this vegetation type one of the most distinctive of tropical areas. Mangrove forests appear to develop best in tropical temperatures, where the average temperature of the coldest month exceeds 68° F (20° C) and the seasonal temperature range does not exceed 9° F (5° C); they favor fine-grained alluvium along deltaic coasts or near river mouths, and shores free from strong wave action—because they cannot tolerate large waves. Mangroves are best developed along the shores of protected bays and estuaries, or back of barrier beaches. Within the general distribution area of mangroves, two major floristic regions are distinguishable—a relatively rich Oriental region stretching from East Africa to the western Pacific, and a relatively impoverished Occidental region comprising tropical America and the west coast of Africa. The more important genera, notably *Rhizophora* and *Avicennia,* are distributed through both these regions and it is doubltful whether the division into Oriental and Occidental is significant geomorphically, but the greater availability of species in the Oriental region is worth bearing in mind when making spatial comparisons.

In tropical America and the Atlantic coast of Africa, four major genera form the bulk of the mangrove trees: *Rhizophora* (red mangrove) *Avicenna* (black mangrove), *Laguncularia* (white mangrove), and *Conocarpus* (sometimes called "buttonwood"). Around the shores of the Indian Ocean and in Southeast Asia there are, at least, 15 genera and more than 20 species. The

MAP 4-12   MANGROVE VEGETATION OF THE WORLD

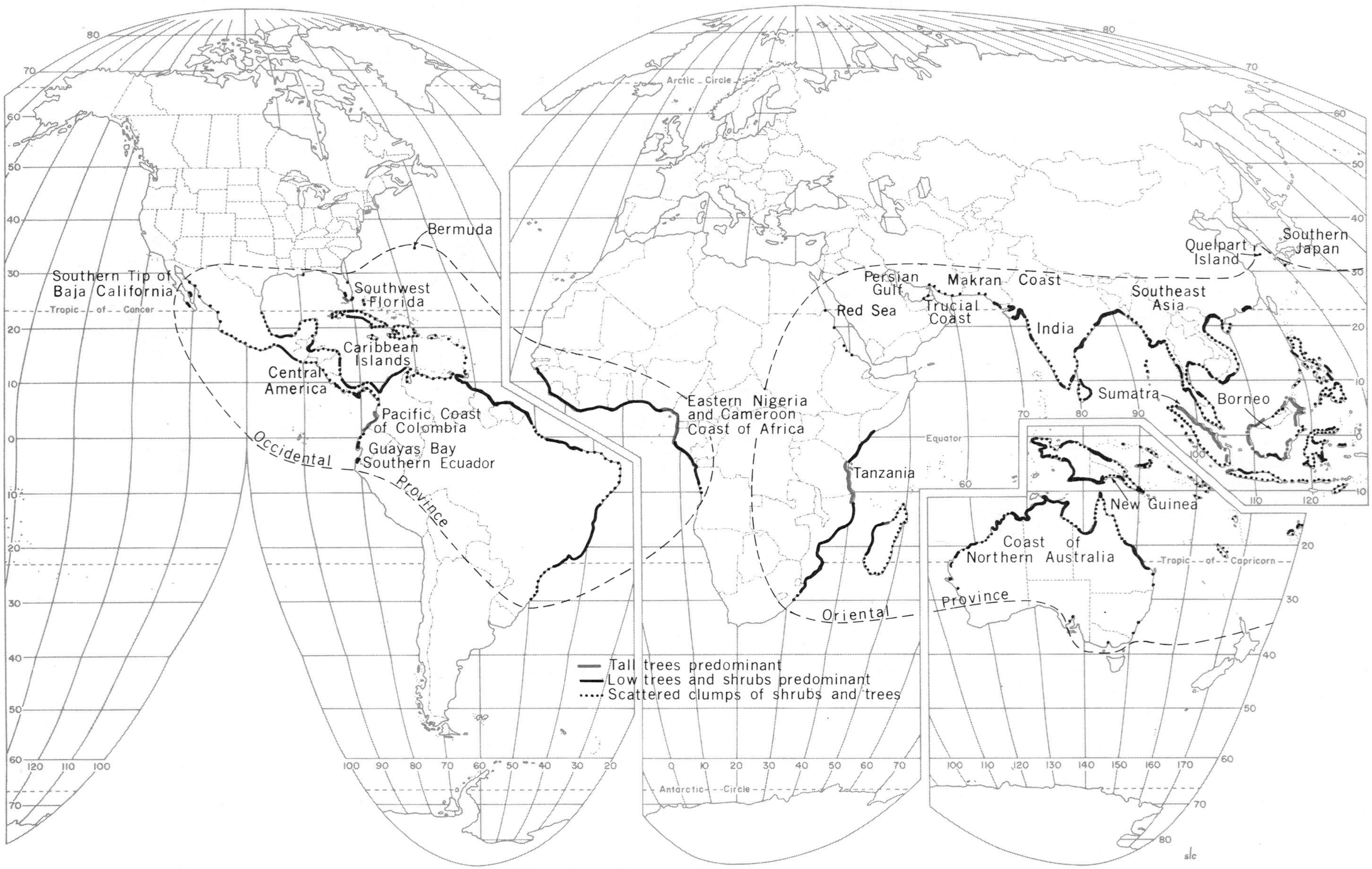

Southern
Japan

Quelpart
Island

Southeast
Asia

Borneo

New Guinea

Sumatra

India

Makran Coast

Persian
Gulf

Trucial
Coast

Red Sea

Coast of
Northern Australia

Province

Tropic of Capricorn

slc

Equator

Tanzania

Oriental

Eastern Nigeria
and Cameroon
Coast of Africa

Antarctic Circle

Arctic Circle

Bermuda

Southwest
Florida

Caribbean
Islands

Pacific Coast
of Colombia

Guayas Bay
Southern Ecuador

Province

Central
America

Occidental

Southern Tip of
Baja California

Tropic of Cancer

——— Tall trees predominant
——— Low trees and shrubs predominant
········· Scattered clumps of shrubs and trees

GOODE HOMOLOSINE EQUAL–AREA PROJECTION
Based on Goode Base Map, copyright
by the University of Chicago; used by
permission of University of Chicago
Press.

0   500   1000   1500   2000 Miles
0 1000 2000 3000 Kilometers
(True distances on mid–meridians and
parallels 0° to 40°)

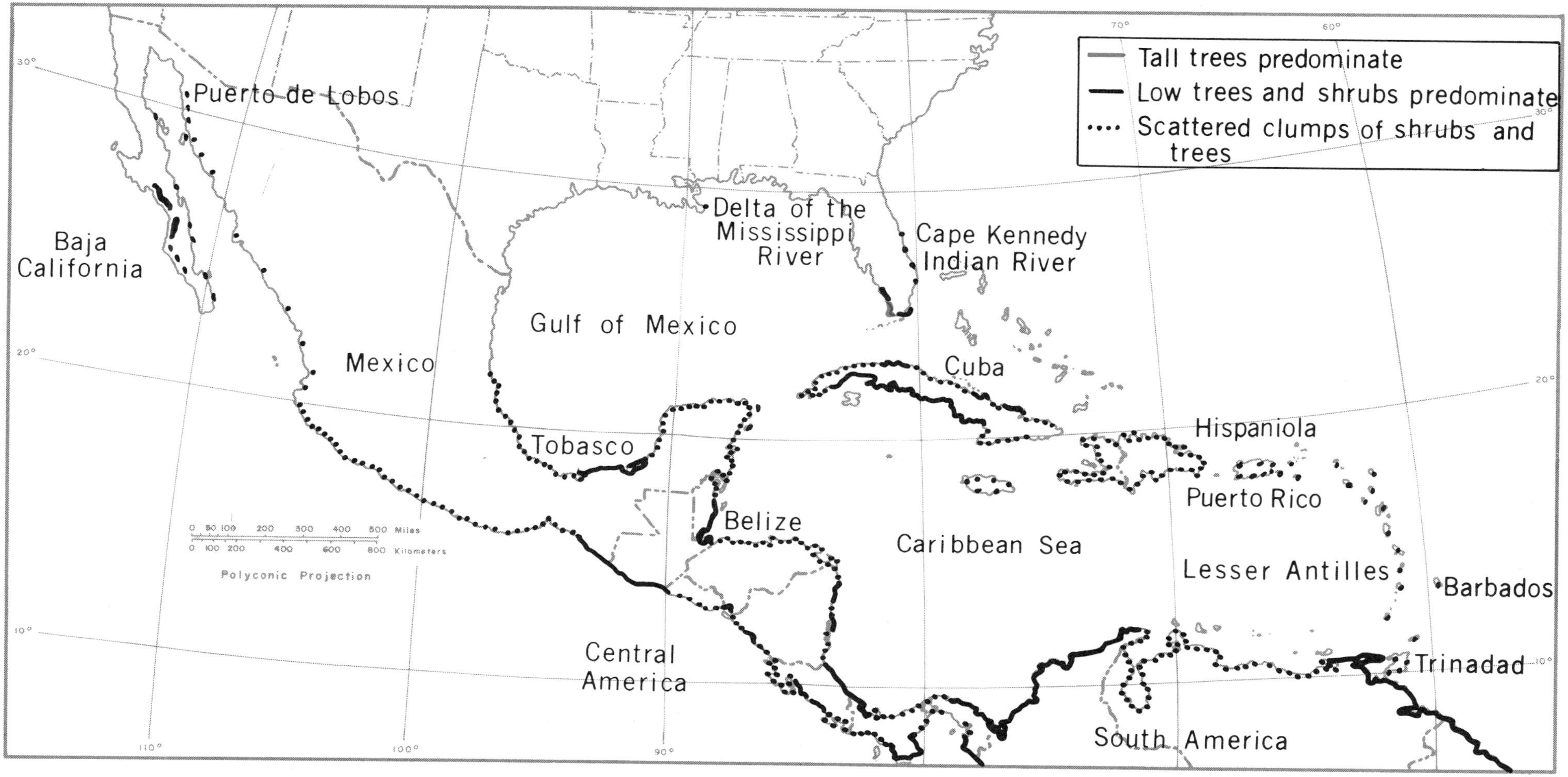
Tall trees predominate
Low trees and shrubs predominate
Scattered clumps of shrubs and trees
Puerto de Lobos
Baja California
Mexico
Gulf of Mexico
Delta of the Mississippi River
Cape Kennedy
Indian River
Cuba
Tobasco
Belize
Caribbean Sea
Hispaniola
Puerto Rico
Lesser Antilles
Barbados
Trinadad
Central America
South America
0 50 100 200 300 400 500 Miles
0 100 200 400 600 800 Kilometers
Polyconic Projection
30°
20°
10°
110°
100°
90°
70°
60°

MAP 4-13 MANGROVE VEGETATION OF MIDDLE AMERICA

tallest mangrove trees in the Americas are found along the Pacific coast of Colombia, where red mangroves grow to heights of more than 100 feet (30 meters), with huge butt resses at their base. In this section of coastal Colombia are some of the most luxuriant forests found anywhere in the world (West, 1956). Only two other mangrove forests in the American tropics are characterized by such luxuriance: one is on the eastern shore of Guayas Bay, southern Ecuador, near the poleward limit of mangrove on the Pacific coast of South America; and the other is a small section in the Ten Thousand Islands area of southwestern Florida, also near the poleward limit of mangrove. Tall mangrove forests, comparable to the ones of the Colombian Pacific coast, are found on the shores of the Cameroons and eastern Nigeria in the Gulf of Guinea, West Africa, along the coast of Tanzania, East Africa, and in various parts of Southeast Asia (for example, the north coast of Borneo) where mangrove forests probably reach their maximum development (Morskoi Atlas, 1950). The poleward limits of mangroves in the Americas is Bermuda. They also occur as low clumps north of Cape Canaveral in Florida, in the delta of the Mississippi River in Louisiana, and along the arid coast of the Gulf of California. In Asia the poleward limit is extreme southern Japan and the island of Quelpart, south of Korea (Putnam et al, 1960). Along the very arid coasts of the Red Sea and Persian Gulf, dark-leaved mangroves, although only 10 to 20 feet tall (3 to 6 meters), are in striking contrast to the barren desert environment. These trees and shrubs are struggling to survive in a harsh environment.

Sources: West, R.C., "Mangrove Swamps of the Pacific Coast of Colombia," *Annals of the Association of American Geographers,* Vol. 46, 1956, pp. 98-121, and Davies, J.L., *Geographical Variation in Coastal Development,* Hafner Publishing Co., New York, 1973, pp. 62-66.

MAPS
4-14
4-15

CORAL REEFS

Hermatypic or reef-forming corals maintain a symbiotic relationship with minute dinoflagellate algae called zoaxanthellae and it is the life requirements of the zoaxanthellae which are critical for reef distribution. Other coral forms extend into cold waters, but are of little geomorphic significance and their distribution are not included on the maps. The corals themselves are animal organisms whose skeletons are formed from secretions of calcium carbonate. The corals and algae live together in large colonies. As the corals die, new organisms grow on top of them, creating a reef made up of strongly cemented calcium carbonate skeletons. Coral growth is inhibited by large rivers, which bring in fresh or muddy waters, and by leeward locations, which lack moving water to bring food to the sedentary polyps, to prevent silting, and to provide a proper oxygen-carbon dioxide balance. Because of the above factors reef development is often more vigorous on the windward sides of islands. However, too much wave action may prohibit or inhibit reef development by preventing establishment of polyp planules or by destroying established coral at too high a rate. Recent volcanic activity also prohibits their growth, as does the excessive turbidity created by the many diatoms found in the ocean waters of the eastern Pacific and Atlantic.

The most important factor in the horizontal distribution of coral reefs is temperature. Live corals are widespread throughout the world, but they are most often found between 30° north and 30° south of the equator, where the ocean temperature is above 68°F (20°C). Although the outside limits are from about 60.8°F to 96.8°F (16°C to 36°C), the band of optimum growth is much narrower and may be taken as 77°F to 84.2°F (25°C to 29°C). Apart from limiting reefs to warm water areas and helping to concentrate development in the western parts of the oceans, a result of the temperature factor is to separate two major distribution provinces. As in the case of mangroves,

MAP 4-14 CORAL REEFS OF THE WORLD

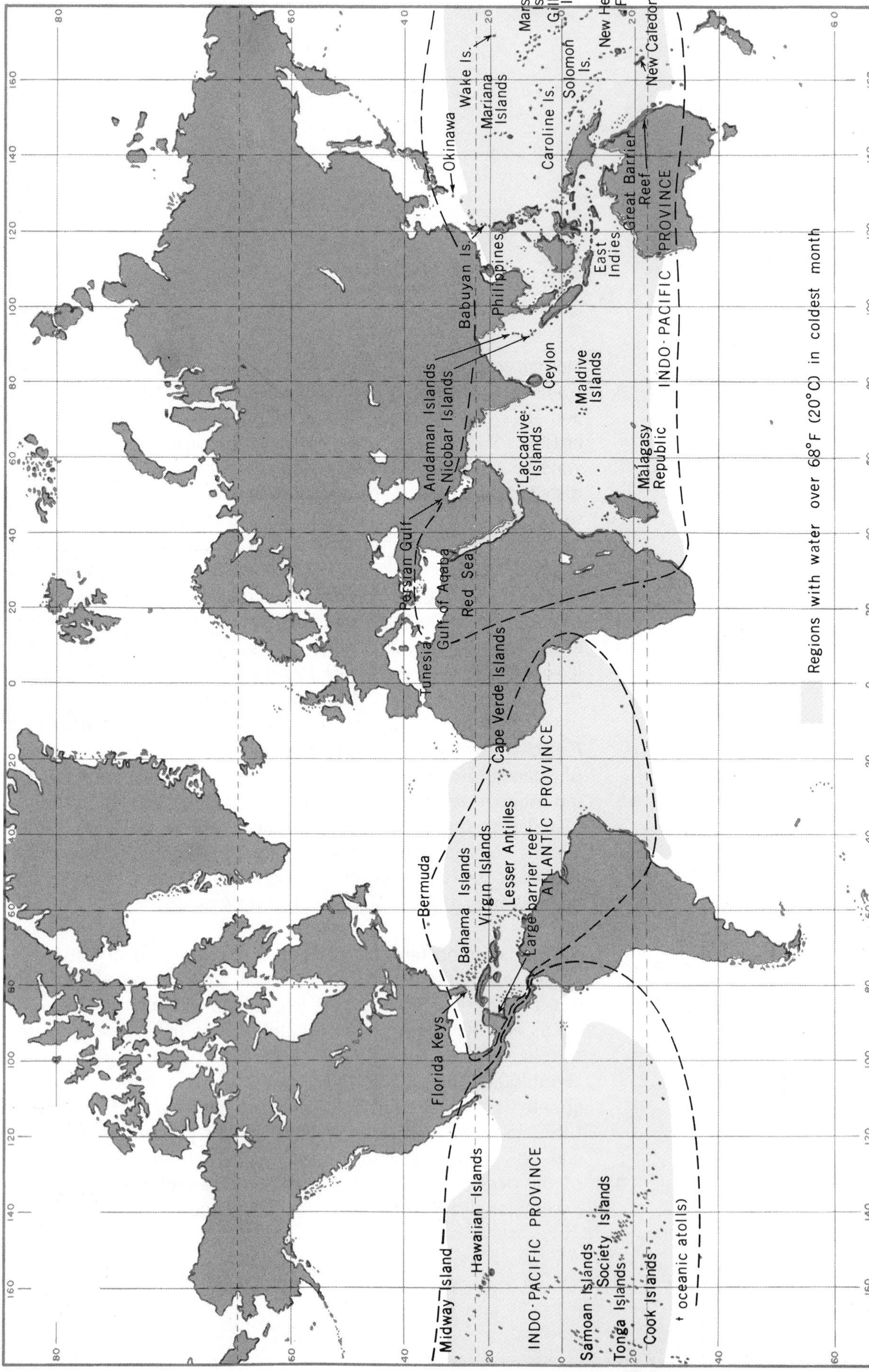

Regions with water over 68°F (20°C) in coldest month

INDO-PACIFIC PROVINCE

Marshall Is.
Gilbert Is.
New Hebrides
Fiji
New Caledonia
Wake Is.
Mariana Islands
Caroline Is.
Solomon Is.
Okinawa
Great Barrier Reef
East Indies
Babuyan Is.
Philippines
Ceylon
Maldive Islands
Malagasy Republic
Andaman Islands
Nicobar Islands
Laccadive Islands
Persian Gulf
Gulf of Aqaba
Red Sea
Tunisia
Cape Verde Islands
ATLANTIC PROVINCE
Bermuda
Bahama Islands
Virgin Islands
Lesser Antilles
Large barrier reef
Florida Keys

INDO-PACIFIC PROVINCE

Midway Island
Hawaiian Islands
Samoan Islands
Society Islands
Tonga Islands
Cook Islands
(oceanic atolls)

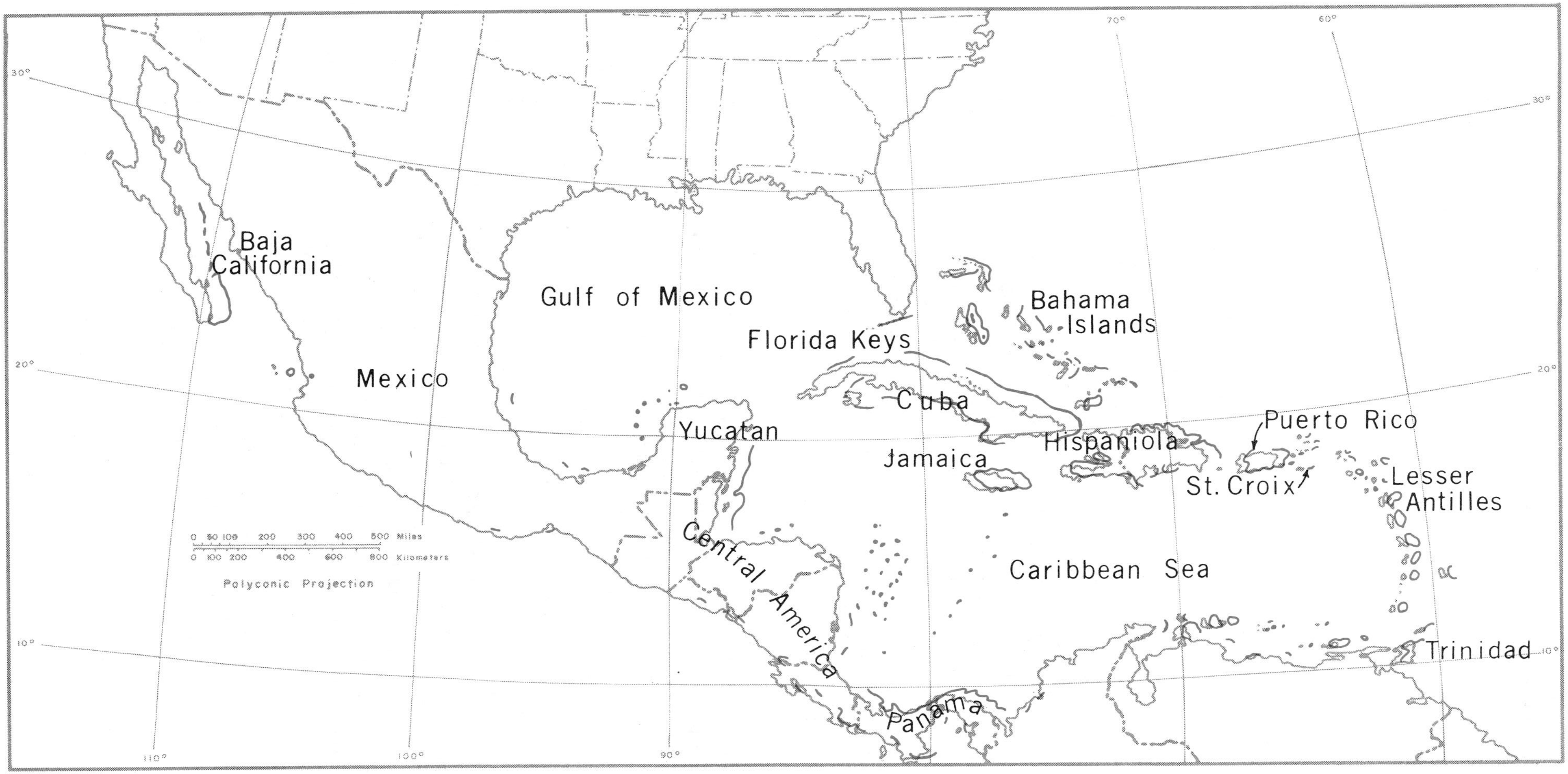

Baja California

Mexico

Gulf of Mexico

Florida Keys

Cuba

Bahama Islands

Hispaniola

Puerto Rico

St. Croix

Lesser Antilles

Trinidad

Jamaica

Yucatan

Central America

Panama

Caribbean Sea

60°
70°
90°
100°
110°
30°
20°
10°
30°
20°
10°

0 50 100   200   300   400   500 Miles
0   100   200   400   600   800 Kilometers

Polyconic Projection

MAP 4-15   CORAL REEFS OF MIDDLE AMERICA

there is an extensive Indo-Pacific Province stretching from East Africa to the Pacific and much less extensive tropical Atlantic Province from the Caribbean to West Africa. Not only are the outside limits of reef development wider in the Indo-Pacific, but so are the limits of optimum development. As the world map shows the zone of corals is most narrow on the western coasts of continents, where cool ocean currents flow equatorward. It is widest on eastern coasts of continents, where warm ocean currents diverge from the equator. The Indo-Pacific Province is noticably richer in species of coral and of assorted organisms than is the Atlantic Province and this is reflected to some extent in the profuseness and variety of the structures which have been produced. It is interesting the way in which the coral fauna increases in luxuriance and diversity as one goes from east to west in the Pacific. Reefs are absent from the Galapagos and Easter Island in what otherwise appear favorable habitats, in contrast, there is most profuse coral development in the Malaysian-Australian region in the extreme west (Davies, 1973).

A coral reef has a very rough surface, with pinnacles and deep wells. The coral formations may take the form of *fringing reefs, barrier reefs,* or *atolls* (Lobeck, 1958). Fringing reefs are built as limestone platforms attached to the shore. Fringing reefs can be found in the Bahama Islands and the islands of the Lesser Antilles in the West Indies (Joubin and Wells, 1954). They are also found in Hawaii and around some of the islands of the Pacific. On Oahu, Hanauma Bay, about 15 miles (24 kilometers) from Honolulu, is a drowned volcanic cone. Inside the cone is a beautiful fringing coral reef that is accessible to swimmers. A second type is the barrier reef, which lies offshore and is separated from the mainland by a lagoon ranging in width from one-half mile to 10 miles (16 kilometers) or more. The most famous and largest barrier reef extends along the northern Australian coast for 1,200 miles (1,930 kilometers) and is separated from the land by a lagoon 20 to 30 miles (32 to 48 kilo-meters) wide. Much of the Great Barrier Reef is actually "patch reefs'' bank or hammock reefs found on the continental shelf of Queenslands, inside the barrier reef in the north and making up the whole system in the south. There is a barrier reef along the Florida Keys. At John Pennekamp Coral Reef State Park, off Key Largo, one can rent boats and swim through one of the most luxuriant coral reefs found anywhere in the world. Great numbers of smaller barrier reefs partially or completely sur-round islands in the West Indies and the Pacific and Indian oceans (Wells, 1957). Buck Island Reef National Monument is a new park near St. Croix in the Virgin Islands. The third type of reef is an atoll. This is a circular or horseshoe-shaped ring of reef rock rising from a submerged platform and encircling a shallow lagoon. The reef ring is usually wider and more con-tinuous on the side from which prevailing winds blow. The principal channels through the atolls are found on leeward sides. The atolls are most widespread in the Pacific, where they number in the hundreds. Wake and Midway Islands are two of the most famous. Nearly all of the Caroline, Gilbert, Marshall, and Mariana islands are coral reef atolls. Other island groups around the Pacific that are famous for coral reefs are Society, Samoa, Tonga, New Hebrides, Fiji, and New Caledonia. Many of these islands have barrier reefs. The East Indies have rich corals, especially the coasts of Celebes and Sumatra. Reefs occur around the islands of the central Philippines, as well as around Okinawa. Most of the islands of the Indian Ocean have coral reefs. Reefs surround much of the Malagasy Republic and nearby islands, as well as the east coast of Africa deep into the Red Sea (McGill, 1960). The Asiatic coast is mostly free of growing coral because of the detritus from large rivers. In Ceylon there is evidence of ancient coral reefs.

Section Five
Glaciation

MAXIMUM EXTENT OF PLEISTOCENE GLACIATION

About one million years ago the whole of the earth's surface became cooler, presaging the first of four stages of glaciation. In polar latitudes and at high elevations, even near the equator, a series of glaciers began to form and move out from source regions. A glacier is a perennial mass of ice that develops where accumulated annual snowfall exceeds the amount annually dissipated. Over a period of years the snow is recrystalized into firn (ice granules) which in time and with intense pressure becomes transformed into plastic or glacial ice which then moves from its accumulation center. Glaciers from a number of centers in North America and Europe moved hundreds of miles as great continental ice sheets. As they advanced and retreated, they created the variety of landforms presented in this series of maps (Daly, 1934; Wright and Frey, 1965). Glacial landforms are created not only when moving ice scrapes, plucks and scours material from an area but also when this material is redeposited by the glacier itself or by meltwater streams. Although each glacier creates both erosional and depositional landforms, erosional forms (cirques, horns, aretes, steps, and troughs) are probably the more striking features of mountain glaciers, whereas depositional features (moraines, eskers, and drumlins) are the products of continental glaciers (Lobeck, 1939, 1958). Ice-free areas were also affected by glaciation. Drainage patterns were disrupted, and streams were overloaded around the edges of the glaciers (Antevs, 1928). Strong winds picked up silts and clays and carried them for miles, dropping them as loess deposits. Along coastal regions, sea level dropped when great quantities of water were stored in the ice; the seas later rose when the glaciers melted. The weight of the ice caused land masses to sink and then rebound with release of pressure. This map shows the maximum extent of glaciation during the Pleistocene. Wherever there is ice today in the world, there was more during the period of the Ice Ages (see Map 5-6, 5-7, and 5-8). Seventeen-thousand foot mountains near the equator today have snow and ice above 14,000 feet (4,267 meters); but during the Pleistocene these same mountains were covered from 6,500 feet (1,981 meters) upward. The equator had no "ice age". However, the equatorial region may have had a much cooler climate than at present.

In North America, ice started to accumulate in the higher, northern, colder areas and then gradually spread. It radiated from several centers as great ice sheets that eventually coalesced and covered virtually the whole of Canada and the northernmost part of the United States. From time to time, one center became more active and spread its ice farther than usual. On at least four occasions the ice started as separate ice flows and expanded to become one continuous cover. Between glacial advances the climate became sufficiently warm for nearly all the ice to melt except in the higher latitudes and on the highest mountains (Embleton and King, 1968). The four classical periods of glaciation recognized in North America are (from the oldest to the most recent): Nebraskan, Kansan, Illinoian, and Wisconsin, named after the states where their deposits are best preserved (Zeuner, 1945). This map depicts the maximum limits of each of these ice sheets. Within the western United States, south of the limit of continuous ice, there were, at least, 75 separate areas of glaciers centered in highlands (mountain ranges, group of ranges, or plateaus). The largest single area of former glaciation lies in the Yellowstone-Teton-Wind River highlands, which include many mountain ranges and plateaus. The second largest is in the Sierra Nevada. In the southern Rocky Mountain region, glaciers occurred as far south as northern New Mexico, but generally these glaciers were at higher elevations than the glaciers to the north and were not so extensive because of their distance from the source of moisture. Even the three highest mountains in Mexico, the volcanic cones of Ixtaccihuatl, Popocatepetl, and Orizaba, had several overlapping glacial advances down their slopes to about 10,000 feet (3,048 meters). Extensive glaciation occurred in the Canadian Cordillera and Alaska Ranges. The coastal ranges in Alaska had the largest mountain glaciers, and glaciation also occurred throughout the Alaskan Peninsula and the volcanic Aleutian island chain. Inland, the glaciers of the Brooks Range were not extensive (Hunt, 1967:422).

The two main centers of glaciation in continental Europe were Scandinavia and Switzerland-Austria. The latter really amounted to an enlargement of the conditions that still exist in the Alps where, in a sense, the Ice Ages are by no means finished. Evidence of five separate periods of glaciation have been found in the Alps; but only three are known definitely in northern Europe. The other two almost certainly occurred, but the "clues" belonging to the older glaciations are very complex and thus far have not been completely untangled. In every case, the northern center of the ice was to the east of the highest mountains in Sweden. The Gunz, Mindel, Riss, and Wurm Ice Ages of the Alps correspond in age to the Nebraskan, Kansan, Illinoian, and Wisconsin of North America. In northern Europe the Elster (Kansan), Salle (Illinoian), and Wurm (Wisconsin), are the three stages from oldest to youngest. The last glaciation was the Alpine Wurm. There is no general name for it in northern Europe, but it is referred to as the "newer glaciation" in Britain. Apart from the Scandinavian and Alpine ice sheets, many glaciers occupied highlands in continental Europe and islands

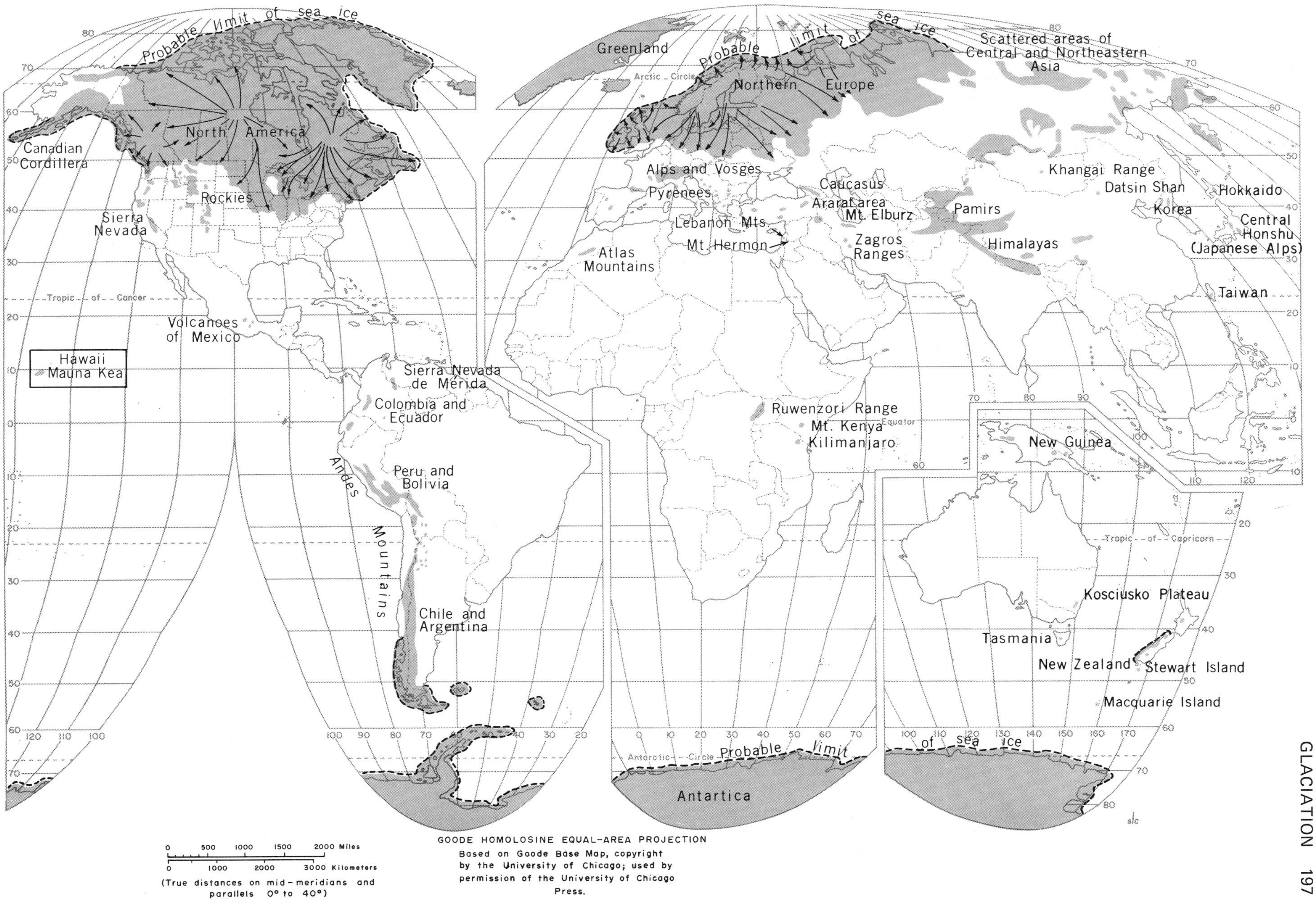

GOODE HOMOLOSINE EQUAL–AREA PROJECTION
Based on Goode Base Map, copyright
by the University of Chicago; used by
permission of the University of Chicago
Press.

MAP 5-1　MAXIMUM EXTENT OF PLEISTOCENE GLACIATION IN THE WORLD

MAP 5-2 MAXIMUM EXTENT OF PLEISTOCENE GLACIATION IN THE UNITED STATES

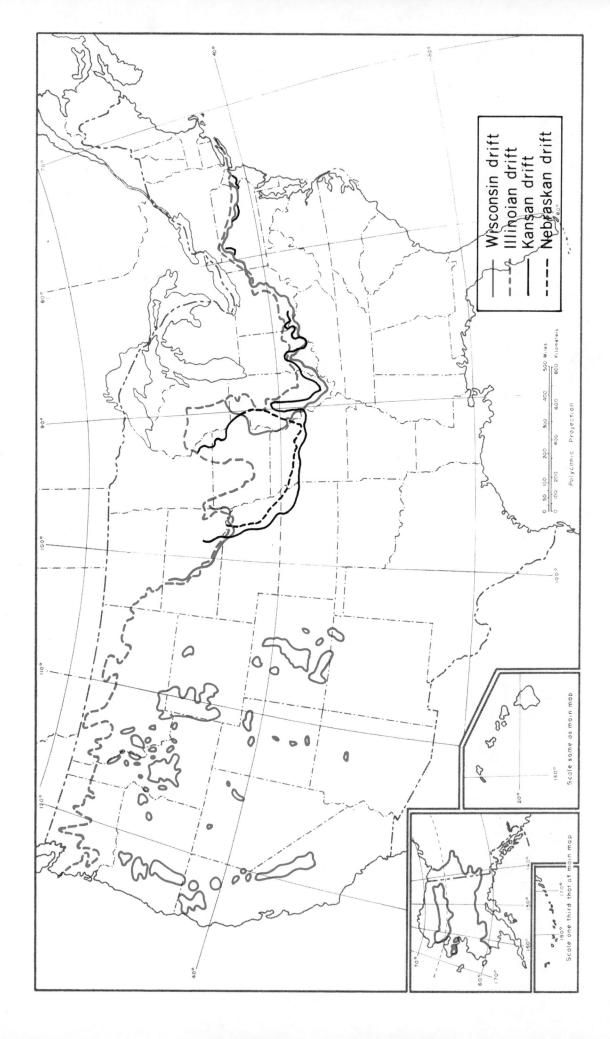

in the north Atlantic and Arctic Oceans, a large area of no fewer than seven centers or groups of centers with radial outflow, each located on a highland (Charlesworth, 1957). During maximum glaciation, 90 percent of Iceland was covered by ice sheets. There were also glaciers in the mountains of Spain, the Pyrenees of France, the Appennines of Italy, the Dinaric Alps and Pindus mountains of the Balkans, and the Transylvanian Alps. The former glaciers of Turkey, Syria, Iran, and the Caucasus were confined entirely to mountains. At one time the main chain of the Caucasus was ice covered almost continuously for a distance of 400 miles (644 kilometers). Mt. Elburz had glaciers down to 6,000 feet (1,820 kilometers). The high peaks of the Zagros Ranges, west of Isfahan were glaciated (Powers, 1966). The high mountains of central and northeastern Asia, as well as those of China, Mongolia, and Korea, probably had extensive glaciers, but much work still needs to be done there to determine the maximum extent of Pleistocene glaciation. As new evidence is found, there will probably be greater changes in the glacial map of central and eastern Asia than in those of other areas of the world. Glaciation was intense in the Pamirs and Himalayan Ranges, but its extent is not known. Pleistocene glaciation in Africa was confined to very high mountains, chiefly the Atlas Mountains in Morocco and the highlands in eastern Africa close to the equator. In South America, glaciers formed in the high mountains close to the equator. In southern Chile and Argentina they extended to the point of spreading out and coalescing on the plains east of the Andes. Ice tongues also moved to the west and cut the now picuresque fiord region of Chile. In northern South America there were glaciers on the Sierra Nevada de Merida in Venezuela. The Kosciusko Plateau in southeastern Australia had only small valley glaciers but the high region of central Tasmania had small ice caps. Only two high volcanoes (Ruapehu and Egmont) on the North Island of New Zealand were glaciated during Pleistocene time, but in the mountainous part of the South Island of New Zealand, glaciation was widespread and intense. The Antarctic ice sheet was much larger than it is today; it spread out across the surrounding sea floor. Submarine ridges extending across the mouth of the Ross Sea are thought to be a series of submerged terminal moraines (not shown on the world map). Glaciers once occurred in areas where today little ice exists. Mauna Kea (13,796 feet/4,206 meters) and Mauna Loa (13,680 feet/4,149 meters), on Hawaii, were glaciated above 10,500 feet (3,200 meters) by small ice caps believed to date from the Wisconsin Age. The Japanese Alps, west of Tokyo, had large valley glaciers on this northern island of Japan. The highest mountains of Taiwan and even the high peaks in New Guinea held glaciers (Flint, 1953 and 1957).

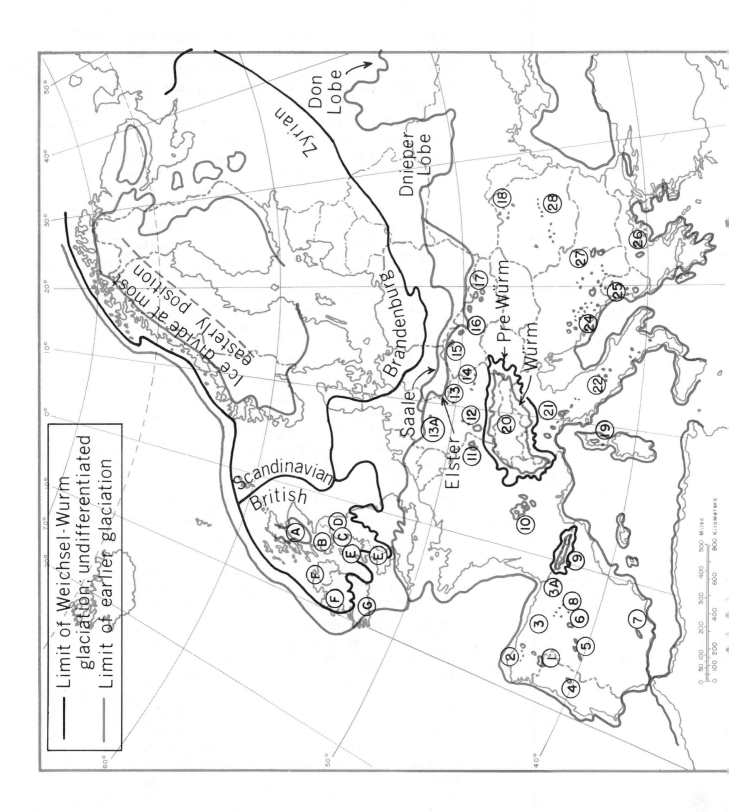

MAP 5-3   MAXIMUM EXTENT OF PLEISTOCENE GLACIATION IN EUROPE

KEY TO LETTERS SHOWING CENTERS OF OUTFLOW OF GLACIER ICE IN BRITAIN AND IRELAND

A   Scottish Highlands
B   Southern Uplands of Scotland
C   Cumberland Highlands
D   Pennine Chain
E   Mountains of Wales
F   Mountains of Connemara and Donegal
G   Mountains of Southern Ireland

KEY NUMBERS SHOWING SEPARATE AREAS OF INDEPENDENT GLACIATION IN CONTINENTAL EUROPE

SPAIN AND PORTUGAL

1   Sierra de Pena Negra
2   Sierra de Picos
3   Penas de Europa
3A   Sierra de Aralar
4   Sierra de Estrela
5   Sierra de Gredos
6   Sierra de Guadarrama
7   Sierra Nevada
8   Sierra Cebollera

FRANCE

9   Pyrenees Mts.
10   Auvergne and Cevennes Mts.
11   Vosges Mts.

GERMANY

12   Schwarzwald
13   Bayerische Wald
13A   Harz Mts.
14   Bohmer Wald
15   Riesen Gevirge

SWITZERLAND AND ADJACENT COUNTRIES

20   The Alps

ITALY

21   Etruscan Apennines
22   Roman Apennines
23   Southern Apennines

BALKAN COUNTRIES

24   Dinaric Alps
25   Pindus Mts.
26   Mt. Olympus
27   Rhodope Mts.
28   Transylvanian Alps

**MAP
5-4**

DEGLACIATION OF THE LAURENTIDE ICE SHEET

This map depicts the retreat of the Laurentide ice sheet from the United States across Canada. It can be compared with Map 5-2 which shows the maximum extent of ice in the United States during four glacial ages with the last being the Wisconsin. This map continues the retreat of the Wisconsin ice sheet into central Canada until it disappears altogether around 7,500 BP. From the beginning of deglaciation (18,000 to 15,000 BP, depending on the sector) to the final disappearance of the Laurentide continental ice as an ice sheet, the elapsed time was not much more than about 12,000 years. Measured along a radius extending from the Hudson Bay region southward nearly to central Illinois, where deglaciation began early, this time can be conveniently divided into four units. From the Late-Wisconsin maximum to Lake Erie is about 4,000 years; from Lake Erie through much of the Great Lakes history is about 3,000 years (see Map 5-5); from the northern edge of the Great Lakes to the two remnants of glacial ice, east and west of Hudson Bay, is about 3,000 years; and then to the final disappearance of the ice sheet proper is about 2,000 years.

It can be seen that deglaciation was an accelerating process in terms of area uncovered. Deglaciation involved not only the abandonment of territory by retreating ice margins but also thinning of the ice bodies. Numbers on the map—17.0, 12.5, 8.0, etc.—are a shortened version of 17,000 BP; 12,500 BP; and 8,000 BP (BP is the abbreviation for Before Present). A more detailed account of the phases of the glacial Great Lakes at the time of deglaciation is present on Map 5-5.

Source: R.F. Flint, *Glacial and Quaternary Geology,* John Wiley and Sons, Inc., New York, 1971 Fig. 18-12, p. 492.

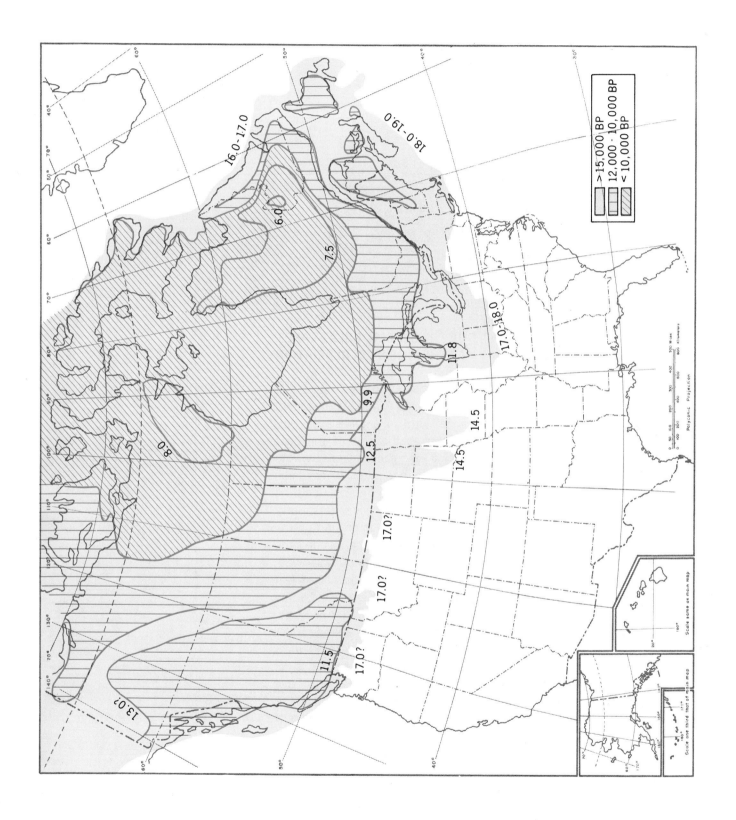

MAP 5-4   DEGLACIATION OF LAURENTIDE ICE SHEET

MAP
5-5

PHASES OF THE GLACIAL GREAT LAKES
AT THE TIME OF DEGLACIATION

The five maps on page 205 show the changes in the Great Lakes at the end of the Wisconsin glacial age. Map A depicts the extent of several stages of glacial advance with an emphasis on Wisconsin drift. When the ice retreated at the end of the Late Wisconsin, a magnificent sequence of glacial lakes, whose successors are the modern Great Lakes, occupied former large stream valleys. These valleys were enlarged by glacial scouring and isostatically depressed under the weight of ice sheets. Early lakes in this sequence were ponded between high ground on the south and the ice sheet on the north. As the ice receded the lakes widened and with glacial readvances the lakes narrowed. An early stage in the retreat of the late Wisconsin ice sheet shows the location of Lake Maumee, Lake Chicago, and three other small ice front lakes in New York State. At this stage, the waters from the ice drain by way of the Susquehanna River in the east and by way of the Wabash and Illinois (Des Plaines) rivers in the west. In Map B the Wisconsin ice sheet has retreated far enough to the north that the Finger Lakes have come into existence, but they drain to the east through the Mohawk-Hudson River System. At this time, the Chicago outlet by way of the Des Plaines River to the Illinois must have been occupied by a very large stream. As long as the ice margins, whether retreating or readvancing stood within the Great Lakes region, the glacial lake system was unstable and changes were frequent. But about 10,000 BP deglaciation had freed most of the lakes region of ice and the system became more

MAP 5-5   PHASES OF THE GLACIAL GREAT LAKES AT THE TIME OF DEGLACIATION

stable. The main changes then were brought about by isostatic recovery of the crust which altered altitudes of lake outlets. In diagram C drainage from the Great Lakes is across Canada and directly into the Champlain Sea. The St. Clair, near present day Detroit, is temporarily abandoned. Ocean waters have advanced into the St. Lawrence Valley, flooding all the lowlands and spreading into the basin of Lake Ontario. The sea waters extended southward into the Lake Champlain basin and may have continued at one time, as an estuary to the present site of New York City. In sketch D, the upper Great Lakes are larger than they are today, and the waters of Georgian Bay are more extensive. Both the Ottawa River and the St. Clair River outlets are in use, but the Champlain Sea has shrunk due to uplift of the American continent and the Mohawk outlet has been abandoned. Lakes Erie and Ontario appear about as they do today. Map E depicts the present day drainage lines. These changes include the disappearance of the Champlain Sea, the abandonment of the Ottawa outlet and the shrinkage of some of the lakes. It is interesting that certain of the low passes formerly used as outlets are now occupied by railways, automobile highways, and canals.

Sources: Sketch maps are drawn from R.F. Flint, *Glacial and Quaternary Geology,* New York, Wiley, 1971, pp. 568-569: W.A. Atwood, *The Physiographic Provinces of North America,* New York, Ginn and Company, 1940, pp. 215-221; William C. Alden, "The Quaternary Geology of Southeastern Wisconsin with a chapter on the older rock formation," *U.S. Geological Survey Prof. Paper,* No. 106, 1918, 356 pages, and F. Leverett, and F.B. Taylor, "The Pleistocene of Indiana and Michigan and the History of the Great Lakes," *U.S. Geological Survey Monograph,* No. 53, 1915, 529 pages.

PRESENT-DAY ACTIVE GLACIERS

This map shows areas where glaciers still exist today. About 17,642,126 square miles (45,692,930 square kilometers) of the earth's land area was ice covered at the height of the Pleistocene, but now only 33 percent of the total land area, or 5,860,708 square miles (15,179,175 square kilometers), is glaciated. The total volume of ice during the maximum glacial period is estimated to have been about 19.5 million square miles (50,485,500 square kilometers). Since glaciers drew water mainly from the oceans, the accumulation of ice resulted in a 350 to 400 foot (102 to 122 meters) lowering of sea level during each glacial advance (Flint, 1953, 1957). With the decrease in area covered by glaciers in the last 18,000 to 20,000 years, sea level has been rising. If all the ice shown on this map melted, sea level would rise, at least, another 120 feet (37 meters), drowning nearly all the world's seaports. The two main areas now covered by continental glaciers are Greenland and Antarctica. Greenland has an area of 839,782 square miles (2,175,026 square kilometers), of which 637,000 square miles (1,649,823 square kilometers) are ice covered. Only the mountainous margin is comparatively bare, but even here the deep valleys are usually filled by tongues of ice that extend downward to the sea from the great interior ice mass. Adjacent large islands, such as Ellsmere, have ice caps that are related to the Greenland glacier, although they are not now continuous with it (Bird, 1967). Antarctica is a continent of about six million square miles, of which about 5,019,300 square miles (12,999,936 square kilometers) are covered by ice. In many places the glacier reaches from the interior to the sea and joins the shelf of ice, or frozen seawater. Through the passes in the mountains that fringe parts of the Antarctic continent, great tongues of ice descend to the coast and push out to the sea beyond (Daly, 1963). Numerous large "noncontinental" glaciers are found at other locations around the world. About one-eight of Iceland is covered by glaciers. Very large valley glaciers are found in Pamir region of west central Asia, one of which is 44 miles (71 kilometers) long. There are also numerous large glaciers in the Karakoram, Himalaya, and Tien Shan ranges of central Asia. The coast ranges of the Alaska-Yukon region are the sites of most of the glacier ice and the large glaciers of continental North America. Some of the Alaskan glaciers are very long. Hubbard Glacier extends for 75 miles (121 kilometers); Seward, 40 miles (64 kilometers); and Sustina 25 miles (40 kilometers). The piedmont glaciers along this coast are also large; especially famous are the Malaspina and Bering glaciers. Coastal mountain ranges with extensive glaciers in the Alaska-

208

MAP 5-6   PRESENT DAY ACTIVE GLACIERS IN THE WORLD

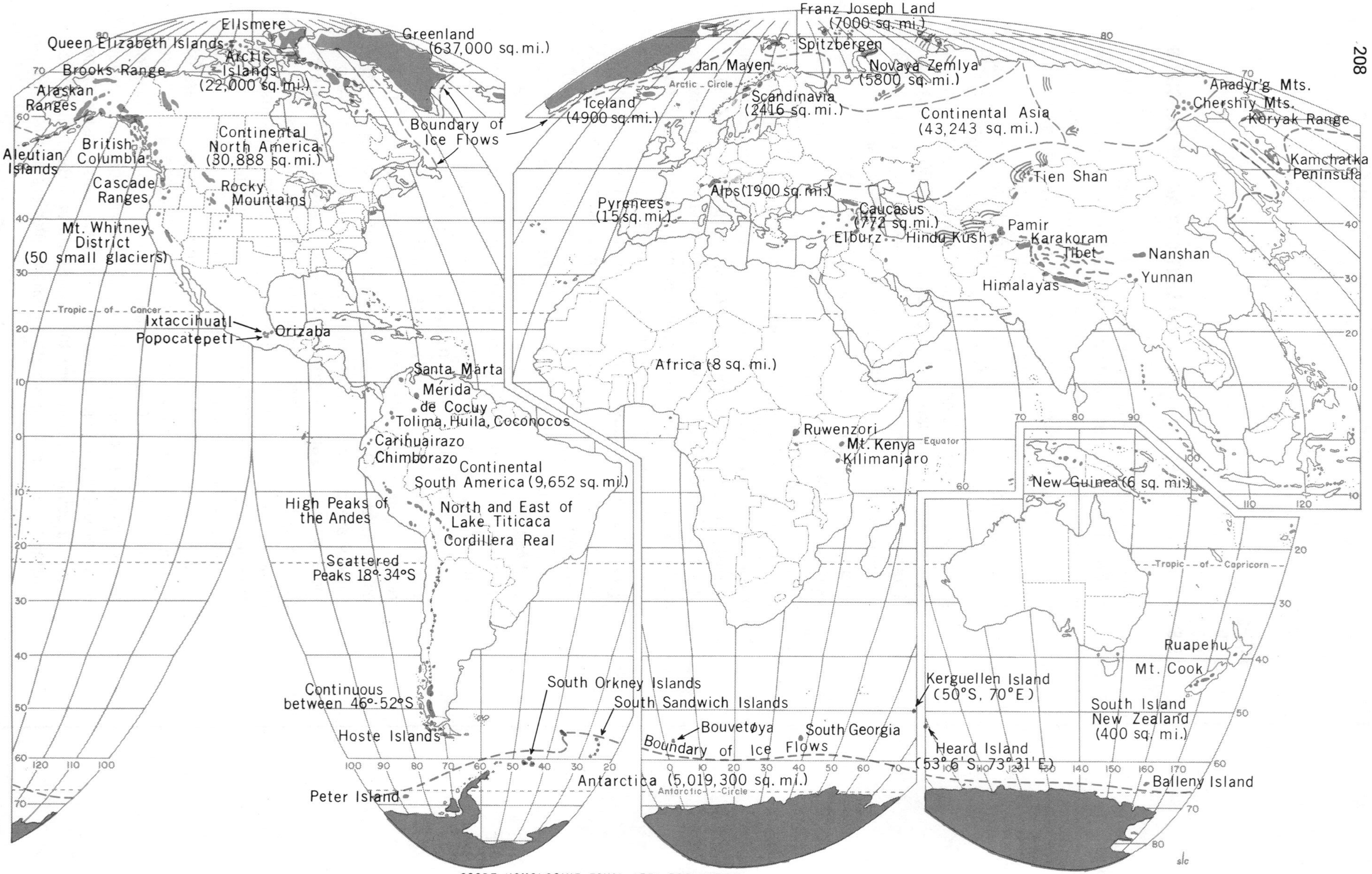

GOODE HOMOLOSINE EQUAL–AREA PROJECTION
Based on Goode Base Map, copyright
by the University of Chicago; used by
permission of the University of Chicago
Press.

British Columbia region include the St. Elias Range (19,850 feet/6,050 meters), The Chugach Range (13,000 feet/3,962 meters), and the Kenai Range (more than 5,000 feet/1,524 meters). Behind these rise the Wrangell Mountains (16,000 feet/4,877 meters), the Alaska Range (20,000 feet/6,096 meters), and the Aleutian Range (10,000 feet/3,048 meters). Plate 12 (centerfold) is a space photograph of the Tustamena glacier moving down the western slopes of the Kenai Range with a delta building out into Tustamena Lake. The Olympic Mountains and high volcanic peaks of the Cascade ranges have small valley glaciers. Mt. Rainier alone has about 26 active glaciers. The Sierra Nevada support about 50 small glaciers, mostly in the Mount Whitney region. Small ice masses largely confined to cirques occur in Glacier National Park and the Wind River and Teton Ranges in western Wyoming. Farther south, the Medicine Bow Range in southern Wyoming and the Front, Sawatch, and Sangre de Cristo ranges in Colorado support small glaciers. The southermost glaciers in the United States are in the Mt. Whitney California area located at 36° 34' north latitude. In Mexico there are a few small glaciers on Ixtaccihuatl at an elevation of 16,000 to 17,000 feet (4,877 to 5,181 meters). In South America glaciers are presently confined to the Andes Cordillera, but they extend from the Caribbean to Cape Horn. Near the equator they do not exist below 18,000 feet (5486 meters); but in the extreme southern part of the continent, in Tierra del Fuego, they occur as low as 3,000 feet (914 meters). In Europe, the Alps are dotted for 350 miles (563 kilometers) with an estimated 1,200 glaciers. The largest individual glacier is the Aletsch, more than 17 miles (27 kilometers) in length. The Scandinavian mountains, north of 60° latitude, have glaciers, and Spitsbergen, Franz Joseph Land, and Novaya Zemlya have ice caps and glaciers. Other areas where there are very small valley glaciers include the northern slopes of the central Pyrenees, the Caucasus, and the Elburz Mountains along the southern coast of the Caspian Sea. Mt. Ararat has several glaciers. Two peaks, Kenya and Kilimanjaro, plus the mountains in the Ruwenzori region, rise to heights of 15,000 feet (4,572 meters) to more than 17,000 feet (5,182 meters) near the equator in East Africa, support the only existing glaciers in Africa. The small Elena Glacier flows from Uganda's highest peak, 16,763 foot (5,109 meters) Mt. Stanley. An unusual place for glaciers is the island of New Guinea. In the central high mountain range at an elevation of 16,000 feet (4,877 meters) there are small glaciers that have formed from snowfields. On the North Island of New Zealand one volcanic peak, Ruapehu, has several small glaciers at 9,000 feet (2,743 meters). South Island, however, has more than 50 valley and intermontaine glaciers extending over some 200 miles (322 kilometers) in the high, nearly continuous, Southern Alps (Field, 1958; Mercer, 1967).

MAP 5-7 PRESENT DAY ACTIVE GLACIERS IN THE UNITED STATES

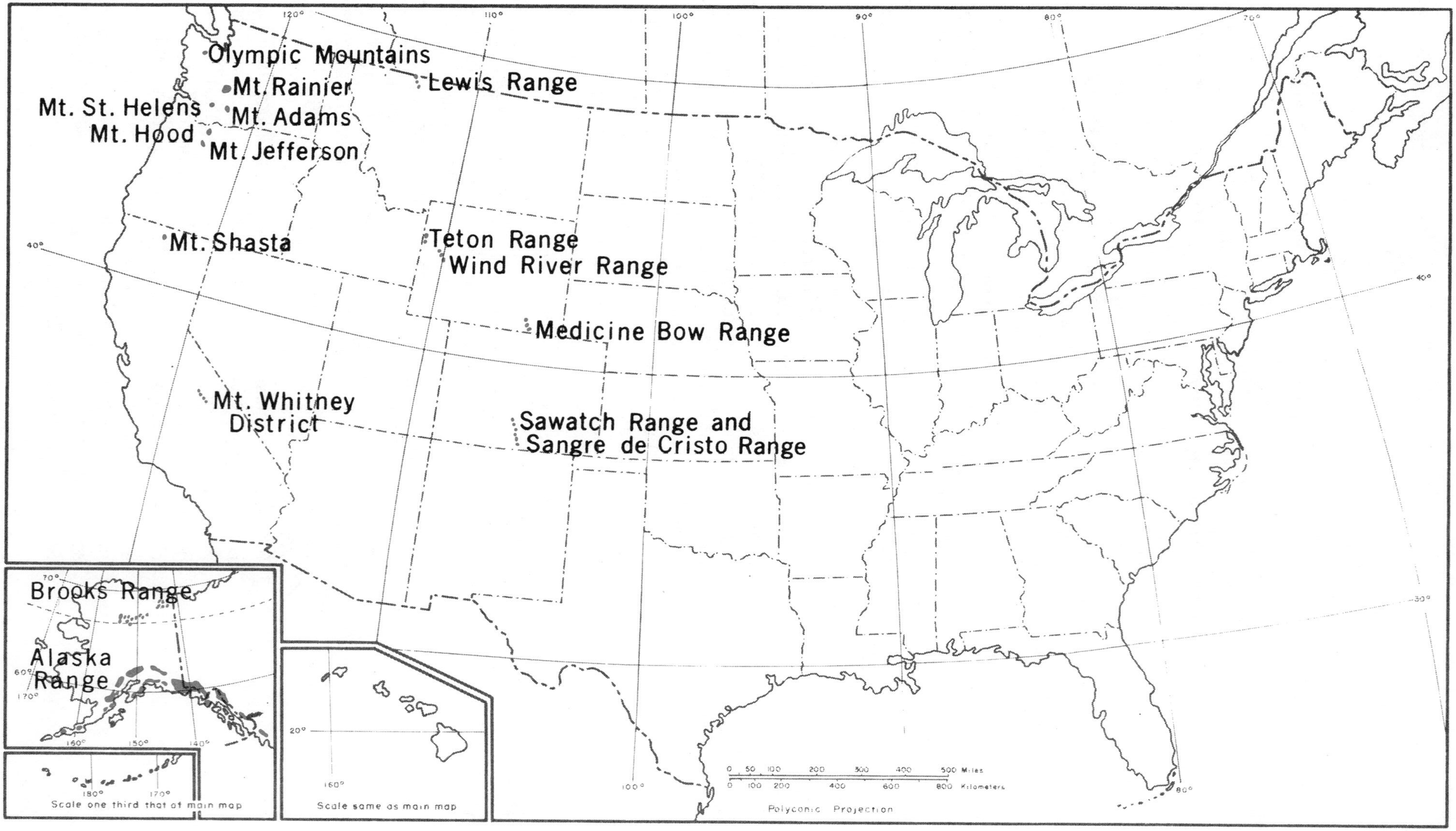
Olympic Mountains
Lewis Range
Mt. Rainier
Mt. St. Helens
Mt. Adams
Mt. Hood
Mt. Jefferson
Mt. Shasta
Teton Range
Wind River Range
Medicine Bow Range
Mt. Whitney
District
Sawatch Range and
Sangre de Cristo Range
Brooks Range
Alaska
Range
Scale one third that of main map
Scale same as main map
0 50 100 200 300 400 500 Miles
0 100 200 400 600 800 Kilometers
Polyconic Projection
120°
110°
100°
90°
80°
70°
40°
30°
20°
160°
150°
140°
170°
180°
60°

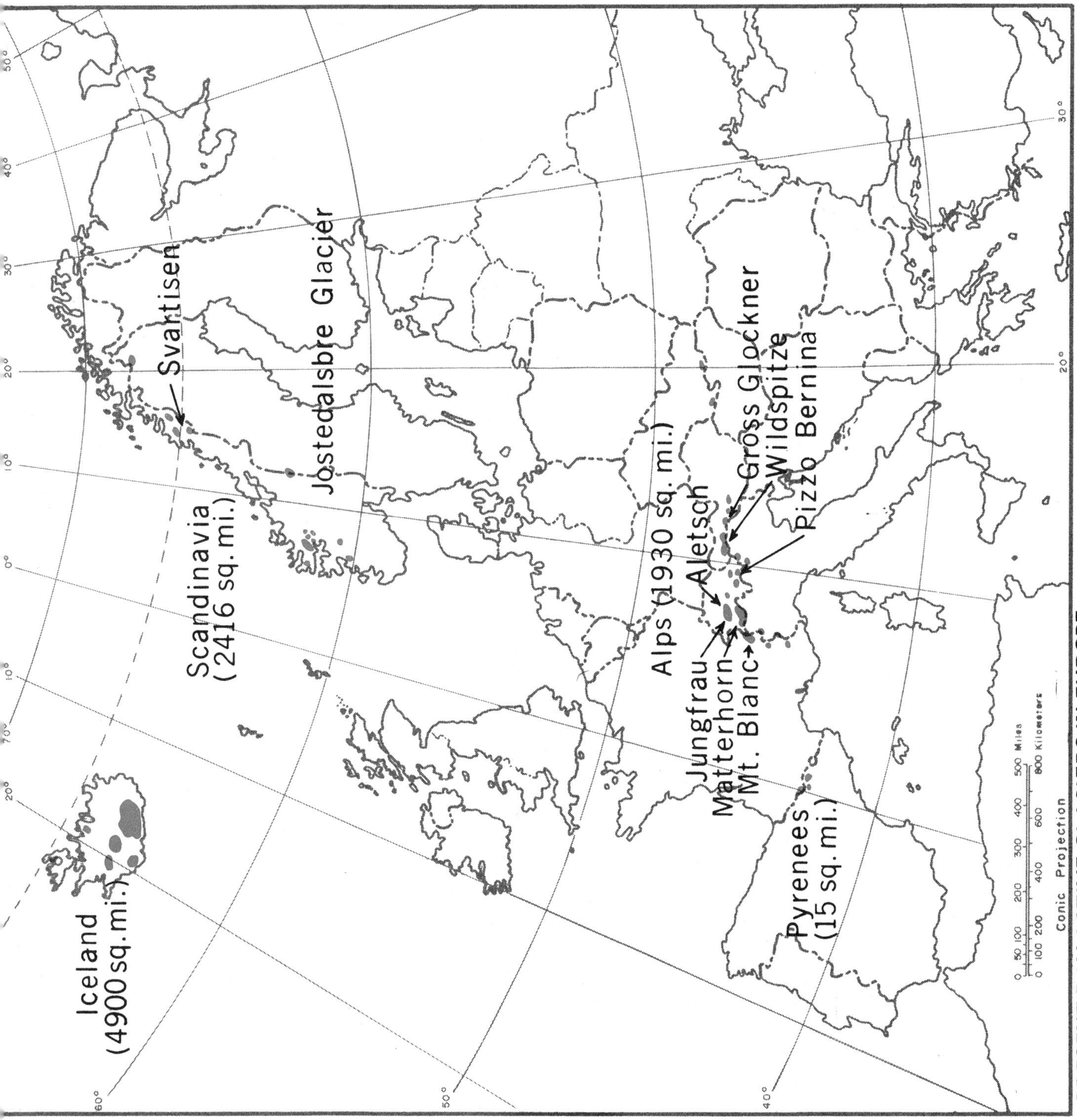
Iceland
(4900 sq. mi.)
Scandinavia
(2416 sq. mi.)
Svartisen
Jostedalsbre Glacier
Alps (1930 sq. mi.)
Aletsch
Gross Glockner
Wildspitze
Pizzo Bernina
Jungfrau
Matterhorn
Mt. Blanc
Pyrenees
(15 sq. mi.)
Conic Projection
0 50 100 200 300 400 500 Miles
0 100 200 400 600 800 Kilometers
60°
50°
40°
30°
20°
10°
0°
70°

MAP 5-8 PRESENT DAY ACTIVE GLACIERS IN EUROPE

**MAP
5-9**

GLACIAL, PERIGLACIAL, AND FROST ZONES

This map delineates regions that are under the influence of ice and freezing conditions during, at least, part of the year. The highly generalized zone of glacial and periglacial conditions has either a permanent ice cover, such as the glaciers of Greenland and Antarctica, or has permafrost layers that restrict the growth of trees. The southern limit of this zone, therefore, approximates the tree line across northern Canada, northern Europe, and the Soviet Union, and ice becomes a significant aspect of the landscape.

A second zone includes areas having seasonal frost. These areas have severe winters, during which the ground and lakes are frozen from one to three months. Ice conditions are significant but more limited than they are north of the tree line. Landforms are influenced by frost-weathering. There is considerable daily and annual variation of temperature which man, animals, and plants must adjust to. In this zone, animals and plants must adapt to frost conditions by either becoming dormant (deciduous trees) or adapting to the cold conditions (animals grow warm fur coats or burrow to depths beyond the cold). A third zone is one of occasional frost. Winters are mild, but frosts do occur. Seldom do the cold spells last long enough for the ground to become frozen or lakes to freeze over solid for long periods. However, specialized crops, such as the oranges and lemons of Florida and California, are hit by an occasional damaging frost. Even occasional freezing conditions bring about animal and plant adjustments to these conditions.

Sources: Butzer, K.W., Figure 21 in *Environment and Archaeology: An Introduction to Pleistocene Geography,* Aldine Publishing Company, Chicago, 1964.

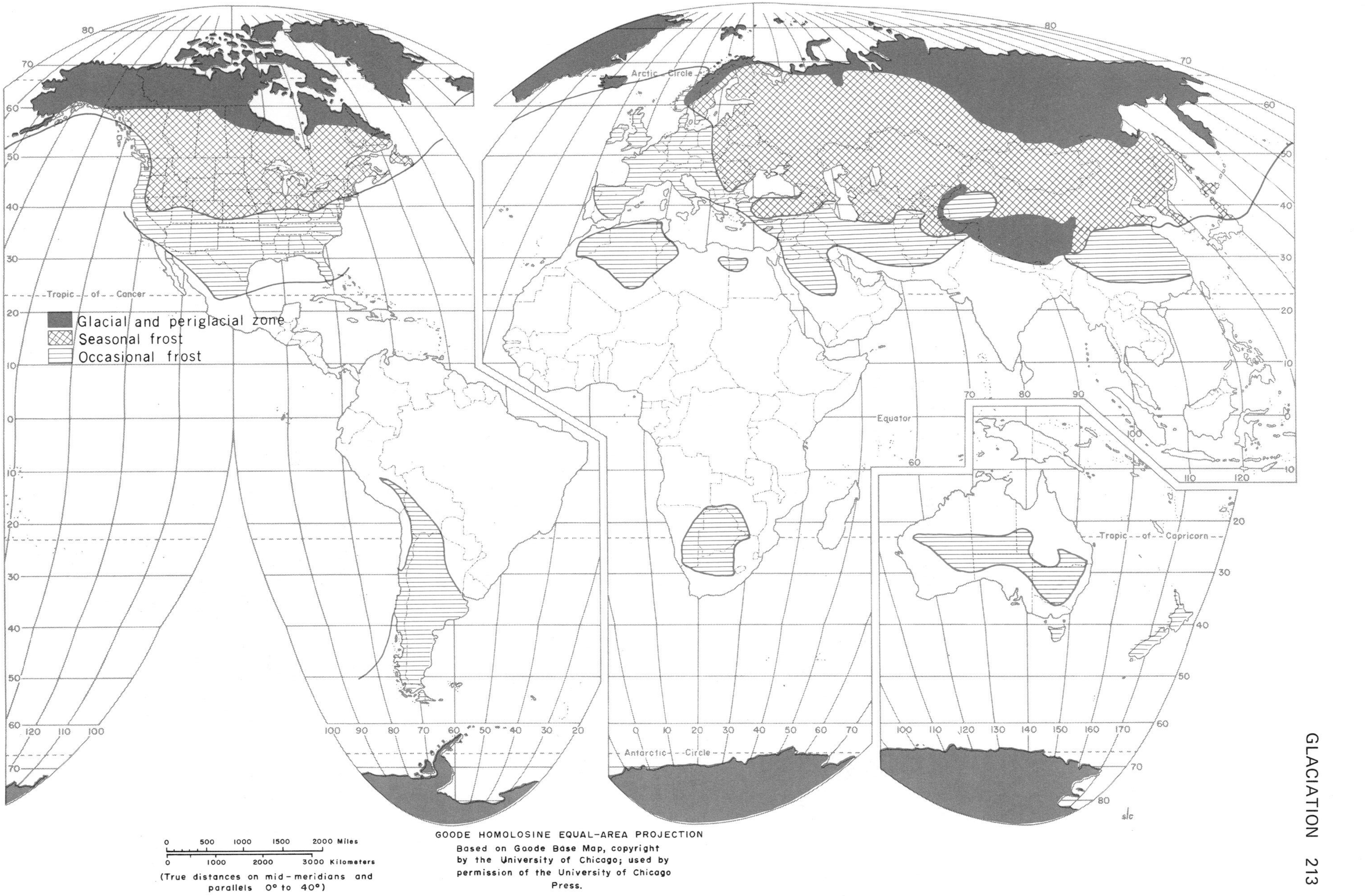

GOODE HOMOLOSINE EQUAL-AREA PROJECTION
Based on Goode Base Map, copyright
by the University of Chicago; used by
permission of the University of Chicago
Press.

Glacial and periglacial zone
Seasonal frost
Occasional frost

MAP 5-9   GLACIAL, PERICLACIAL AND FROST ZONES

**MAP
5-10**

PERMAFROST AND ICE CAVES

Permafrost is ground that has been frozen for many years. It occupies approximately 20 percent of the earth's suface: In the Soviet Union about 3,860,000 square miles (9,997,361 square kilometers) , or nearly 45 percent of its area! In Canada, frozen ground occupies about one-half of the country's area (Jenness, 1949). Although there is a general correlation between the distribution of permafrost and that of glacial, periglacial, and frost zones, one deals with subsurface freezing and the other deals with surface freezing. The undulating layers of permafrost are of different thicknesses; they lie from a few inches to several feet below the surface of the ground. Climate, relief, soil characteristics, and degree of vegetative cover are factors that help to determine the upper limit of permafrost. Its lower limit is the point at which warmth from the earth's interior or circulating ground water raises the temperature above 32°F (0.0°C). Permafrost is usually covered by a layer of soil, called the "active layer," that generally is quite moist and varies in temperature with the seasons. This is known as "wet permafrost." This layer freezes every winter; during a warm summer it will thaw down to the upper surface of the permafrost. There are three main types of permafrost: continuous, discontinuous, and sporadic. In northern areas of greatest cold, "dry" or permanent permafrost is present everywhere close to the surface, except beneath large water bodies, reaching a depth of more than 600 feet (183 meters). It is thickest in Siberia, where it attains a maximum depth of 2,000 feet (610 meters). Farther south the permafrost becomes thinner with the depth variable, and at some sites it may vanish completely, although nearby it may still be 200 feet (61 meters) thick. Under these conditions permafrost is said to be discontinuous. In a third zone, small islands of permafrost are surrounded by thawed ground; hence, this is called sporadic permafrost. There has long been a controversy over the extent to which permafrost is the product of present climatic conditions, and over its age. There is evidence that permafrost

The following is a short list of major ice caves plotted on the world map:

1. Dobsina ice cave, Solvakia, Czechoslovakia
2. Eiriesenwelt cave near Salzburg, Austria
3. Grotte de la Glaciere near Passavant in the Jura
4. Numerous small ice caves in the Alps, usually above (4,757 feet) 1,450 m.
5. Numerous small caves in the Pyrenees
6. Perpetual Ice Cave (The Desert Ice Box), New Mexico
7. Sunset Crater National Monument, Arizona
8. Ice Cave in the Black Hills of South Dakota
9. Four ice caves east of Warren, Montana
10. Crystal Ice Cave, Idaho
11. Boy scout cave, Craters of the Moon National Monument
12. Shoshone Indian Ice caves, Idaho
13. Merrill Ice cave, Lava Beds National Monument, California
14. Lava River Caves State Park, Oregon
15. Paradise Ice caves, Mt. Rainier National Park, Washington
16. Crow's Nest Area, Alberta

first came into existence during the Pleistocene. A comparison of the above map with the map of maximum Pleistocene glaciation shows that the areas of former ice sheets bear no evident relation to that distribution of permafrost at the present time. Instead, it appears that this phenomenon is in balance with existing weather conditions, and if a decided warming trend takes place, sporadic and discontinuous permafrost will decrease substantially, as it may well have done during the interglacial ages. There is permafrost associated with and surrounding most large glacial areas. Thus a base map showing the location of existing glaciers is given.

Ice caves and permanent sheets of ice have been added to this map, but their location is underground where the temperature remains close to freezing all year. The ice in many of these caves was formed during Pleistocene glaciation. There are numerous small Pleistocene ice caves often found in high mountain areas such as the Alps, Rockies, and Pyrenees.

There are many ice caves in the United States, but only a few are in areas that are accessible to the tourist. Ice caves are actually natural storage areas for the cold air of winter. The rock covering of the cave acts as an insulation against the sun and summer air. The duration of the summer in ice-cave areas is not long enough for the temperature to rise above freezing. The result is that each winter and spring more cold air and water replenish the ice perpetually stored within the cave.

Some of the lava tubes in the western United States have large quantities of ice that have formed in the freezing air of the cave. Some of these deposits are thousands of years old. Many of the more delicate ice formations melt in the fall of the year, but they are replaced in the spring as the melting water from the surface drips into the freezing air of the cave.

MAP 5-10 PREMAFROST AND ICE CAVES

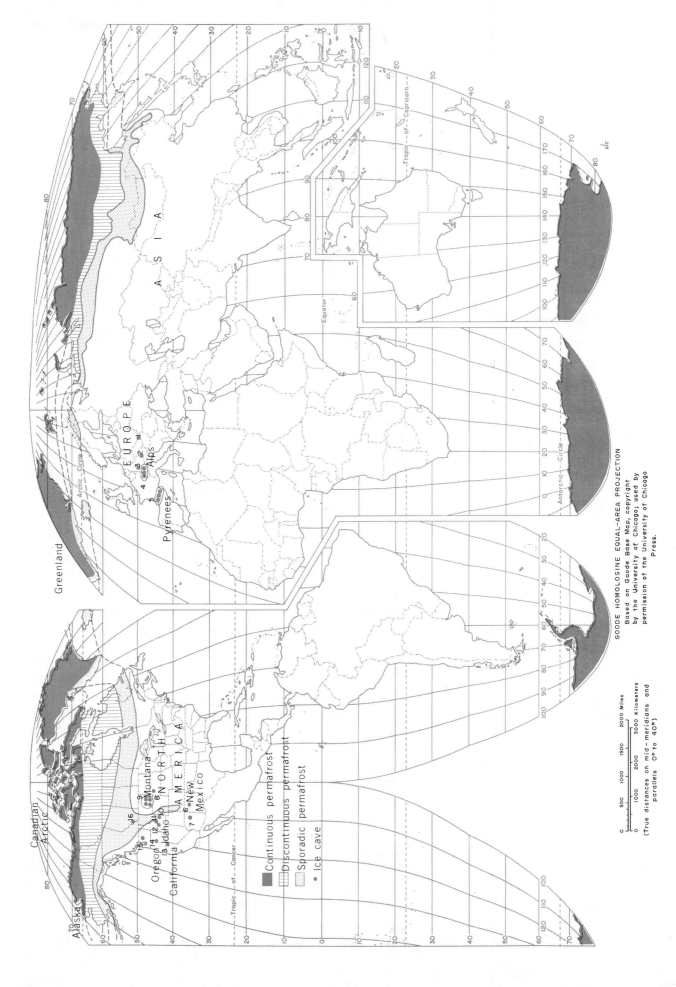

GOODE HOMOLOSINE EQUAL-AREA PROJECTION
Based on Goode Base Map, copyright
by the University of Chicago; used by
permission of the University of Chicago
Press.

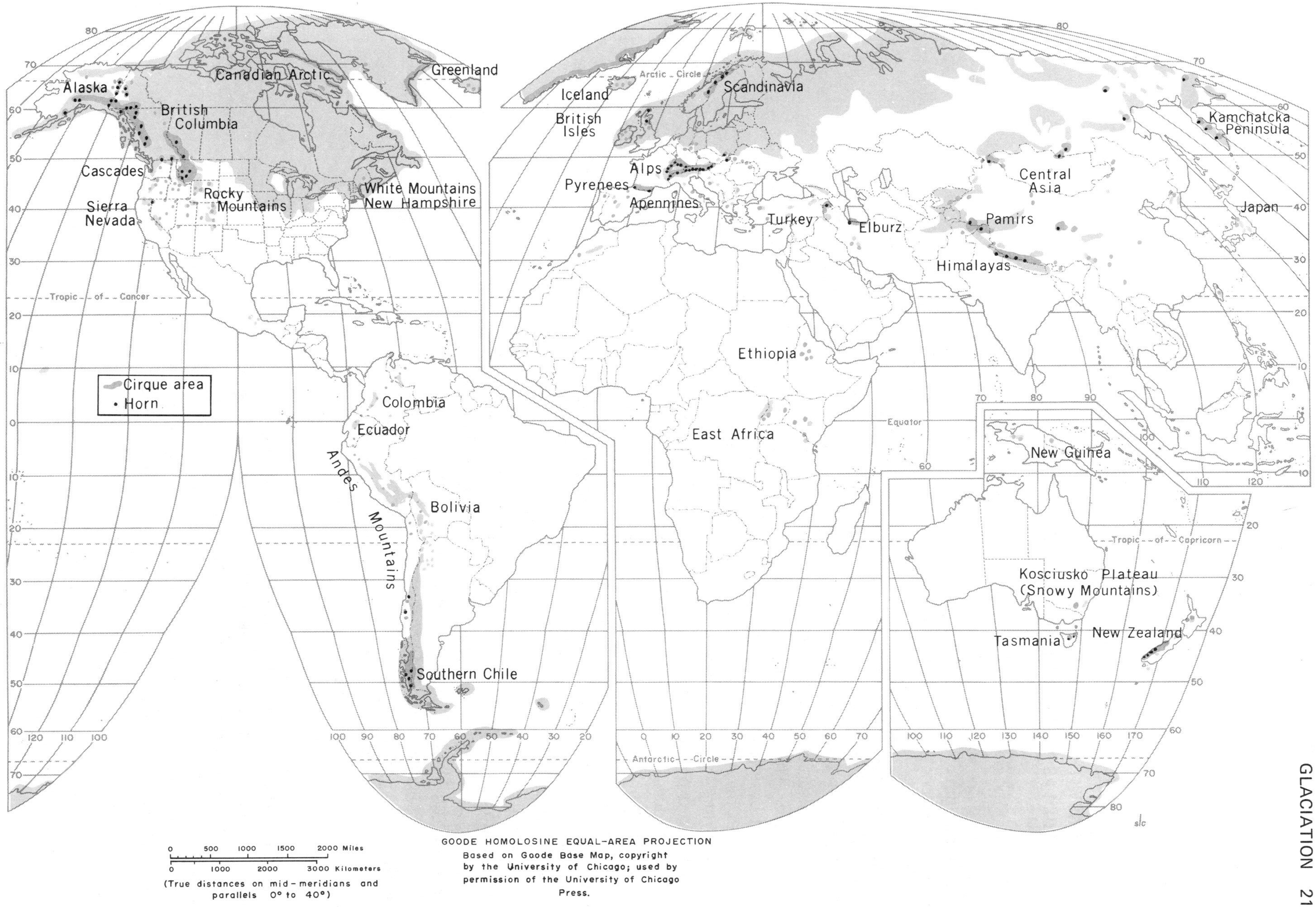

Kamchatka Peninsula
Japan
Central Asia
Pamirs
Himalayas
Elburz
Turkey
Scandinavia
Iceland
British Isles
Pyrenees
Alps
Apennines

New Guinea
Kosciusko Plateau (Snowy Mountains)
New Zealand
Tasmania
Tropic of Capricorn

Ethiopia
East Africa
Equator
Arctic Circle
Antarctic Circle

Greenland
White Mountains New Hampshire
Canadian Arctic
British Columbia
Rocky Mountains
Cascades
Sierra Nevada
Alaska
Tropic of Cancer

Colombia
Ecuador
Bolivia
Andes Mountains
Southern Chile

Cirque area
• Horn

GOODE HOMOLOSINE EQUAL-AREA PROJECTION
Based on Goode Base Map, copyright
by the University of Chicago; used by
permission of the University of Chicago
Press.

0   500   1000   1500   2000 Miles
0 1000 2000 3000 Kilometers
(True distances on mid-meridians and
parallels 0° to 40°)

MAP 5-11   REGIONS WITH PROMINENT MOUNTAIN GLACIAL FEATURES

**MAP
5-11**

REGIONS WITH PROMINENT MOUNTAIN GLACIAL FEATURES SUCH AS; CIRQUES, TARNS, HORNS, ARÊTES, AND COLS

All five of these mountain glacial features are closely related; being formed by active glacial erosion. The more intense and longer a mountain area has been subjected to the action of ice, the sharper will be the features discussed below.

The *cirque* has long been recognized as one of the most characteristic examples and proofs of mountain glacial erosion. In its fully developed form, a cirque is a rounded basin partially enclosed by steep cliffs and sometimes containing a small lake, called a *tarn.* The rounded basin has been scoured out or plucked out by a glacier. In large cirques, the cliffs at the back of the basin may rise to great heights and culminate in sharp mountain ridges called *arêtes.* Cirques can also sculpture a mountain into pyramidal shapes called *horns.* There are many types of cirques. At one extreme, there are the small shallow depressions, only a few feet across, which contain snowfields that are changing into the first stage of glacial ice called *firn,* — ice granules that often represent the beginning of cirque formation because, as firn changes into plastic ice, it may begin to move and cut the depression deeper. At the other extreme, there are the great cirques of Antarctica and the Himalayas whose widths may be measured in terms of miles and whose backwalls attain heights of hundreds or thousands of feet. A large cirque on Mount Everest has a maximum width approaching two and one-half miles and backwall which, if the ice were removed, would probably rise to more than 9,000 feet (2,743 meters). Many of the large cirques are not perfectly formed, and, therefore, may be classified as glacial troughs rather than cirques (Loeback, 1939; Flint, 1953, and 1957).

Since it is difficult to identify true cirques, especially cirques with snow and glaciers still in them, and even more difficult to plot individual cirques on a world map, only those regions having groups of cirques have been mapped. Cirques occur worldwide, confined to areas of present or former glaciation. As a consequence of this broad distribution, they have received numerous local names, such as *corrie,* or *coire* in Scotland, *cwm* in Wales, *kar* in Austria and Germany, *botn* in Norway, and *nisch* in Sweden. Major cirque regions are the Cordilleran Ranges of Canada and the United States (Krinsley, 1965).

British Columbia has a large number of cirques. Alaska is noted for active glaciers presently forming large cirques. There are numerous cirques in the Sierra Nevada of California, the Rocky Mountains, particularly Glacier National Park in Wyoming, and in the Colorado Rockies (Harrison, 1952). The most southern glaciated peak in the United States with one small cirque is Cerro Blanco in New Mexico (Smith and Ray, 1941). Only small cirques are forming today because few active glaciers still exist in these mountains. In the eastern United States, mountains are too low in elevation to have active glaciers, but during the height of Pleistocene glaciation small cirques that can still be seen developed in the Presidential Range of the White Mountains of New Hampshire, the Katahkin Group in Maine, and the Green Mountains of Vermont. Whiteface Mountain and the north slope of Sentinel Peak in the Adirondacks also have remnant cirques. European cirques are prominent in the Swiss and Austrian Alps, but cirques also developed during the Pleistocene in the Lake District and Pennines of Great Britain and Ireland, the Apennines of Italy, and in Turkey. Some of the most magnificent cirques in the world occur in the mountains of Asia, especially in the Himalayas and Pamirs. Many of these cirques are still forming. High volcanic peaks in Ecuador on the equator show the effect of glacial action and have small cirques. In Columbia, rivers take their sources from the small tarns that are found in cirquelike depressions. The highest peaks of New Guinea, reaching over 16,000 feet (4,877 meters), supported small glaciers during the Pleistocene, and cirques have been reported there. In New Zealand the Southern Alps offer a number of excellent mountain glaciated features. In Australia there is one small section of the Snowy Mountains with cirques and the highlands of Tasmania have several small areas.

A *tarn* is a lake which frequently occupies a rockcut basin usually formed by a cirque. In this sense, they are synonymous with *cirque-lake, corrie-lake,* or *cwm.* In northern England the word is frequently used to indicate a small mountain lake but most geomorphologists restrict the term to a rock basin set in a steep-walled amphitheatre in the mountainside. Such basins are the product of erosion by cirque glaciers, while the amphitheatre was largely the result of a freeze-thaw action (Lewis, 1960).

By definition, a *horn* is a high pyramidal-shaped mountain peak that has been under the influence of intense glaciation. It is produced by the encroachment of three or more cirque glaciers that sculpture the preexisting surface to a stage where three serrated ridges converge upward into a pyramid. Since their shape is dependent on the work of cirque glaciers, all mountain areas with horns also have cirques. However, not all cirque regions have horns, because only areas with fully developed cirque glaciation will have pyramidal-shaped peaks. Very few perfectly shaped horns are found in the world. The most famous and classic example is the Matterhorn (14,685 feet/4,531 meters) in Switzerland, visible from Zermatt or Breuil-Cervinia. The Alps of Switzerland, Italy and Austria have many hornlike features, although few of them are perfectly shaped. Other mountain ranges in Europe that have a few horns are the Pyrenees, parts of the Carpathian Ranges, the Caucasus, and jagged crest line of mountains between Sweden and Norway. In North America the higher glaciated mountain ranges exhibit a few examples of horns. The Cordilleran Ranges of Alaska, British Columbia, and Alberta have some of the best examples, as do Glacier National Park and the Grand Tetons. None of the ranges in the eastern United States are high enough or far enough north to have supported glaciers of sufficient size or for a sufficient length of time to form angular glacial features. In South America the high southern Andes of Chile and Argentina present a scenery not unlike that of the Canadian Rockies. Horns are found at the crests of the higher peaks. The glacier-clad Southern Alps of New Zealand furnish an excellent example of glacial landforms, with spectacular horn peaks as well as steep-walled valleys, waterfalls, lakes, and fiords. Of course, the high mountain ranges of Asia, notably the Himalayas, have many horns.

As a general definition a *arête* is any sharp mountain ridge but is more often the abrupt angle or scarp at the upper limit of a cirque. The knife edge or serrated ridge caused by the intersection of two adjacent cirque walls has been given the French name *arête* or *grat* in German (Cotton, 1942).

A *col* is a French word for mountain pass, saddle, or low point on a ridge. It is derived from a Latin *collum,* meaning neck. In geomorphology it has come to mean the meeting point of two opposing cirques or the low place between two aretes that may well culminate in horns.

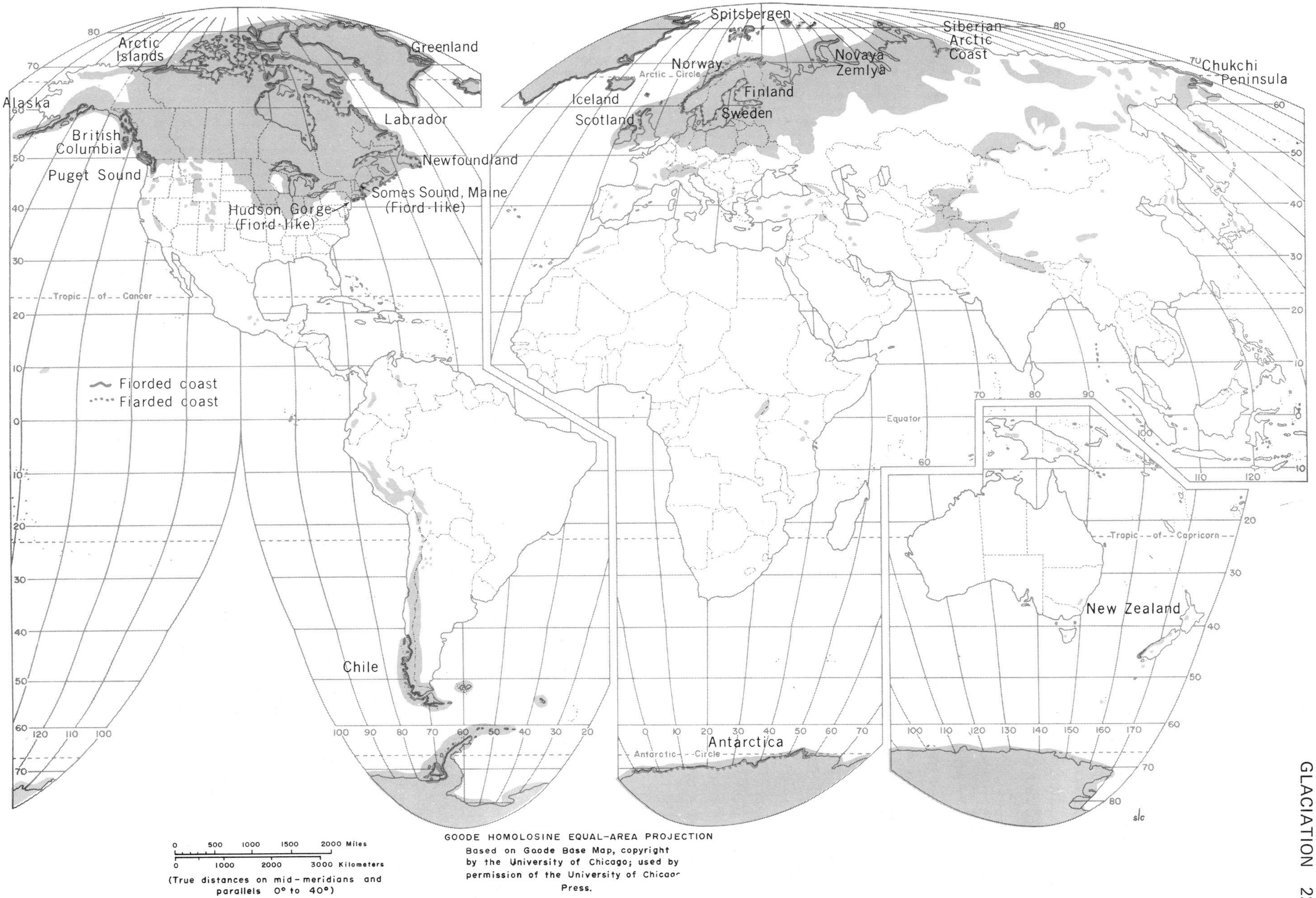

GOODE HOMOLOSINE EQUAL-AREA PROJECTION
Based on Goode Base Map, copyright
by the University of Chicago; used by
permission of the University of Chicago
Press.

MAP 5-12   FIORDS AND FIARDS

MAP
5-12

FIORDS AND FIARDS

A *fiord* (fjord) is a long, deep arm of the sea that occupies a portion of a channel having high, steep walls, and a bottom made uneven by basins and sills; side streams enter from high-level valleys by cascades or steep rapids. Fiords are found along mountainous coasts in latitudes 50 to 70 degrees north and south. The most universal belief now is that fiords are glacial troughs, eroded by the ice below sea level. Since ice will not float until approximately seven-eigths of its volume has been submerged, a glacier 1,000 feet (305 meters) thick will, on entering the sea, continue the eroding until it has worn its channel down to 800 or more feet (244 meters) below sea level (Cotton, 1941). The best examples of fiords are found in five regions (Loebeck, 1939 and 1958). The fiords of coastal Norway are among the deepest and largest in the world; they extend along the entire Atlantic coast but are best developed in the regions of Bergen and Stavanger. Sogne Fiord is the largest, extending inland 115 miles (185 kilometers). A second major region of fiords is along the coast of British Columbia and Alaska. Fiords are found from Puget Sound north to the western tip of the Alaskan Peninsula, and here they are forming at the present time, because several large glaciers, (for example, Muir Glacier) come directly to the coast. A third sea is southern coastal Chile, beginning at Puerto Montt and extending south to Tierra del Fuego. A fourth region of beautiful fiords is the southwest coast of New Zealand, along sections of the South Island. The fifth and largest region of fiords is around most of the island of Greenland, with the most magnificent scenery on the east and south sides. Some of these fiords are more spectacular than the ones in Norway, but they are little visited. The glaciers forming the fiords in Greenland reach the coast and calve off to form large icebergs that float southward as far as the Grand Banks. Several other fiord regions should be mentioned. They are located along the coasts of Labrador north of Hamilton Inlet, around the island of Novaya Zemlya, and along sections of extreme eastern Siberia, especially the Chukchi Peninsula (Gregory, 1913; Suslov, 1961). When a fiord is short and shallow it is called a *fiard*. Fiards are found along the coast of Sweden, Finland, eastern Norway, southern Labrador, Newfoundland, around many of the Arctic Islands, and along the Siberian Arctic coast (Bird, 1967). Although the New England coast is sometimes called a fiorded coast and sometimes a ria coast (a river valley drowned by the sea), it would be best to consider it a fiarded coast. Only Somes Sound on Mount Desert Island, Maine can be called a true fiord, because its walls are steep and high. A stretch of the Hudson River near Peekskill, New York is fiordlike, but it would be best to call it a fiard. Good examples of fiards can be found between Calais and Portland, Maine, while smaller features can be found along the coast of Rhode Island and Connecticut (Embleton and King; 1968).

DRUMLINS OF NORTH AMERICA

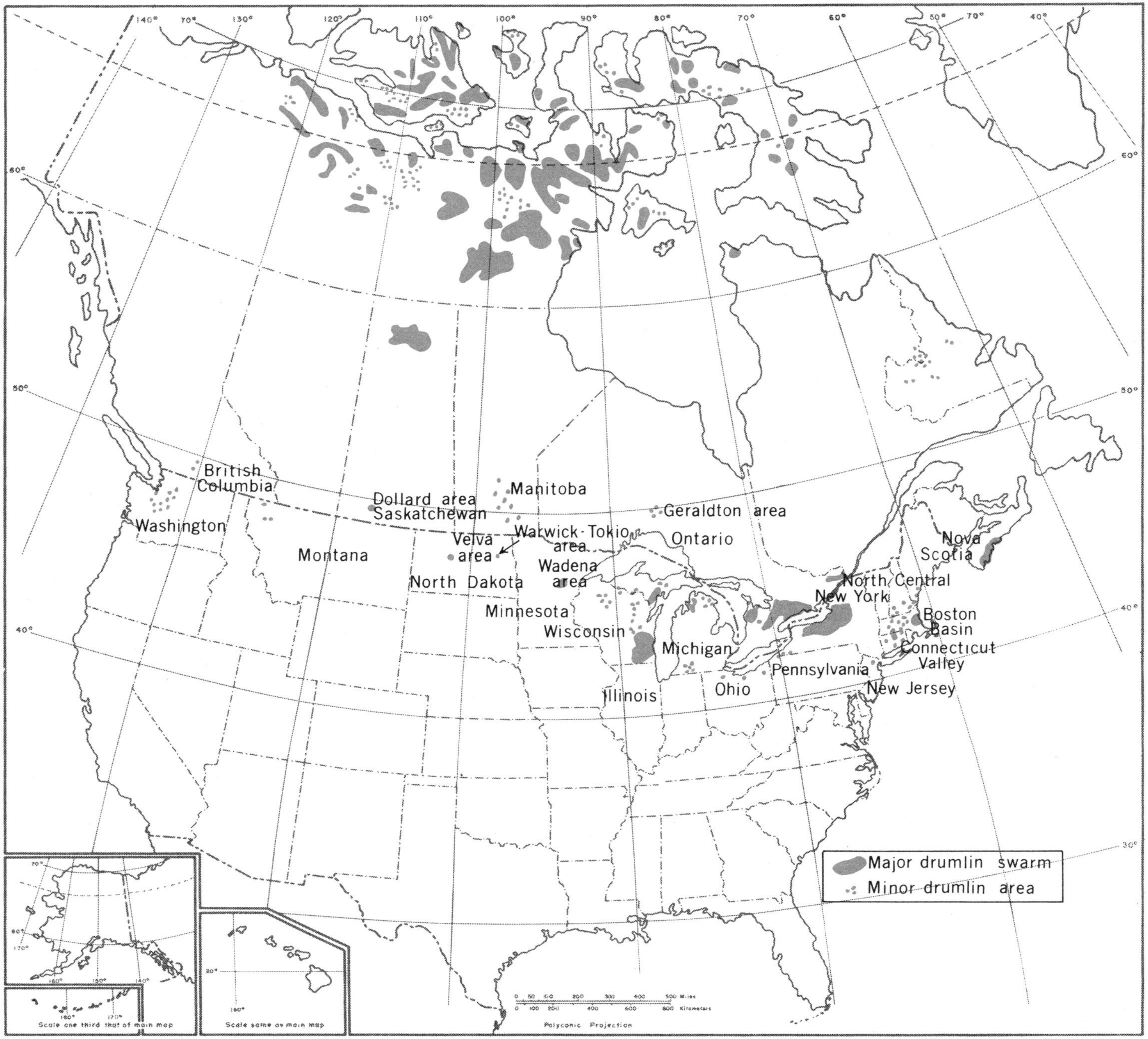

Major drumlin swarm

Minor drumlin area

Nova Scotia

Boston Basin

North Central New York

Connecticut Valley

New Jersey

Pennsylvania

Ohio

Michigan

Illinois

Wisconsin

Minnesota

Geraldton area

Ontario

Warwick-Tokion area

Wadena area

Velva area

North Dakota

Manitoba

Saskatchewan

Dollard area

Montana

British Columbia

Washington

Polyconic Projection

0   100 200 300 400 500 Miles
0   100 200 300 400 600 800 Kilometers

Scale same as main map

Scale one third that of main map

MAP 5-13

**MAPS
5-13
5-14**

DRUMLINS

Drumlins are smoothly rounded, oval hills composed mainly of unstratified boulder-clay (till), but may also include lenslike masses of gravel and sand. It has been suggested that they form where the ice at the base of glacier has become so loaded with rocks and clay that it becomes stagnant and is then overridden by the overlying, cleaner glacial ice. Drumlins are commonly one-half mile to a mile in length and 100 to 200 feet (30 to 61 meters) high with their long axis parallel to the direction of glacial movement. Often they occur in swarms which dot the landscape for miles. Although drumlins are known to occur with mountain glaciers, they are more numerous in areas of continental glaciation (Flint, 1957, and 1971). Only well-known drumlin localities have been plotted. Many scattered drumlins can be found, but it is impossible to identify cartographically the location of all of them.

One of the best-known localities is southeastern Wisconsin, where there is a great swarm of drumlins, possibly as many as 5,000 (Flint, 1971). Another major region is along the south shore of Lake Ontario, in central-western New York state. This is probably the most extensive drumlin belt in the world. Approximately 10,000 drumlins have been counted between Rochester and Rome, New York, with their greatest concentration in the Syracuse area. Along the north shore of Lake Ontario within Canada, there are drumlin swarms near the towns of Peterborough, Guelph, Teeswater, and Geraldton (Chapman and Putnam, 1951). Another drumlin region is the Boston Basin and adjacent eastern Massachusetts, with about 3000 of the oval hills (Davis, 1901). Bunker Hill, Belsson Hill, and Telegraph Hill, all in the Boston area, are drumlins. In Boston Bay, most of the islands are either true drumlins or *drumlinoids* (Lobeck, 1939). A *drumlinoid* (or *rock drumlin*) is a particular type of drumlin whose core is bedrock; glacial till merely forms a veneer on top. Many New England and European drumlins are really drumlinoids. Many of the drumlins that exist as islands in Boston Bay have been eroded by wave action so that their elliptical shapes have been destroyed. In the other harbor, a group of six to eight islands have been tied together to form a complex tombolo, known as Nantasket Beach (see Map 5-9). Drumlins have been cut and partially submerged by the sea near Halifax, Nova Scotia, and several have been connected to the mainland by tombolos. A count of drumlins in Southern Nova Scotia suggests that there may be as many as 2,300 separate forms. Other drumlin fields include the northwest part of the southern peninsula of Michigan, especially south of Charlevolx, and scattered drumlins in the Connecticut Valley between Hartford and Springfield. Large groups of remarkable long, narrow forms with pointed ends have been found in various parts of central and northern Canada; most of them are drumlinoids (Bird, 1967).

In Europe, Northern Ireland has one of the largest and most continuous drumlin fields in the world. Most of the drumlins are arranged in a close set series known as the *basket of eggs* in the county of Down, which is south of Belfast (Holmes, 1945; Charlesworth, 1957; Vernon, 1966). There are also well-developed scattered drumlins across much of Ireland, especially in the counties of Donegal and Mayo. Submerged forms can be found in Clew, Donegal, Bantry Bays, and Lough Erne. Scotland has three areas of scattered drumlins: the region around Galloway, the Midland Valley between Glasgow and Edinburgh, and the Tweed Valley. England has widely scattered drumlins in Wensleydale and in the neighborhood of Kendal, Oxenholme, and the Ribble Valley; they are also common in the Eden Valley, between the Lake District and the northern Pennines (Wright, 1937). Drumlins are not as numerous in continental Europe as in the British Isles, but they are found in north Germany, in the Holstein Valley, in Holland, and in Denmark (Schou, 1949). A few are found in the Alpine valleys of Switzerland and Austria, and in a region to the east and southeast of the Baltic Sea in Lithuania, Latvia, and Estonia. They are also found in a number of districts of Sweden, and at several locations in Finland. Drumlin forms are still being identified in other parts of the world. They have been recognized in the Tienshan ranges of China, in parts of Siberia, and on Novaya Zemlya. (Wright, 1937; Powers, 1966).

MAP 5-14   DRUMLINS OF EUROPE

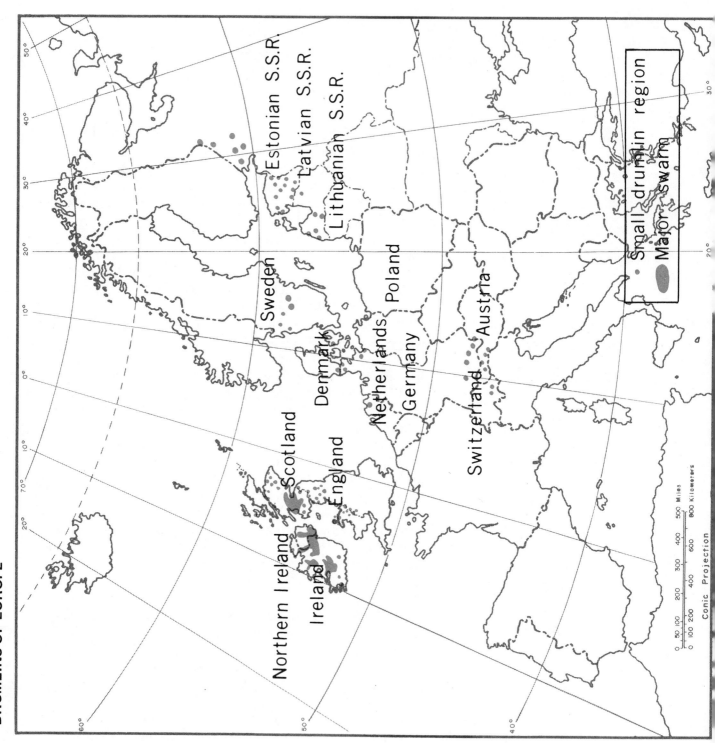

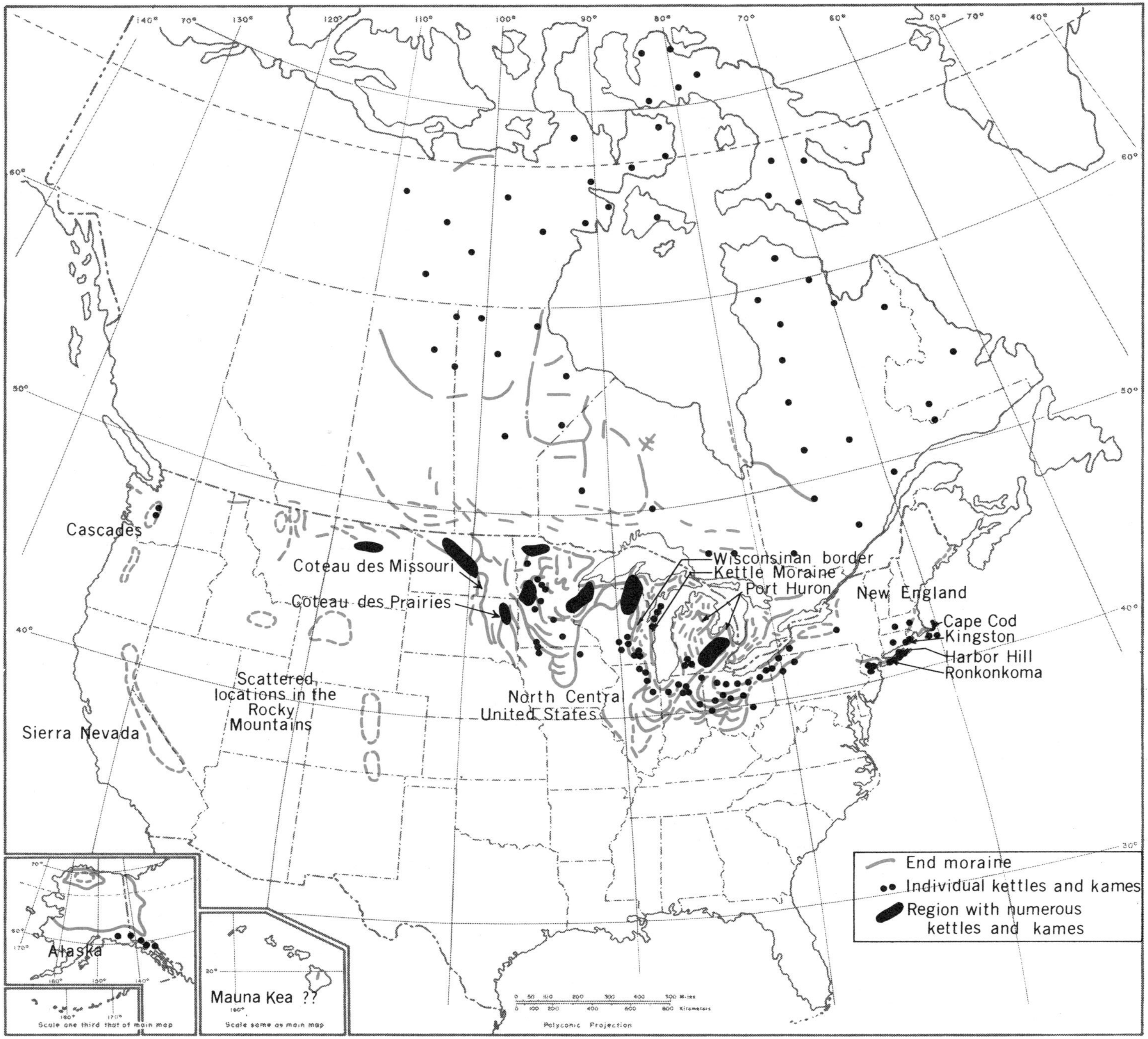

MAP 5-15   END MORAINES, KAMES AND KETTLES OF NORTH AMERICA

**MAPS
5-15
5-16**

END MORAINES, KAMES, AND KETTLES OF NORTH AMERICA

A *moraine* by definition is any rock material, for instance, boulders, *till,* gravel, sand or clay (*drift*), deposited chiefly by direct glacial action, mainly when a glacier margin is retreating. The various associated topographic forms are described according to their position in relation to an ice sheet. They can be *end* or *terminal* moraines (formed at the moraine extent), *recessional moraines* (resulting from temporary halts in glacial retreat), *inter-lobate moraines* (situated between ice lobes, or *ground moraines* (glacial materials scattered across a surface). Moraines can also be designated according to the material composing them: *till moraines* (non-sorted, non-stratified sediment carried or deposited by a glacier), *waterlaid moraines*, *delta moraines*, or *kame moraines* (a conical hill or short ridge deposited in contact with glacier ice) (Flint, 1945, 1953, 1957, and 1971).

It is a difficult task to plot end moraines with any accuracy because of their variety, difference in size and number. Moraines grade from simple smooth ridges with very low slopes to the most complex aggregation of knobs and ridges interspersed with enclosed kettles (a depression in glacial drift). This is the so-called kame and kettle or knob and basin type of topography (see Maps 5-15 and 5-16). The local relief within a moraine may exceed 100 feet (30 meters) or be less than 10 feet (3 meters). Total thickness may be on the order of several hundred feet. Because more recent glaciers have advanced over earlier glacial landforms, it is difficult or impossible to find the surface topography formed during each of the four major ice advances (Totten, 1969). This map shows mainly the end (terminal and recessional) moraines that still have well-defined or significant surface topography today. Most of the end moraines are Wisconsin (Wurm) in age, the exception being where the Illinoian (Riss) glaciation had advanced beyond the limits of Wisconsin glaciers (Ostrem, 1961). Mountains with local glaciation may have small moraines, but they are far too numerous and widespread to show on large scale maps. Only the major well-known end moraines have been plotted. Probably the most famous and

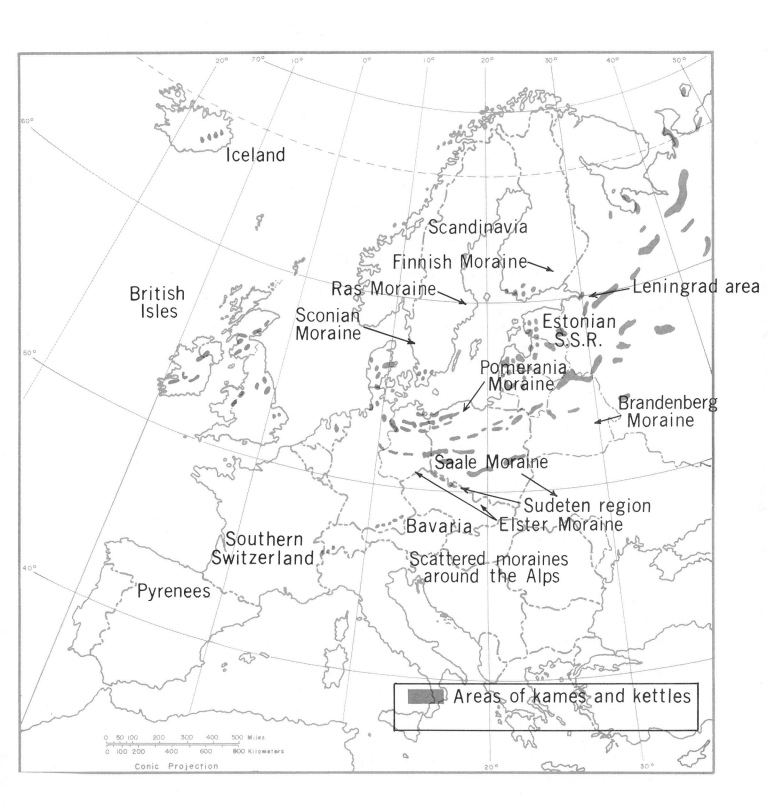

Iceland

Scandinavia

Finnish Moraine

Ras Moraine

British
Isles

Sconian
Moraine

Leningrad area

Estonian
S.S.R.

Pomerania
Moraine

Brandenberg
Moraine

Saale Moraine

Sudeten region
Elster Moraine

Southern
Switzerland

Bavaria

Scattered moraines
around the Alps

Pyrenees

Areas of kames and kettles

0 50 100   200   300   400   500 Miles
0 100 200   400   600   800 Kilometers
Conic Projection

MAP 5-16 END MORAINES, KAMES AND KETTLES OF EUROPE

best developed terminal moraines in the United States are the Ronkonkoma on Long Island are the moraines that make up much of Cape Cod, Martha's Vineyard, and Nantucket (Lobeck, 1958). They consist of rolling hills with areas of kame and kettle topography. Near the moraines are large **outwash fans** (material washed out from glacial ice), veneered with till during several advances of the ice. In the central part of the United States terminal moraines are not continuous. In fact, the outer limit of the youngest or Wisconsin ice sheet in North America from the Atlantic Ocean to the Rocky Mountains is marked by an end moraine along not much more than one-half of its total length. In most cases, moraines were never built, either because melt water was so abundant and flowed so rapidly that the drift was washed away as fast as it was delivered, or because the glacier did not stay at its farthest line of advance long enough to build an end moraine.

However, recessional moraines are numerous in the north central portion of the United States. Their number is too large to map, but a few areas where moraine systems occur have been plotted. Ohio, Michigan, and sections of eastern and northern Indiana and Illinois have well-developed moraines (Taylor, 1912). The Fort Wayne moraine can be followed from near Fort Wayne, Indiana to north of Detroit, Michigan, and east into Ohio around the south side of former glacier lake Maumee (now lake Erie) (see Map 5-5). Much of Michigan is covered by moraines. Two of the largest are the West Branch Moraine and the Port Huron Moraine in the north-central part of the state (Wooldridge and Morgan, 1937). Wisconsin has well-developed moraines adjacent to the **driftless area** (region not covered by a glacier) in the southwest part of the state. The largest of these moraines is the Johnstown. The Kettle Moraine, which extends from west of Milwaukee to the Door Peninsula, is another well-developed landform. In Minnesota, Iowa, and North and South Dakota there are long extensive morainic chains, which suggests that many oscillations of the ice front occurred during the period of ice retreat. Many of these forms run in a northwest-to-southeast direction. One of the largest is associated with the

Coteau de Missouri, which extends across central North Dakota into South Dakota. There are indications of morainic features on Mauna Kea in Hawaii.

In Europe the largest and most extensive end moraines extend from Denmark across Germany into Poland and the Soviet Union. There are numerous small moraines; but the two largest series, the Pomeranian and Frankfort moraines are well-defined in northern Germany and Poland (Holmes, 1945). Across southern Finland there is a recessional moraine that has helped to create some of the large lakes for which Finland is famous. The morainic material acts as a long irregular dam. In Ireland and England small end moraines are associated with several ice advances (Embleton and King, 1968).

Moraines plus kames and kettles are not shown on a world map because their greatest concentrations are in North America and Europe. However, these features do occur near the outer limits of Pleistocene glaciation in Chile and Argentina, the Himalayas, the Pamirs, Tibet, northern Siberia, and New Zealand.

In Tasmania, there are only scattered short moraines and in southeastern Australia there is only one small area in the Snowy Mountains which have indications of these landforms.

A *kame* is a conical hill, mound, or short irregular ridge composed of gravel, sand, or stratified drift deposited in contact with glacier ice, usually at the edge of a retreating glacier. The term is of Scottish origin. Kames originate in, at least, two principal ways. Some are masses of sediment deposited in *cavaties* and *crevasses* (a fissure in the ice) in or on the surface of nearly stagnant glacier ice that later melts away, leaving the accumulated sediment in the form of an isolated or semi-isolated mound. Another type of kame consists of small deltas or fans built outward from ice, or inward against ice, that later melts, collapses, and isolates the mass of sediment to form an irregular mound — a kame. These features are often difficult to recognize in the field because true conical-kamic hills are often associated with or grade into kame terraces, eskers and *terminal, recessional,* and *interlobate moraines*. The counterparts of kames are the steep-sided depressions, pits, or basins called *kettles,* or kettle-holes. These features represent the *ablation* (combined processes by which a glacier wastes) of a former mass of glacier ice that was wholly or partially buried in *drift.* As the ice melts away a depression results, and this creates a kettlehole. *Kame* and *kettle* topography, which is also called hillswale and knob-basin topography, occurs where there is a maze of these mounds and depressions (Flint, 1953; 1957 and 1971; Embleton and King, 1968).

Some of the best examples of kames and kettles exist in Minnesota and southeastern Wisconsin, where the kettles are quite large, particularly on the 150-mile (241 kilometers) long interlobate moraine between Green Bay and Lake Michigan. A belt of moraines with kame and kettle topography can be followed from Wisconsin through the north-central states into New York. This topography loops around the western and southern ends of the Great Lakes basin. Such hill-swale topography is most characteristic of the Battle Creek-Jackson of Michigan. Two other regions with well-known kame and kettle country are near Whitewater, Wisconsin and Vergas, Minnesota. The terminal moraine near Franklin Furnace, New Jersey has kame and kettle topography, as do the Long Island moraines and the Kingston moraine across Rhode Island (Atwood, 1940). There is similar knob and basin topography in Massachusetts on Cape Cod, especially on the Mashpee plain, in the Ayer Region, and near Mt. Toby, where there is a *delta kame*. Kettle holes are in the making on the margins of Alaskan glaciers (for example, in the Tasnuna Valley) where they number as many as 100 or more per square mile.

In Europe, kame and kettle topography is common on the plains of northern Germany, especially in the Mecklinburg and Bradenburg regions. North of Poznan, Poland there is a region with about 30,000 kettle holes (Charlesworth, 1957).

MAP 5-17   ESKERS OF NORTH AMERICA

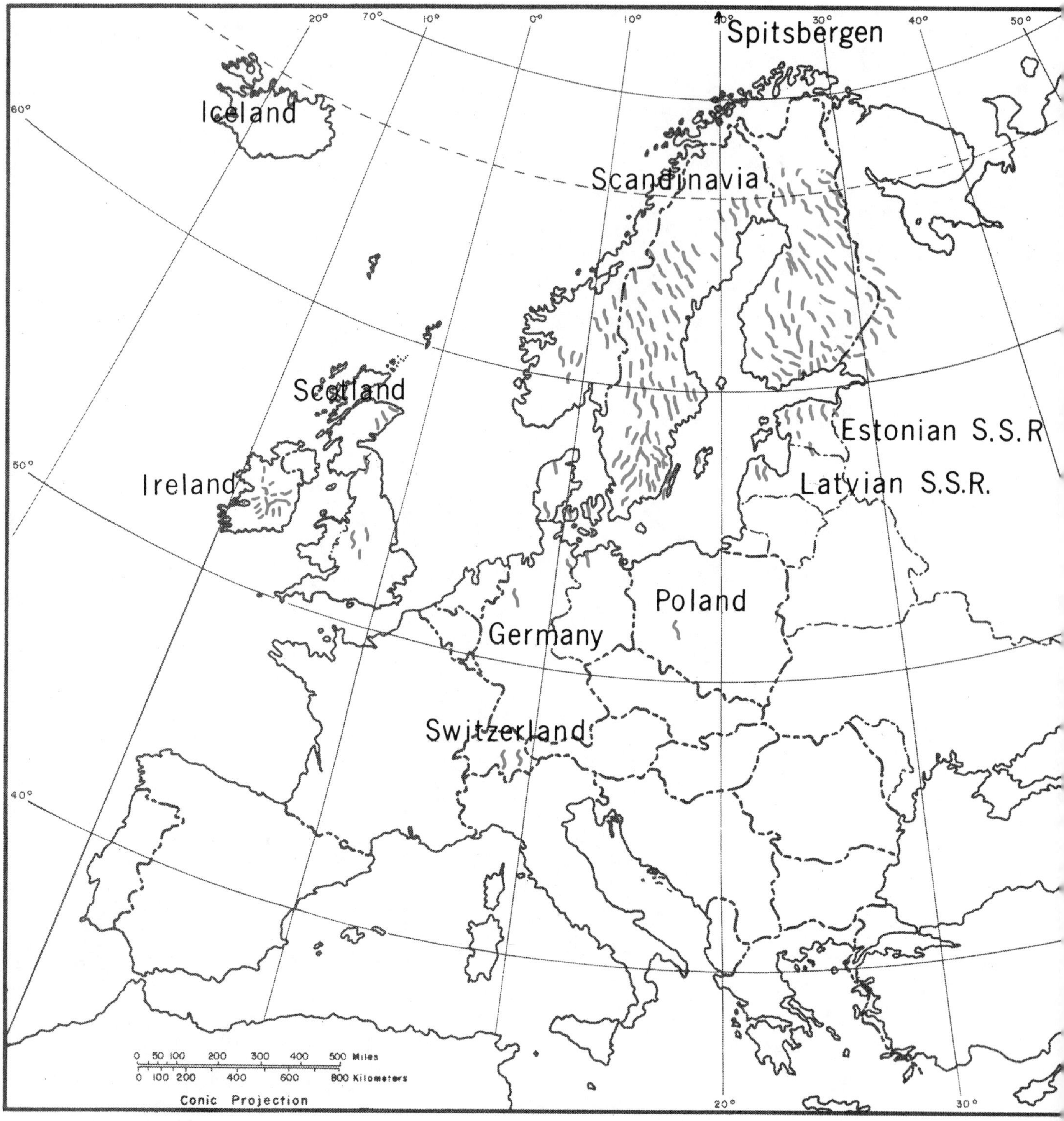

Iceland

Spitsbergen

Scandinavia

Scotland

Ireland

Estonian S.S.R

Latvian S.S.R.

Poland

Germany

Switzerland

0   50 100   200   300   400   500 Miles
0   100 200   400   600   800 Kilometers
Conic Projection

ESKERS OF EUROPE

MAP 5-18

An *esker,* also called Os, plural Osar, from the Swedish, is a sinuous gravelly ridge, often with flat crests, that is deposited by a glacial stream in an ice tunnel. In many cases an esker represents a drainage channel under a decaying ice sheet through which a stream washes out much of the finer drift, leaving the sand and coarser gravel between the ice walls in a somewhat stratified form. It is a long, thin ridge that often extends for many miles across the countryside. Eskers are most commonly found on low, swampy plains, but they do lie on all kinds of surfaces and disregard the underlying topography. Since they often develop during the wastage stages of ice, they generally are found near end moraines; and at their lower ends, they may merge with deltas, kames, and outwash plains. They vary greatly in size, ranging up to 150 feet (46 meters) or more in height and extending, sometimes in a series separated by short gaps, for more than 150 miles (241 kilometers). They are often winding and branching, like braided streams (retriculate) (Flint, 1971). Eskers are most common across central, eastern, and northern Canada (Bird, 1967). Two great concentrations are found: literally thousands of eskers radiate away from former ice sheets in northern Quebec and Labrador, and a second in the eastern part of the Northwest Territories, west of Hudson Bay and east of Great Slave and Bear Lakes (Raisz, 1949-1950; Atlas of Canada, 1957). Most eskers indicate drainage conditions at the end of the Wisconsin ice period but there are a few Illinoian age eskers. Only a few of these forms are plotted on the map. In the United States the best eskers are found in Maine, with striking features at the head of Penobscot Bay. Some extend for more than 40 miles (64 kilometers), rising 40 feet (12 meters) above the adjacent tamarack swamps. Roads follow the crests. Eskers are thus found in scattered locations in Massachusetts, Connecticut, central and northern New York, and in northern New Jersey (Lobeck, 1939). Other states with eskers are Wisconsin, Minnesota, Michigan, and Illinois. There are excellent eskers near Fort Ripley, Minnesota, in the region of Sun Prairie, Wisconsin, and the Webberbille, Michigan area. A few are found in the high mountain regions of the western United States, for example, the Wind River Range of Wyoming (Wright and Frey, 1965). In places, eskers are forming. At Malaspina Glacier in Alaska, a detailed study is being made of an esker in the process of formation. The fullest European development of eskers is in Finland and in central and southern Sweden, where there are large plains with hundreds of eskers, and they can even be traced across the floor of the Baltic Sea. Some of the largest eskers in Finland average 60 to 100 (18 to 30 meters) high and 150 miles (241 kilometers) long. Roads and railroads follow their crests across swamp lands. Much smaller eskers are found in Denmark and in several localities of northern Germany. In the Soviet Union they are found between Lake Ladoga and the White Sea and east of the Baltic Sea in the Latvian and Estonian Soviet Socialist Republics (Somme, 1960).

**MAPS
5-19
5-20
5-21**

PLEISTOCENE LAKES

During Pleistocene glaciation large sections of North America and Europe were covered with huge continental ice sheets. With the increase of ground moisture, both through the melting of the enormous ice masses and as a result of the increased storminess in the vicinity of the glaciers, a series of very large lakes were formed in large basins. This map shows the extent of these major lakes. Remnants of the lakes are shown in black. Because thousands of lakes were left as continental glaciers receded, only a few of the major ones can be plotted here.

The largest of the late Pleistocene ice marginal lakes in North America was Lake Agassiz, whose remnant is now Lake Winnipeg in southern Manitoba (Powers, 1966). Lake Agassiz at different times occupied large portions of the Red River Basin of North Dakota and Minnesota plus extensive adjacent areas in Manitoba, Saskatchewan, and Ontario. A lacustrine plain (former lake bed) of more than 100,000 square miles (258,999 square kilometers) marks its site. The Great Lakes were also at one time much larger than they are today. For example, Lake Maumee was the much larger, earlier form of present-day Lake Erie. The history of the Great Lakes is complex, and much work still needs to be done before the sequence of events if fully understood (Clark and Stearn, 1960; Daly, 1963; Thornbury, 1965). One of the most famous of the Pleistocene Lakes was Lake Bonneville, today represented by the Great Salt Lake in Utah. Great Salt Lake averages about 30 feet (9 meters) in depth with an area of 2,000 square miles (5,180 square kilometers). During the Pleistocene its predecessor was 1,000 feet (305 meters) above its present level and covered more than 20,000 square miles (51,799 square kilometers)! In the Great Basin of Nevada and California several large lakes formed during the Glacial period.

Pyramid and Winnemucca Lakes, northeast of Reno, were part of a huge lake of very irregular outline that spread into several fault troughs. This huge lake, known as Lahonton, had a depth

PLEISTOCENE LAKES

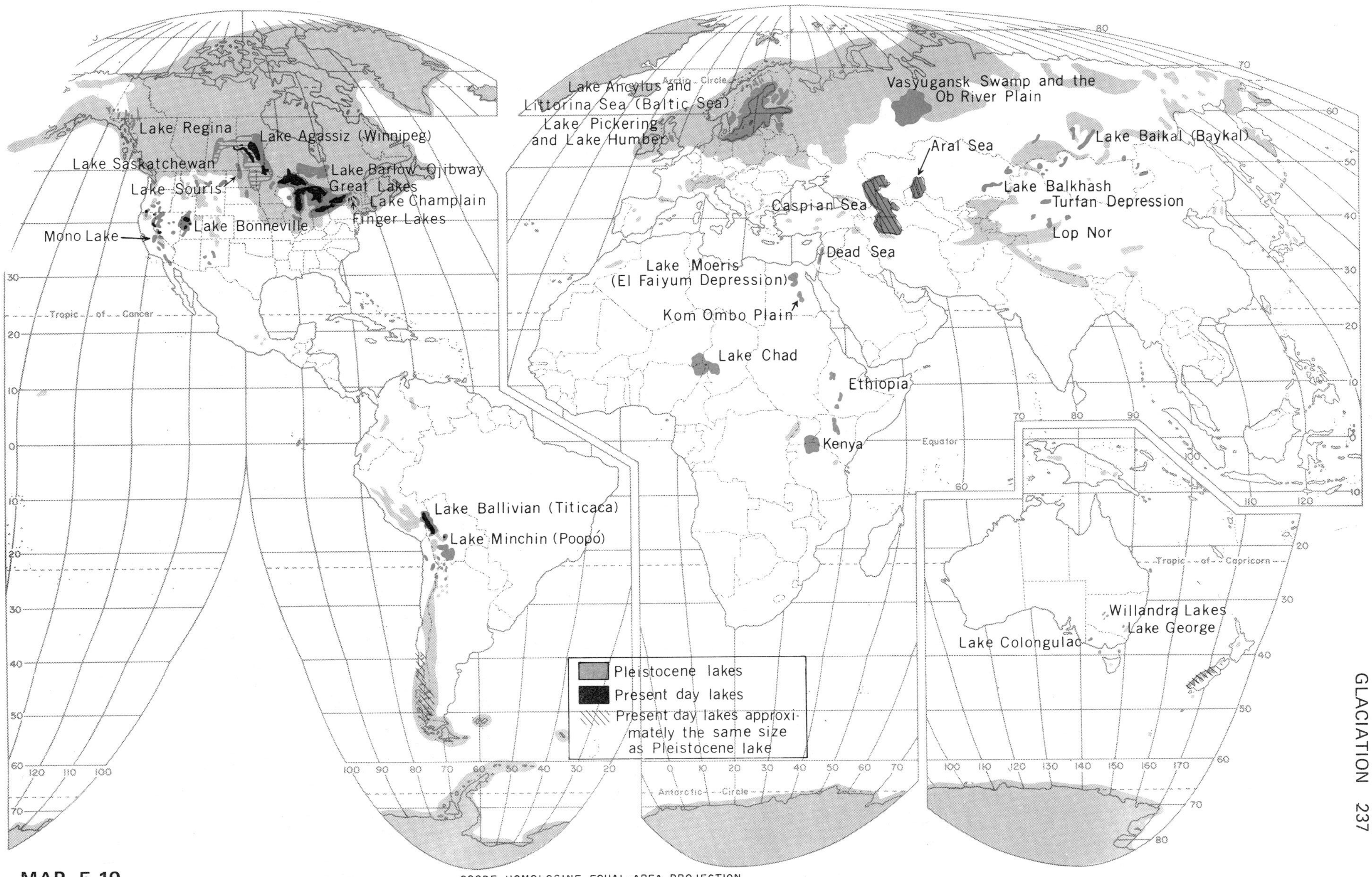
Lake Regina
Lake Agassiz (Winnipeg)
Lake Saskatchewan
Lake Barlow-Ojibway
Great Lakes
Lake Souris
Lake Champlain
Finger Lakes
Lake Bonneville
Mono Lake
Tropic of Cancer
Lake Ballivian (Titicaca)
Lake Minchin (Poopó)
Lake Ancylus and
Littorina Sea (Baltic Sea)
Lake Pickering
and Lake Humber
Arctic Circle
Vasyugansk Swamp and the
Ob River Plain
Aral Sea
Lake Baikal (Baykal)
Lake Balkhash
Turfan Depression
Lop Nor
Caspian Sea
Dead Sea
Lake Moeris
(El Faiyum Depression)
Kom Ombo Plain
Lake Chad
Ethiopia
Kenya
Equator
Tropic of Capricorn
Willandra Lakes
Lake George
Lake Colongulac
Antarctic Circle
Pleistocene lakes
Present day lakes
Present day lakes approximately the same size as Pleistocene lake

MAP 5-19

0 500 1000 1500 2000 Miles
0 1000 2000 3000 Kilometers
(True distances on mid-meridians and parallels 0° to 40°)

GOODE HOMOLOSINE EQUAL-AREA PROJECTION
Based on Goode Base Map, copyright by the University of Chicago; used by permission of the University of Chicago Press.

of about 500 feet (152 meters). Mono Lake, just east of the Sierra Nevada in California, was 900 feet deep and much larger than today (Flint, 1953 and 1957). In northwest Europe there developed a series of similar ice-marginal water bodies. However, an oscillating ice front in Scandinavia created alternating fresh and marine waters, depending on whether the connection between the Baltic basin and the North Sea was open or closed (Woodridge and Morgan, 1959; Daly, 1963). On the watershed between the Ob and Irtysh lies the enormous Vasyugansk swamp, the site of a proglacial lake (formed in front of the ice) similar to Lake Agassiz in North America. Blocked by ice to the north, the lake drained to the southwest through the Lurbgi corridor. The Pripet marsh region south of Minsk, Russia was a glacial lake now covered by vegetation. In other areas of central Asia there were numerous lakes, identified now by the sediments they left. The Caspian and Aral seas, Lake Balkhash, the swampy lakes in the Turfan depression, and Lop Nor in the Sinkiang region of western China were formerly much larger (Mosley, 1937). In Israel and Jordan, the Dead Sea expanded over a large section of the plain bordering the Judean Hills. In Australia, there is evidence that large lakes occurred during the Pleistocene, but today most of the old lake basins are completely dry. Ancient Lake Dieri, with an area of about 40,000 square miles, flooded several contiguous basins that are now dry. Present Lake Eyre is one of the remnants (Daly, 1963). Around Mexico City, former lakes were much larger and deeper than the present ones. In South America, lakes were formed in the southern Andes. Lake Titicaca, the former Lake Ballivian, was much larger and flowed into Lake Poopo, the former Lake Minchin. The lakes in Africa, particularly in Ethiopia and Kenya, also increased and decreased in size during the Pleistocene. Today many users of groundwater in arid zones are living on the capital provided by the rainier periods of the Pleistocene. To know the full extent of these once much larger lakes would be of benefit to the nations that are faced with severe water shortages, but much research must be carried out before the size and volume of these former lakes can be accurately established.

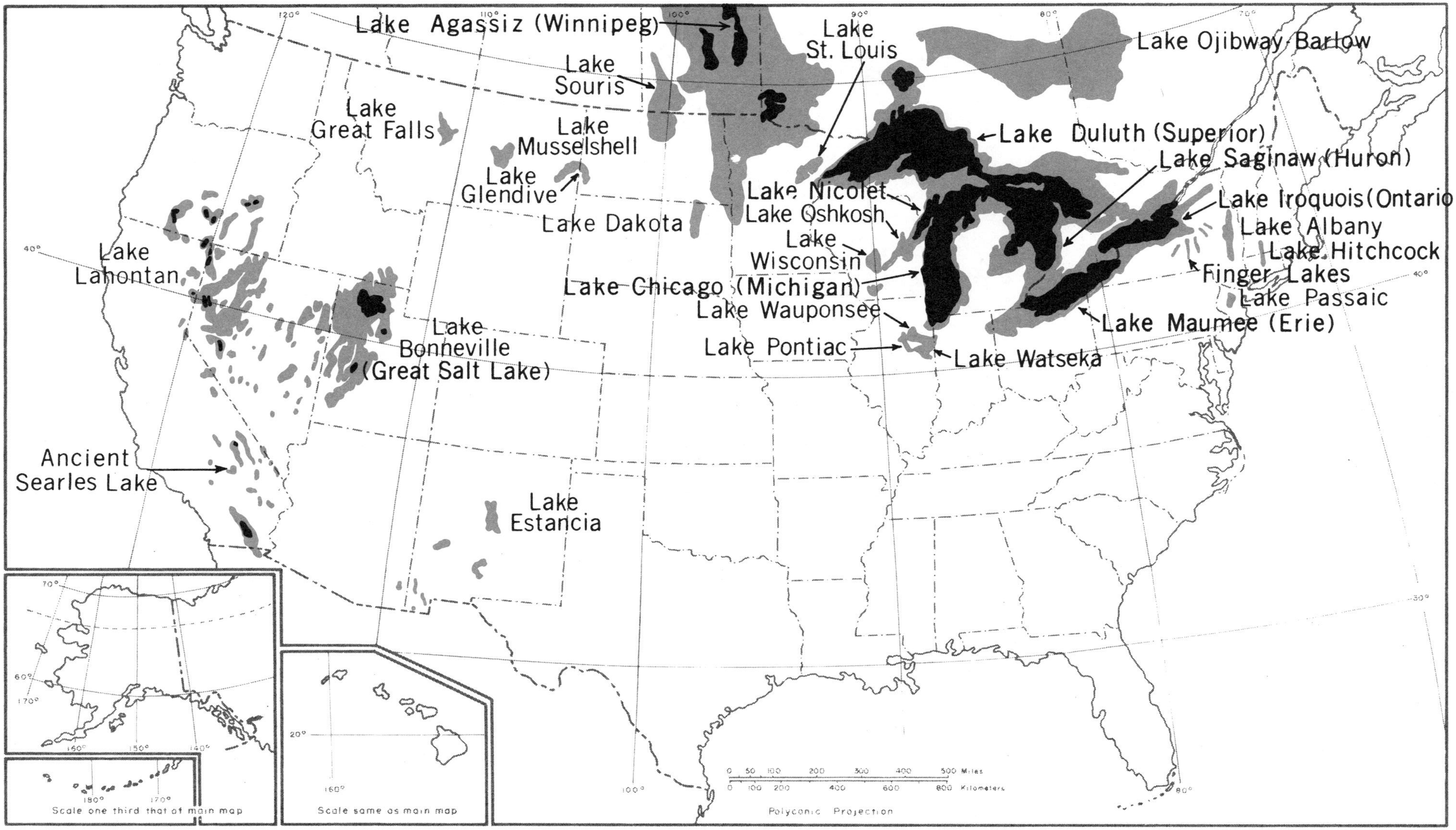

MAP 5-20   PLEISTOCENE LAKES OF THE UNITED STATES

Lake Ojibway Barlow
Lake Duluth (Superior)
Lake Saginaw (Huron)
Lake Iroquois (Ontario)
Lake Albany
Lake Hitchcock
Finger Lakes
Lake Passaic
Lake Maumee (Erie)
Lake Watseka
Lake St. Louis
Lake Nicolet
Lake Oshkosh
Lake Wisconsin
Lake Chicago (Michigan)
Lake Wauponsee
Lake Pontiac
Lake Agassiz (Winnipeg)
Lake Souris
Lake Musselshell
Lake Glendive
Lake Dakota
Lake Great Falls
Lake Bonneville (Great Salt Lake)
Lake Estancia
Lake Lahontan
Ancient Searles Lake

Polyconic Projection

0   100   200   300   400   500 Miles
0   100 200 300 400 500 600 Kilometers

Scale one third that of main map

Scale same as main map

MAP 5-21 PLEISTOCENE LAKES OF EUROPE

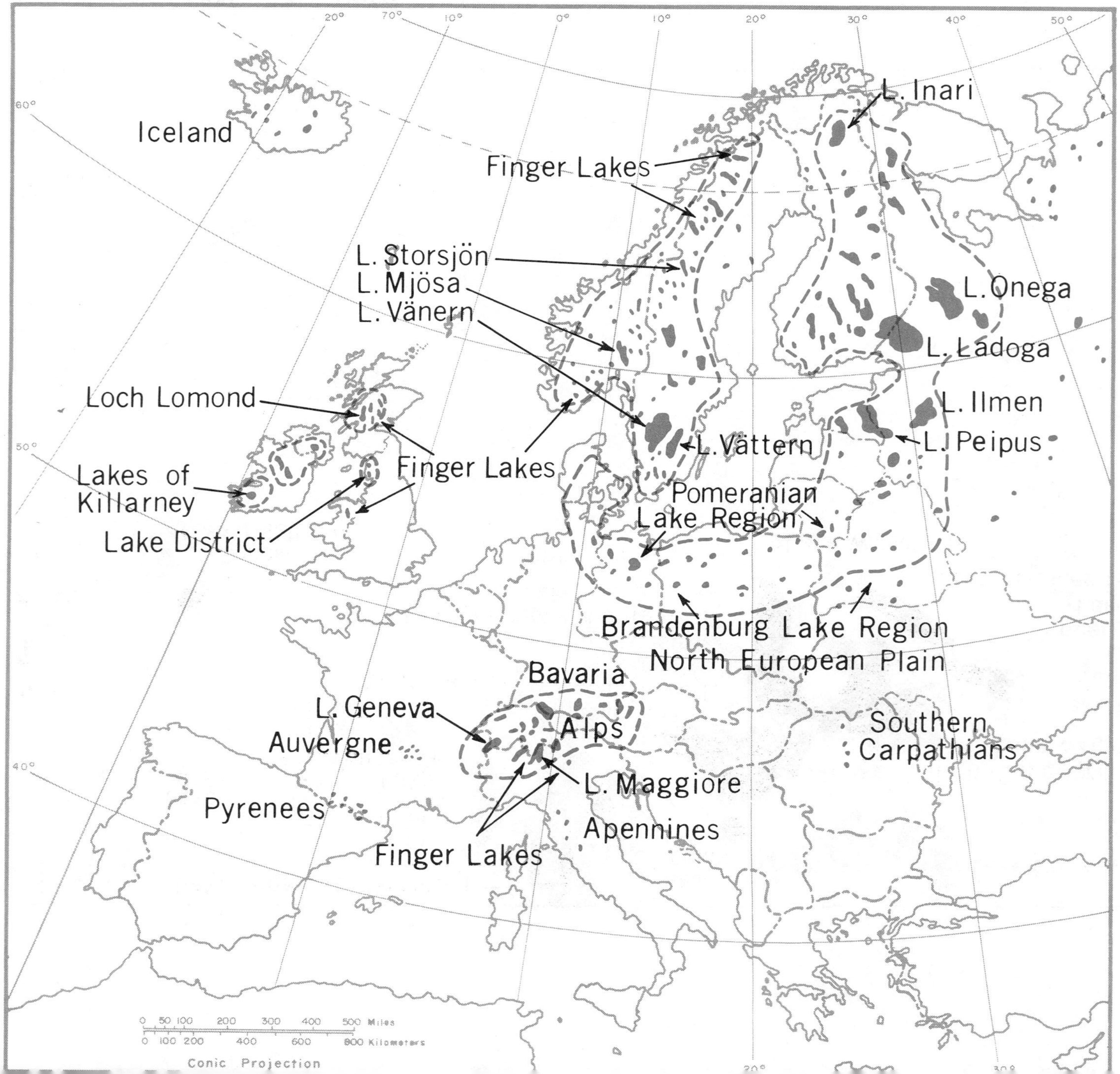
Iceland
L. Inari
Finger Lakes
L. Storsjön
L. Mjösa
L. Vänern
L. Onega
L. Ladoga
L. Ilmen
L. Peipus
L. Vättern
Loch Lomond
Finger Lakes
Lakes of Killarney
Lake District
Pomeranian Lake Region
Brandenburg Lake Region
North European Plain
Bavaria
L. Geneva
Alps
Auvergne
L. Maggiore
Southern Carpathians
Pyrenees
Apennines
Finger Lakes
0 50 100 200 300 400 500 Miles
0 100 200 400 600 800 Kilometers
Conic Projection
60°
50°
40°
20°
70°
10°
0°
10°
20°
30°
40°
50°

Section Six

Wind Created Landforms

ACTIVE AND STABLE SAND DEPOSITS

Sandy regions are widespread throughout the world, but true active sand dunes are confined to a few main regions. Most of the sand found in deserts and along coastal plains is derived from four main sources: bedrock sandstone subjected to weathering, river beds, glacial and periglacial deposits, and beaches and lake shores. Inland or desert dunes and sand flats are the chief dune regions. Nearly every desert has sand flats and scattered isolated dunes, most of which are covered with low scrub vegetation. However, sand dunes and deserts are not necessarily synonymous concepts, because sand dunes are often found outside dry regions and some deserts lack dune forms.

On the world map three symbols have been used to indicate where excellent examples of longitudinal, barchan (transverse), and star dunes can be found. A longitudinal dune (sometimes called a seif dune) is a very general term given to various types of linear dune ridges, generally more or less symmetrical in cross winds and growing in length during periods when the prevailing wind is parallel to the trend of the seif chain. Only the location of large, well formed longitudinal dunes are shown on the world map. Some of these forms are 200 miles (322 kilometers) long and 400 to 500 feet (122 to 152 meters) high. The *seif* may well be a modification of the second type of dune form, called a barchan (Bagnold, 1941). A barchan is a dune having a crescentic shape with the convex side facing the wind. The profile of such a dune is asymmetric, with the steeper slope on the concave or leeward side. The horns of the crescent mark the lateral advance of the sand. These dunes when lined up perpendicular to the wind, form transverse dunes. Barchans are most frequently found in an open area where vegetation is largely absent. This type of dune has wide distribution, but in size it is generally smaller than a seif. The third dune form is called a star dune because of its pyramidal shape. These unusual dunes have radiating ridges of sand from a central point. They form because wind, blowing from all four compass directions, piles the sand up against a central core. The best example can be found in the Rub-al-Khali of southern Saudi Arabia and in the Egyptian Sand Sea, where swarms of these dunes reach heights of 300 feet (91 meters).

The largest dune deposits are in the deserts of Africa, Asia, and Australia. In Africa, one-ninth of the entire area of the Sahara has stable and active dunes, or about 300,000 square miles (776,997 square kilometers). The Sahara is noted for its long

MAP 6-1 ACTIVE AND STABLE SAND REGIONS OF THE WORLD

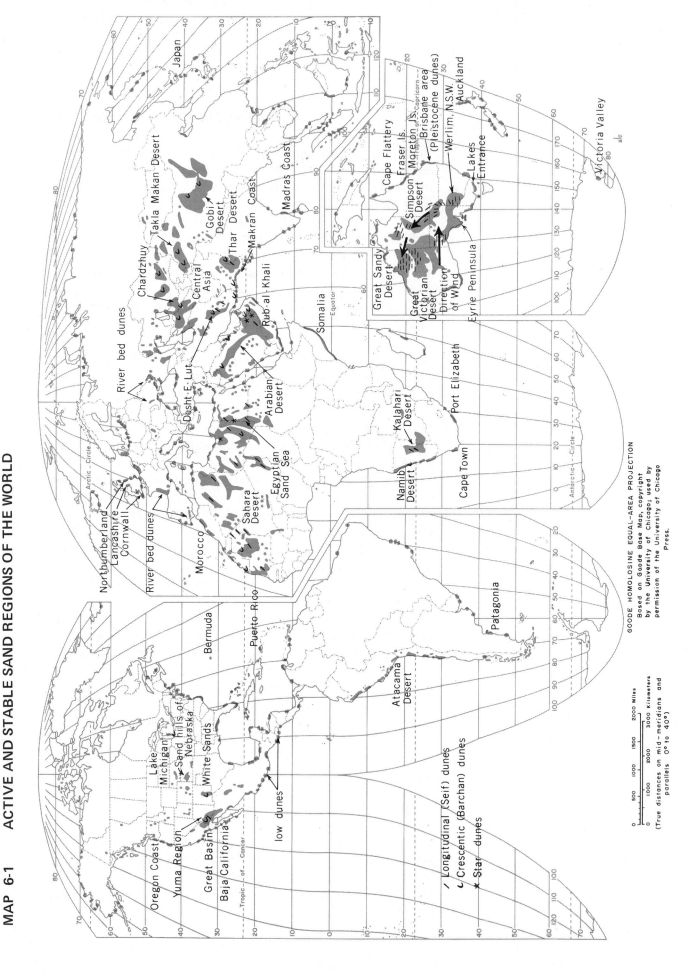

GOODE HOMOLOSINE EQUAL-AREA PROJECTION
Based on Goode Base Map, copyright
by the University of Chicago; used by
permission of University of Chicago
Press.

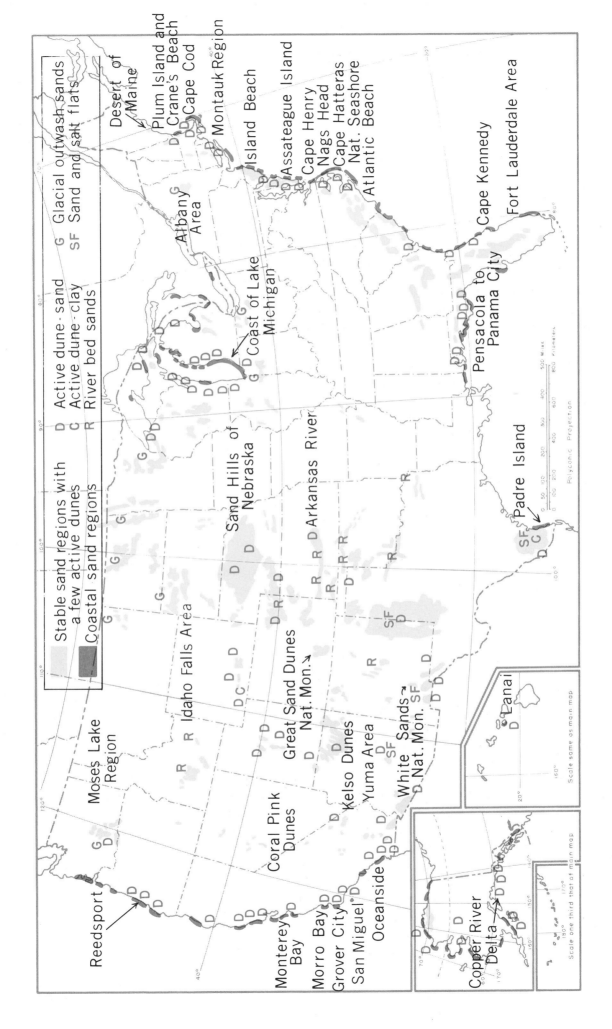

MAP 6-2   ACTIVE AND STABLE SAND REGIONS OF THE UNITED STATES

seif dunes, sometimes called sword dunes, its large whaleback sand accumulations which are shaped like a rounded levee, and its star-shaped or pyramidal dune forms. In central Algeria can be found the Occidental and Oriental Ergs (sand regions), consisting of vast fields of sand extending 400 miles (644 kilometers) from west to east and 100 miles (161 kilometers) from north to south. Plate 13 (centerfold) shows 75 mile (121 kilometers) long seif dunes in the Erg Iquidi a section of the Occidental Erg region of western Algeria. The Libyan Desert and western Egypt have vast areas of complex barchan and seif dunes. Some individual seifs are several hundred miles long. Many appear to be controlled by underlying structures and do not migrate. In south Africa the Kalahari Desert has fossil dunes that are aligned with former wind directions. There are large active transverse dunes in the Namid Desert. One-third of Saudi Arabia, approximately 400,000 square miles (1,035,996 square kilometers), is covered with dunes. In the Nefud Desert and in the great "empty quarter" of south Saudi Arabia (the Rub-al-Khali Desert), dunes may be 700 feet (213 meters) high. Vast tracts of sandy desert exist in central Asia extending from Israel and Syria eastward through Iran, the Baluchistan part of western Pakistan, into the Thar Desert of India to eastern Turkestan and Mongolia, as well as in Soviet Central Asia and southern Afghanistan. The Dasht-E-Lut of eastern Iran, 30 by 100 miles (48 by 161 kilometers), has some seif dunes that reach as much as 600 feet (183 meters). The Thar (Rajasthan) Desert of India has 400 to 500 foot (122 to 152 meters) high sand ridges that extend parallel to the strong southwest winds. In the northern part of the desert the dunes are lower and transverse to the less intense winds. The Tarim Basin, in the heart of Central Asia, its center the Takla Makan Desert, has dune ridges known as "Davans" that rise 300 feet (91 meters) above the general level of the country. In the Turkestan region, sandy deserts occur in wide areas on either side of the Amu Dava River. Here barchan forms appear to be prevalent. Australia has dune areas mainly in its western and central portions. The dunes in the Great Sandy, Simpson, and Victoria Deserts are extensive. Much of the sand making up the dunes is derived from very old sandstone rocks. Barchan dune forms are practically lacking in Australia, but seif or longitudinal dunes are widespread. For example, the Simpson Desert has many active dunes with extensive north-south trending seif ridges. In South America non-coastal dunes cover much of the great Atacama Desert of Chile and Peru. Where streams have deposited sand on the coastal plain, the wind has reworked the deposits to form large barchans. Some of the dunes are 300 feet (91 meters) high and 500 feet (152 meters) wide. They tend to migrate up the Andean piedmont. Most of Patagonia lacks dunes.

In North America, small continental sand dune areas are found in the Great Basin in Nevada, Colorado, and southern California. In southern California the kelso dunes are south of Baker, and a small group of intersecting transverse ridges occupy the central part of Death Valley National Monument. There is a small, developed barchan dune area on the east side of Imperial Valley north of Yuma, Arizona. There are also active dunes at Great Sand Dunes National Monument in the San Luis Valley of southern Colorado, and small fields of barchan dunes west and south of Albuquerque, New Mexico. One of the most unusual dune sites is White Sands National Monument near Alamogordo, New Mexico, where an area of 500 square miles (1,295 square kilometers) is covered with snow-white gypsum dunes. River-bed dunes are found along many of the large rivers of the western United States. Dunes border the eastern or leeward sides of the Arkansas River and its tributaries. One of the most extensive river dune areas in North America is that of the Columbia and Snake rivers in Oregon and Washington.

River-bed dunes are found in the dry regions of southern France and Spain. The valley of the Garonne River in southern France has dunes 30 to 40 feet (9 to 12 meters) high; 100-foot-high (30 meters) dunes are found along the banks of the Guadalquivir River in Andalusia, Spain. In southern Russia, river dunes occur along the banks of the Dnieper, Con, Donetz, and Volga rivers. River valleys of Asia have extensive dune deposits.

Fixed interior dunes can be found in the sand hill region of Nebraska and nearby states. This entire area comprises approximately 18,000 square miles (46,620 square kilometers) of stable dunes largely covered with vegetation. Sand is mainly derived from large river beds, such as the North and South Platte. Other sand regions in the Great Plains have occasional shifting dunes when the vegetation is removed. A small region of fixed dunes is south of Saratoga, New York, Here knobby hills are covered with evergreen trees. Like the stable sand regions in parts of Europe, the sand is derived from reworked glacial and alluvial deposits.

Sand dunes found along coastal regions are discussed under the topic, Depositional Coastal Features (Maps 4-8, 4-9, and 4-10).

**MAPS
6-3
6-4
6-5**

LOESS DEPOSITS

Sand is not the only earth material carried by the wind. A much finer material, composed of angular particles of silt and clay size, is called *loess,* and it is generally found on the leeward side of coarser sand deposits because prevailing winds have been able to carry the lighter materials farther from their sources. Loess is a buff-colored, nonindurated, calcareous eolian (wind-carried) deposit composed typically of subangular to angular silt-size particles (0.05 — 0.01 mm in diameter) with associated sands and/or clay. It generally has a highly varied mineralogical composition with a distinct vertical structure but no or weak horizontal stratification. The origin of loess is not completely understood, but there appear to be at least four or five modes of origin. The most important is the finely-ground glacial materials that were laid down thousands of years ago in the marginal areas of the huge continental glaciers across the North American and European continents. The largest amount of loess was derived from glacial outwash or from old glacial lake beds. Wide valley floors, such as the Mississippi River, were glacial spillways that provided an abundant source of fine materials that could be picked up by the wind. A narrow band of thick loess deposits exists on the eastern side of the Mississippi Valley as far south as Mississippi and Louisiana (Fisk, 1951). In some areas, loess consists of materials removed from desert regions by deflation; or it may also be derived from disintegrated Cretaceous and Tertiary alluvium (sediments laid down in river beds) like that which mantles much of the great Plains of the central United States. The greatest concentrations of loess extend across the central part of the United States and the central part of Europe (with a wide band into the Ukraine), and in north eastern China (Kriger, 1965). Deposits more than 16 feet thick cover much of western Kansas, central and eastern Nebraska, and extend into Iowa (Flint, 1971; Lugn, 1962). Most of these deposits are probably wind-carried primarily from the flood plains of the large rivers. This is well demonstrated in the rapidly decreasing loess thickness away from the buffs of the large flood plains. Less thick but extensive deposits in Illinois, Iowa, and Wisconsin appear to be derived from old glacial lakes or outwash plains. Patches of this fine silty material can

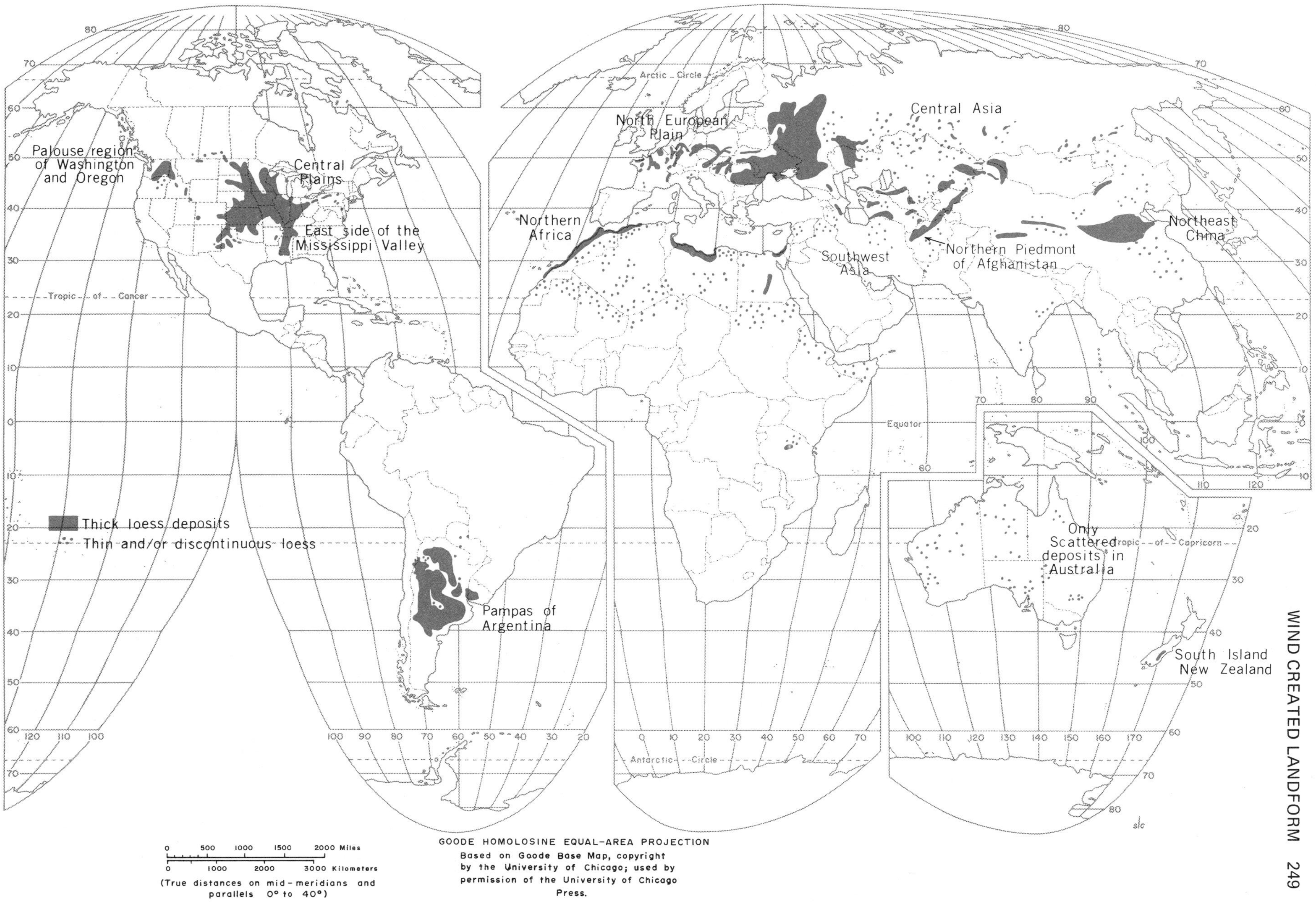

GOODE HOMOLOSINE EQUAL–AREA PROJECTION
Based on Goode Base Map, copyright
by the University of Chicago; used by
permission of the University of Chicago
Press.

MAP 6-3   LOESS DEPOSITS OF THE WORLD

MAP 6-4   LOESS DEPOSITS OF THE UNITED STATES

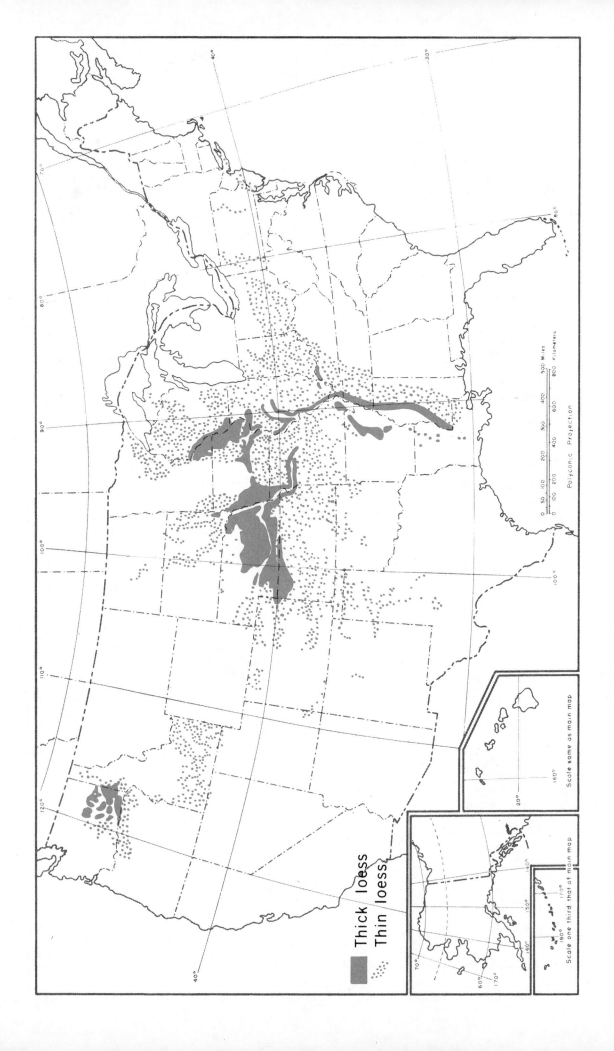

Thick loess

Thin loess

Scale same as main map

Scale one third that of main map

Polyconic Projection

Miles

Kilometers

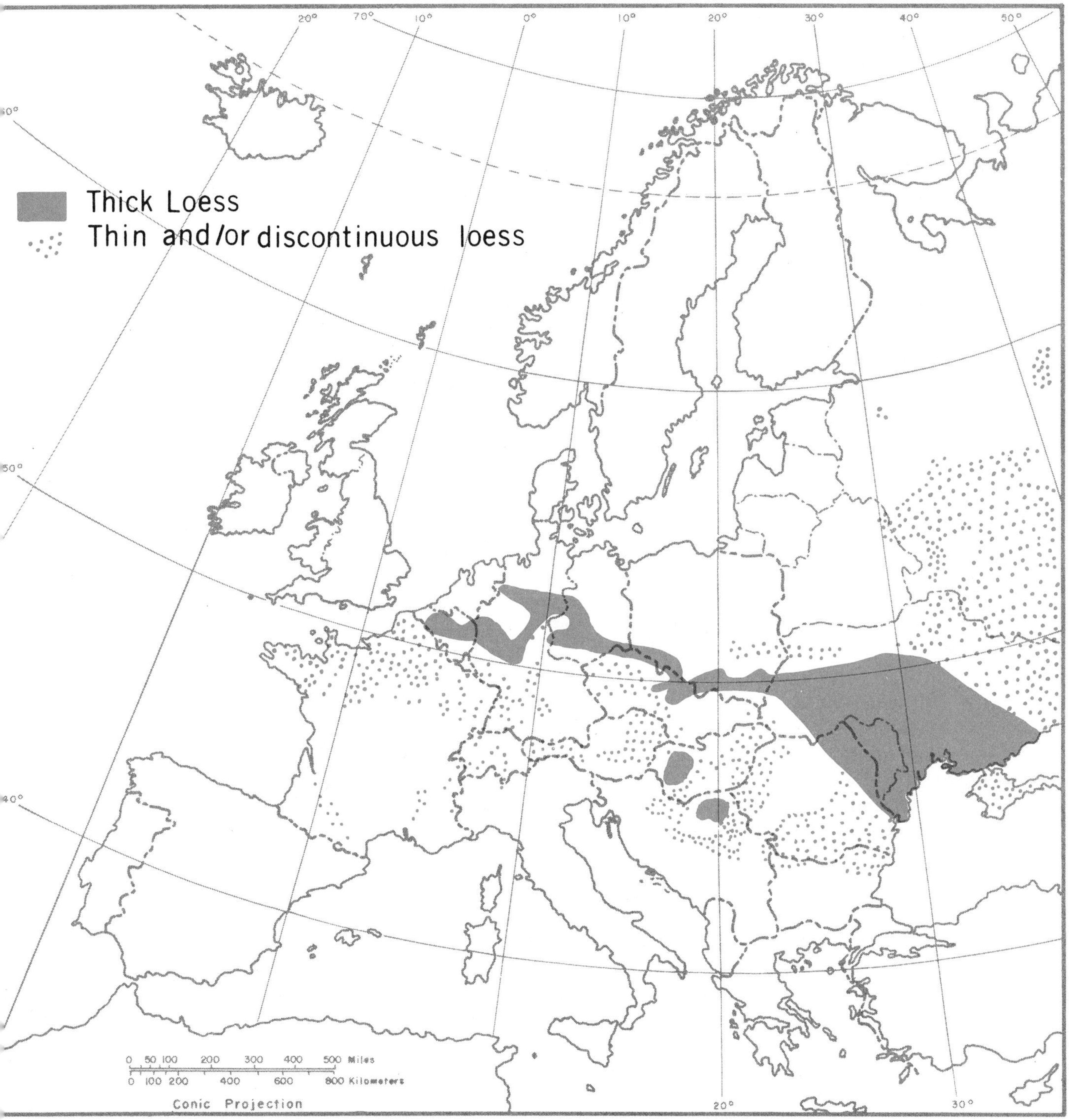

Thick Loess
Thin and/or discontinuous loess

LOESS DEPOSITS OF EUROPE

MAP 6-5

0 50 100 200 300 400 500 Miles
0 100 200 400 600 800 Kilometers
Conic Projection

be found as far east as Ohio (King, 1962). Generally loess is found as a mantle over preexisting topography, and it therefore is difficult to distinguish from residual soils. When it is eroded, it may form a few ridges or knolls that often express the underlying topography. Its most noticeable characteristics are its buff color, unstratified profile, and capability of being formed into steep bluffs and cliffs, such as those near Council Bluffs, Iowa. An area of thick loess is found in the Palouse region of Washington, Oregon and Idaho. Plate 14 (centerfold) is a satellite photograph showing this Palouse region of central Washington State. These deposits are largely derived from old glacial outwash plains and disintegrated volcanic lavas. Other deposits of loess are found in the northeastern United States (King, 1962). An unusual area is the Connecticut Valley, where there is a band of loess derived from glacial outwash plains and dried-up lake beds. Some of the best-known and thickest deposits of loess are in northeast China, most of it derived from the Ordos and Gobi Deserts to the northwest and from the flood plains of the Huang Ho and its tributaries rather than from glacial origin (King, 1962, Flint, 1971). Deposits reach a depth of 300 feet (91 meters) locally, but average about 100 to 200 feet (30 to 61 meters) in thickness over most of the area. The loess deposits that cover much of the Pampas of northern Argentina, extending into Uruguay and southern Brazil, are windblown silts and clays from the arid regions farther west in the lee of the Andes. The meager glacial deposits to the south do not appear to be a major source. Like the Chinese loess, much of the windblown material has been reworked and redeposited by streams. There are well-formed loess deposits on the east side of the South Island of New Zealand especially on the plains of Canterbury, the port hills of the Banks Peninsula, subdued hills about Oamaru and Timaru, and in southland. This material is formed from glacial outwash of the former large mountain glaciers in the southern Alps. Some geomorphologists believe no real loess has been found in Australia. Clay dunes (lunettes) were deposited as pellets rather than dust. The Parna of the Riverina in southeast Australia is close to being loess but it is thought to be mainly former leached soil material blown away as dust. The dots shown for central and western Australia represent possible scattered loess deposits from old Pleistocene lake beds (see Maps 5-19, 5-20, and 5-21) (Schultz and Frye, 1968).

Section Seven

Water Created Landforms

MAP
7-1

CANYONS

Only a United States map is shown for canyons, because select-ing true deep canyons for the world is difficult. A canyon is a valley that is notably deep in proportion to its width. Most canyons are much wider than they are deep. The Grand Canyon of the Colorado River, for example, is as much as 10 miles (16 kilometers) wide at many points but averages only about one mile deep. Yet this is one of the most stupendous chasms in the world. In rare case, gorges are deeper than they are wide, but this is the exception. The Royal Gorge and the Black Canyon of the Gunnison River in Colorado are examples of very deep gorges that are much deeper than they are wide. Also, the average idea of most people regarding the depth of canyons is much exaggerated. I visited Chaco Canyon in northwestern New Mexico only to find that it is not a true canyon but only a very wide valley. The canyons shown represent in most cases a youthful stage of river erosion in which a river has been busy cutting deep into a gorge and has not had time to widen its valley. Most of the canyons and gorges shown can be con-sidered very young in geologic age.

The Grand Canyon of the Colorado River (5,700 feet/1,737 meters) is not the deepest canyon in the United States. Hells Canyon, part of the Snake River between Idaho and Oregon, is the deepest gorge on the North American continent, its depth averaging 6,600 feet (2,012 meters) with a width of 4 to 9 miles (6 to 14 kilometers). But the Grand Canyon is much more spectacular because of its variegated rock colors and its numerous tributary gorges. Plate 15 (centerfold) is a satellite view of this world famous canyon. Mesa Verde, Colorado is shown on the map because its many small deep canyons and streams cut into a large tablelike mesa, and Flathead River Canyon in Glacier National Park, Montana, is shown because it is a 2,000-foot (610 meters) deep glacial trough.

MAP 7-1   CANYONS OF THE UNITED STATES

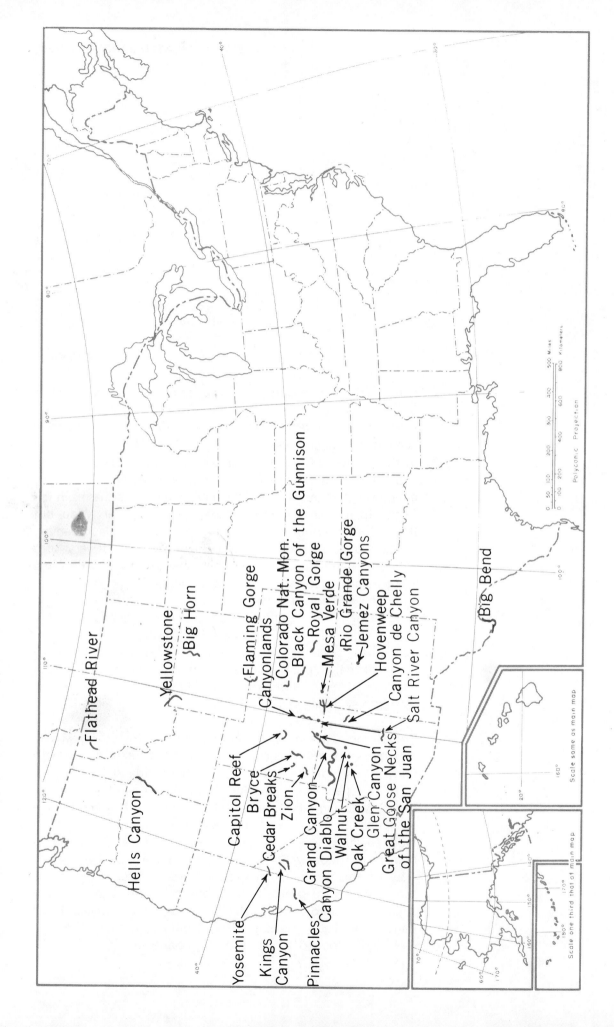

Flathead River

Yellowstone

Big Horn

Flaming Gorge
Canyonlands
Colorado Nat. Mon.
Black Canyon of the Gunnison
Royal Gorge
Mesa Verde
Rio Grande Gorge
Jemez Canyons
Hovenweep
Canyon de Chelly
Salt River Canyon

Big Bend

Hells Canyon

Yosemite

Kings
Canyon

Pinnacles
Canyon Diablo
Walnut
Oak Creek
Glen Canyon
Great Goose Necks
of the San Juan

Capitol Reef
Bryce
Cedar Breaks
Zion

Grand Canyon

Polyconic Projection

0   50   100   200   300   400   500 Miles
0   100   200   400   600   800 Kilometers

Scale same as main map

Scale one third that of main map

MAP
7-2

LARGE NATURAL BRIDGES AND ARCHES

A United States map showing the location of large natural bridges and arches has been included because these features are in some cases spectacular landforms. Natural bridges and arches are openings through a rock mass; they can be formed in a number of ways. The largest ones known are formed by the cutting action of streams, but most natural bridges occur in limestone regions and result from solution along joints and bedding planes. (It has been suggested that natural bridges result when a cave collapses and only a small part of the roof remains. But very rarely do caves collapse). One of the most famous bridges is Natural Bridge of Virginia. It was formed as follows: water seeped through a joint or fissure across a stream, then the water followed along a bedding plane until it emerged under a fall or rapid farther downstream, and the channel thus formed was gradually enlarged until all the water of the stream was diverted from the stream bed below the point of ingress, leaving a bridge. The height of the arch of a natural bridge above a stream will naturally depend on the amount of cutting subsequent to the formation of the bridge and to the weathering of the underside of the arch. A number of the large arches in the western part of the United States represent the weathering of the weak part of a mass of rock, finally forming a hole right through the rock mass. In Arches National Monument, Utah, nearly 90 arches in all stages of development and decay have been discovered.

Most of these remarkable forms were created when water entered the sandstone, weathering away softer areas and leaving perforations, or windows. In time the windows are cut through the entire rock mass, leaving arches. Continued thinning of the arches by weathering eventually leads to their collapse. Arches and natural bridges are also formed when a river creates a very tight meander and then breaks through its own narrow neck, thus abandoning the meander and flowing through the arch. Rainbow Bridge in southern Utah was formed in this manner. These landforms can be formed on a coast where caves are worn by marine erosion on either side of a rock projection, leaving an arch or natural bridge. These features are shown on the map of erosional coastal features (Maps 4-5, 4-6, and 4-7).

MAP 7-2   NATURAL BRIDGES AND ARCHES OF THE UNITED STATES

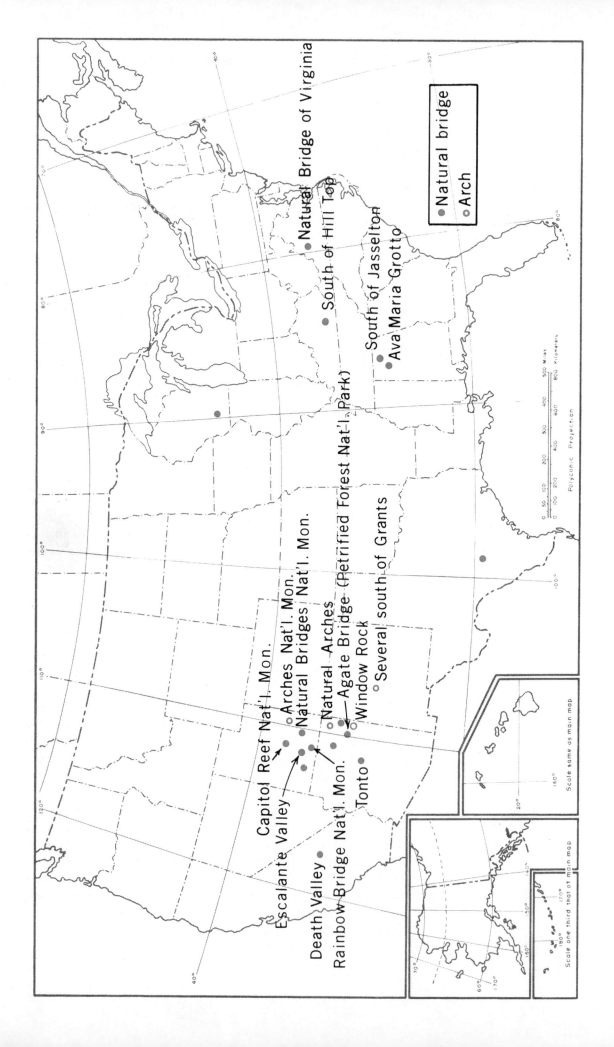

Natural bridge
Arch

Natural Bridge of Virginia
South of Hill Top
South of Jasselton
Ava Maria Grotto

Capitol Reef Nat'l. Mon.
Arches Nat'l. Mon.
Natural Bridges Nat'l. Mon.
Natural Arches
Agate Bridge (Petrified Forest Nat'l. Park)
Window Rock
Several south of Grants
Death Valley
Rainbow Bridge Nat'l. Mon.
Escalante Valley
Tonto

Polyconic Projection
0 50 100 200 300 400 500 Miles
0 100 200 400 600 800 Kilometers

Scale same as main map

Scale one third that of main map

MAP
7-3

LIMESTONE AND KARST REGIONS

The word *karst* is a comprehensive term applied to limestone rock areas largely modeled by surface and underground solution. Non-crystalline and amorphous limestones, such as chalk, are mostly too soft to give rise to a well-developed karst relief. Underground drainage and karst landforms also develop upon soluble rocks such as dolomite, gypsum and salt, but these rocks are not widely distributed. Hence, a karst area normally means an area of pure and massive limestone. The word came from the name of a narrow strip of limestone plateau in Yugoslavia and Italy bordering the Adriatic Sea, where there is a remarkable assembly of earth features dependent on subsurface solution. Many of the names of individual karst features have come from Yugoslavia, but other countries also have their karst terms (Fairbridge, 1968). Climate is a very important factor in the development of karst landscapes. Surface karst features are not well-developed where there is a lack of rainfall. For example, much of the central Sahara is underlain by limestone rock, but there are few karst features because the region is too arid. However, underground drainage in desert limestone areas can form large caverns such as Carlsbad Caverns in southern New Mexico. Some of these landscapes appear to have been formed during earlier, rainier periods (Thornbury, 1969). In humid regions mechanical and chemical erosion is more rapid and in places only remnant rounded hills are left. In northern Puerto Rico these rounded features are called pepino hills; in Cuba, they are called mogotes; and in Jamaica, a series of limestone depressions form the so-called cockpit country. The location of the water table influences the development of karst features. Large circular lakes in central Florida have formed in a limestone region where the water table is close to the surface. Plate 16 (centerfold) is a satellite photograph over eastern Florida depicting the numerous karst lakes. The map of limestone rock regions shows that this rock type is widespread

MAP 7-3   LIMESTONE AND KARST REGIONS

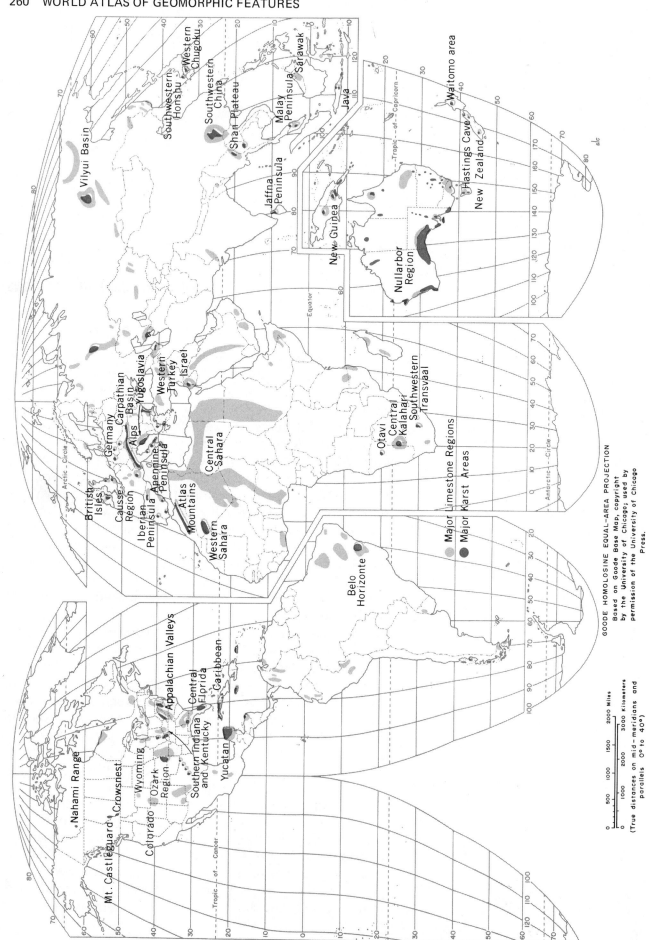

GOODE HOMOLOSINE EQUAL-AREA PROJECTION
Based on Goode Base Map, copyright
by the University of Chicago; used by
permission of the University of Chicago
Press.

0   500   1000   1500   2000 Miles
0   1000   2000   3000 Kilometers
(True distances on mid-meridians and
parallels 0° to 40°)

Major Limestone Regions
Major Karst Areas

Mt. Castleguard
Nahami Range
Crowsnest
Colorado
Wyoming
Ozark Region
Appalachian Valleys
Southern Indiana and Kentucky
Central Florida
Caribbean
Yucatan
Belo Horizonte

British Isles
Causse Region
Iberian Peninsula
Germany
Alps
Carpathian Basin
Yugoslavia
Apennine Peninsula
Western Turkey
Israel
Atlas Mountains
Western Sahara
Central Sahara

Vilyui Basin
Southwestern Honshu
Western Chugoku
Southwestern China
Shan Plateau
Jaffna Peninsula
Malay Peninsula
Sarawak
Java
New Guinea

Otavi
Central Kalahari
Southwestern Transvaal

Nullarbor Region
Hastings Cave
Waitomo area
New Zealand

around the world, but the distribution of karst landscapes is restricted to a relatively small number of localities. In some regions, the limestones maybe too permeable or too impermeable. In gypsum and chalk areas one does not find well-formed karst features because these are mechanically weak rocks and may be too permeable. On the other hand, a dolomitic rock might be so massive that water cannot percolate through it. Another criterion for good karst development is that the rock must either be well jointed or have bedding planes that enable the water to pass freely through. The limestone regions of central Kentucky and Missouri, southern Indiana, and the classic karst region of Yugoslavia all have thick series of limestone strata that are well jointed (Hunt, 1967). On the surface of these karst regions one finds a broken terrain with jagged, etched, pitted rocks (called lapies) and a series of solution sinks that take the form of rounded depressions (dolines). Beneath the surface can be found extensive caverns. Other areas of karst togography are the large limestone valleys of the Appalachians, which extend from Pennsylvania to central Alabama and include the Shenandoah Valley of Virginia. Little known karst regions in other parts of the world include the large Nullarbor Plain which fringes the Great Australian Bight in southern Australia. This is really a plateau area 150 to 600 feet (46 to 183 meters) above sea level, up to 200 miles (322 kilometers) wide, and over 400 miles (644 kilometers) long (Jennings, 1971). In southern China and the northern part of Vietnam there is a karst region with numerous sinkholes, caverns, and remnant karst features. The area is particularly known for the towers which appear in Chinese paintings. The region covers over three hundred thousand square miles (600,000 square kilometers) of Yunnan, Kwangsi, and Kweichou Provinces the world. Throughout southeast Asia, particularly in Malay Peninsula, Sumatra, Java, Celebes, and the Moluccas, there are erosional karst features similar to the pepino hills of Puerto Rico (Sweeting, 1973).

MAP 7-4 CAVES AND CAVERNS OF THE WORLD

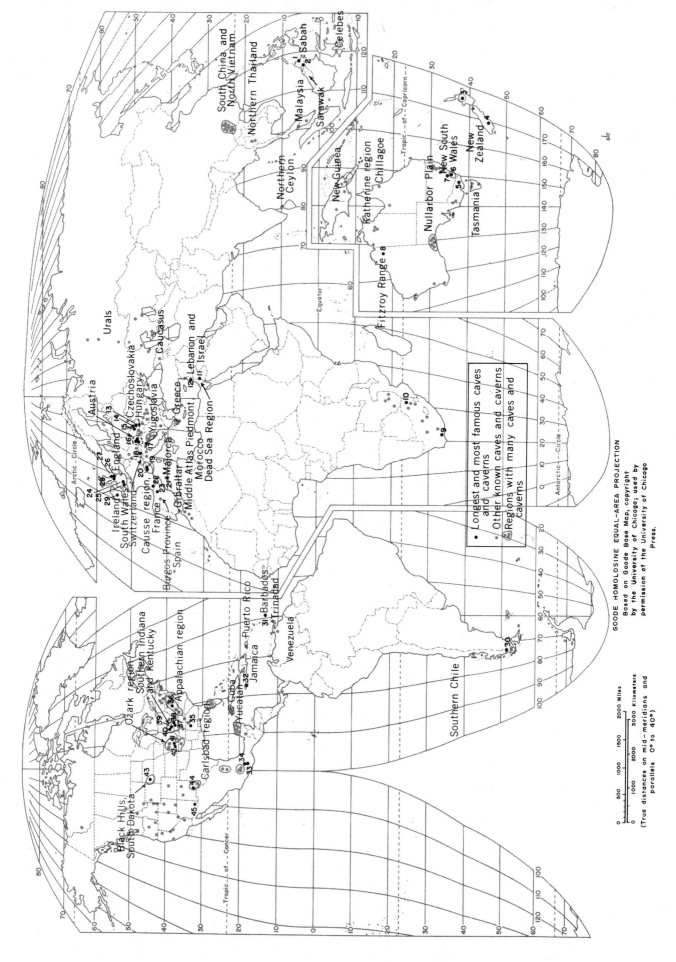

GOODE HOMOLOSINE EQUAL-AREA PROJECTION
Based on Goode Base Map, copyright
by the University of Chicago; used by
permission of the University of Chicago
Press.

Black Hills,
South Dakota

Ozark region
Southern Indiana
and Kentucky
Appalachian region
Carlsbad region
Yucatan
Cuba
Jamaica
Puerto Rico
Barbados
Trinidad
Venezuela

Southern Chile

Austria
Urals
Czechoslovakia
Hungary
Caucasus
Yugoslavia
Greece
Lebanon and
Israel
Dead Sea Region
Morocco
Middle Atlas Piedmont
Gibraltar
Mallorca
France
Spain
Burgos Province
Causse region
Switzerland
South Wales
England
Ireland

South China and
North Vietnam
Northern Thailand
Malaysia
Sarawak
Sabah
Celebes
Northern
Ceylon
New Guinea
Katherine region
Chillagoe
Fitzroy Range
Nullarbor Plain
New South
Wales
New
Zealand
Tasmania

Longest and most famous caves
and caverns
Other known caves and caverns
Regions with many caves and
caverns

0 500 1000 1500 2000 Miles
0 1000 2000 3000 Kilometers
(True distances on mid-meridians and
parallels 0° to 40°)

MAPS
7-4
7-5

LARGEST AND MOST FAMOUS CAVES AND CAVERNS

Closely associated with limestone and karst regions are caves and caverns, because nearly all well-developed karst regions possess underground passageways. The major karst regions of Kentucky, southern Indiana, central Missouri, and the Great Valley of Virginia - as well as the karst regions of south China, the Nullarbar Plain of Australia, and the classic karst region of Yugoslavia, have large underground caverns. This map depicts the major regions of the world where numerous caves and caverns exist. The numbers indicate the largest and best known caves found. A cave is a hallowed-out chamber in the earth while a cavern, although having all the characteristics of a cave, is much larger in size, often with indefinite extent. The United States map, shows caves and caverns open to the public. Lava caves and tunnels are treated in conjunction with the map on lave flows and remnant volcanic features. The Ozark Plateau of Missouri and Arkansas has the largest number of commercial caves. More than 25 caverns are open to the public. One of the most spectacular is Meramec Caverns, 50 miles (81 kilometers) southwest of St. Louis. The largest cave system in the world is the Mammoth Flint Ridge system of caverns about 155.3 miles (250 kilometers) long. Mammoth Cave is 44.5 miles (71.2 kilometers) long. There are several levels to this cavern; at least five are well defined. One great room, known as Mammoth Dome, is 400 feet (122 meters) long, 150 feet (46 meters) wide, and 250 feet (76 meters) high. This huge cavern system is one of numerous caves found from southern Indiana south into Tennessee. The caves in Indiana are famous for such unusual features as subteranean stream piracies that result in lost rivers.

A very famous cavern is Carlsbad in the Guadalupe Mountains of southeastern New Mexico. This cavern has an entrance 90 feet wide and 40 feet (12 meters) high. There are at least three principal levels, the lowest of which is 1,320 feet (402 meters) below the surface. More than 30 miles (48 kilometers) of galleries and rooms have been explored and many more are known to exist. The Big Room in Carlsbad may be the largest chamber in the world.

MAP 7-5 CAVES AND CAVERNS OF THE UNITED STATES

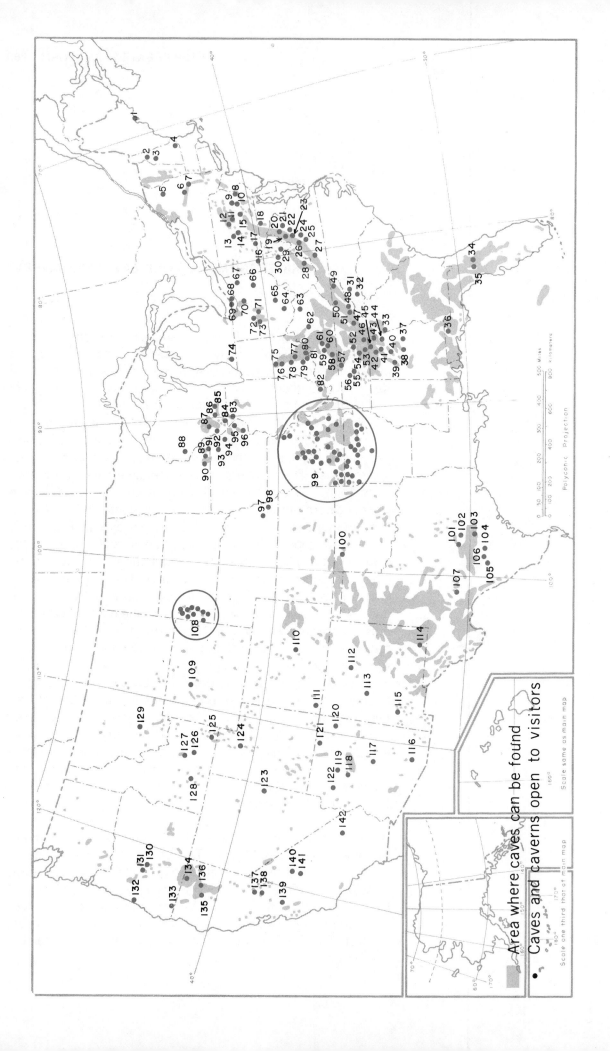

Area where caves can be found

Caves and caverns open to visitors

The following is a list of some of the major caves and caverns found around the world:

| | | where known | |
		mi.	km.
1.	Sandakan Caves - Sabah - Malaysia		
2.	Niah Caves - Northern Sarawak - Malaysia		
3.	Waitomo Caves - North Island - New Zealand		
4.	TeAna - Au Caves (Glow-worm Caves) - South Island - New Zealand		
5.	Buchan Caves - Victoria - Australia		
6.	Bungonia Caves - New South Wales - Australia		
7.	Jenolan Caves - Blue Mountains - New South Wales - Australia		
8.	Tunnel Cave - Napier Range - Western Australia		
9.	Congo Caves near Oudtshoorn - South Africa	2.0	9.7
10.	Makapan Caves - Transvall - South Africa		
11.	Dead Sea Caves - Dead Sea Region - Israel		
12.	Jeita Cave - Lebanon		
13.	Domica Cave - Hungary and Czechoslovakia	13.7	22.0
14.	Demanova Cave - Czechoslovakia	8.7	14.0
15.	Postojnska Cave - Czechoslovakia	9.4	15.0
16.	Eisriesenwelt Cave - near Salzburg, Austria	26.3	42.0
	Tantalhohle Cave - near Salzburg, Austria	10.0	16.0
	Dachstein Mammuthohle - Austria	8.7	14.0
17.	Gro Ha Gigante - near Trieste		
18.	Hollach Cave - Muotatal - Switzerland	48.7	78.0
19.	Gouffre Berger - near Grenoble, France		
20.	Dent de Crolles System - Isere, France	11.2	18.0
21.	Cuavas del Drach - Majorca, Spain		
22.	Gouffre de la Pierre St. Martin - Basses - Pyrenees - France		
23.	Cueva Palomera - Burgos Provinee - Spain	7.5	12.0
24.	Agen Allwedd Cave - South Wales - British Isles	9.0	14.4
25.	Dan yr Ogof Cave - South Wales - British Isles		
26.	Lost John's Cave and Gypsum Cave - N.W. Yorkshire - British Isles		
27.	Gaping Gill Cave and Peny-y Ghent Caves - British Isles		
28.	Ease Gill Caverns - Lancaster area - British Isles		
29.	Vigo Cave - Clare County - Ireland		
30.	Cave of Mylodon - Puerto Natales - Chile		
31.	Cole's Cave and Animal Flower Cave - Barbados		
32.	Windsor Caves - Jamaica		
33.	Juxlahuaca Cave - Zihuatanejo near Tixtla - Mexico		
34.	Gruntas de Cachuamilpa - near Taxco - Mexico		
35.	Anvil Cave - 10 miles south of Decatur, Alabama	12.0	19.2
36.	Greenbrier Caverns - Greenbrier State Forest - West Virginia	15.2	24.3
37.	Mammoth Cave - Central Kentucky	44.5	71.2
38.	Flint Ridge Cave System - next to Mammoth Cave - Kentucky	40.5	64.8
39.	Wyandotte Caverns - Wyandotte, Indiana	23.0	37.0
40.	Sullivan Cave - near town of Sullivan, Indiana	8.5	13.6
41.	Carroll Cave - Camden County - Missouri	8.3	13.3
42.	Marvel Cave (Powell's Cave) near Bransom, Missouri	8.2	13.2
43.	Jewel Cave - Black Hills - South Dakota	13.1	21.0
44.	Carlsbad Caverns - Southern New Mexico	7.4	11.8
45.	Colossal Cave - east of Tucson, Arizona	6.0+	9.7+

The following is a list of the major caves and caverns in the United States open to visitors:

1. Anemone Cave	Maine	26. Melrose Caverns	Virginia	52. Cummberland Caverns	Tennessee
2. Lost River	New Hampshire	27. Dixie Caverns	Virginia	53. Crystal Cave	Tennessee
3. Polar Caves	New Hampshire	28. Organ Cave	West Virginia	54. Wonder Cave	Tennessee
4. Mystery Hill Caves	New Hampshire	29. Seneca Caverns	West Virginia	55. Ruskin Cave	Tennessee
5. Natural Stone Bridge & Caves	New York	30. Smoke Hole Cavern	West Virginia	56. Jewel Cave	Tennessee
6. Secret Caverns	New York	31. Linville Caverns	North Carolina	57. Dunbar Cave	Tennessee
7. Howe Caverns	New York	32. Chimney Rock Park	North Carolina	58. Diamond Caverns	Kentucky
8. Lost River Caverns	Pennsylvania	33. Cave Springs Cave	Georgia	59. Mammoth Cave Nat'l Park	Kentucky
9. Crystal Cave	Pennsylvania	34. Ocala Caverns	Florida	60. Lost River Cave	Kentucky
10. Onyx Cave	Pennsylvania	35. Fire Maker's Cave	Florida	61. Mommoth Onyx Cave	Kentucky
11. Woodward Cave	Pennsylvania	36. Florida Caverns	Florida	62. Daniel Boone's Cave	Kentucky
12. Penn's Cave	Pennsylvania	37. Crystal Caverns	Alabama	63. Carter's Cave	Kentucky
13. Indian Caverns	Pennsylvania	38. Kymulga Cave	Alabama	64. Seven Caves	Ohio
14. Lincoln Caverns	Pennsylvania	39. Rickwood Caverns	Alabama	65. Hocking Hills State Parks	Ohio
15. Indian Echo Caverns	Pennsylvania	40. Guntersville Caverns	Alabama	66. Devil's Den Park	Ohio
16. Laurel Caverns	Pennsylvania	41. Cathedral Caverns	Alabama	67. Nelson's Ledges	Ohio
17. Wonderland Caverns	Pennsylvania	42. Sauta Cave	Alabama	68. Perry's Cave	Ohio
18. Crystal Grottoes	Maryland	43. Sequoyah Caverns	Alabama	69. Crystal Cave	Ohio
19. Shenandoah Caverns	Virginia	44. Manitou Cave	Alabama	70. Seneca Caverns	Ohio
20. Battlefield Crystal Caverns	Virginia	45. Russell Cave Nat'l Monument	Alabama	71. Olentangy Caverns	Ohio
21. Skyline Caverns	Virginia	46. Ruby Falls	Tennessee	72. Zane Caverns	Ohio
22. Luray Caverns	Virginia	47. The Lost Sea	Tennessee	73. Ohio Caverns	Ohio
23. Endless Caverns	Virginia	48. Alum Cave	Tennessee	74. Bear Cave	Michigan
24. Massanutten Caverns	Virginia	49. Bristol Caverns	Tennessee	75. Porter's Cave	Indiana
25. Grand Caverns	Virginia	51. Tuckaleechee Caverns	Tennessee		

76. McCormick's Creek State Park	Indiana	
77. Cave River Valley Park	Indiana	
78. Spring Mill State Park	Indiana	
79. Marengo Cave	Indiana	
80. Wyandotte Caverns	Indiana	
81. Little Wyandotte Cave	Indiana	
82. Cave-In-Rock	Illinois	
83. Mississippi Palisades State Park	Illinois	
84. Badger Mine	Wisconsin	
85. Cave of the Mounds	Wisconsin	
86. Lost River Cave	Wisconsin	
87. Eagle Cave	Wisconsin	
88. Crystal Cave	Wisconsin	
89. Mystery Cave	Minnesota	
90. Minnesota Caverns	Minnesota	
91. Niagara Cave	Minnesota	
92. Kickapoo Caverns	Wisconsin	
93. Wonder Cave	Iowa	
94. Spook Cave	Iowa	
95. Crystal Lake Cave	Iowa	
96. Maquoketa Caves	Iowa	
97. Robber's Cave	Nebraska	
98. John Brown's Cave	Nebraska	

99.
Big Spring Onys Caverns	Missouri
Bluff Dwellers Cave	Missouri
Boone Cave	Missouri
Bridal Cave	Missouri
Cameron Cave	Missouri
Civil War Cave	Missouri
Crystal Caverns	Missouri
Fairy Cave	Missouri
Fantastic Caverns	Missouri
Fisher's Cave	Missouri
Honey Branch Cave	Missouri
Jacob's Cave	Missouri
Mark Twain Cave	Missouri
Marvel Cave	Missouri
Meramec Caverns	Missouri
Missouri Caverns	Missouri
Mystic River Cave	Missouri
Old Spanish Cave	Missouri
Onondaga Cave	Missouri
Ozark Caverns	Missouri
Ozark Wonder Cave	Missouri
Rebel Cave	Missouri
Round Spring Cavern	Missouri
Stark Caverns	Missouri

99
Truitt's Cave	Missouri
Big Hurricane Cave	Arkansas
Blanchard Springs Caverns	Arkansas
Buffalo River State Park	Arkansas
Bull Shoals Caverns	Arkansas
Cave City Cave	Arkansas
Civil War Cave	Arkansas
Devil's Den State Park	Arkansas
Diamond Cave	Arkansas
Mystery Cave	Arkansas
Mystic Cavern	Arkansas
Old Spanish Treasure	Arkansas
Onyx Cave	Arkansas
Ozark Mystery Cave	Arkansas
Petit Jean State Park	Arkansas
Rowland Cave	Arkansas
Shawnee Cave	Arkansas
Wonderland Cave	Arkansas

100. Alabaster Caverns	Oklahoma	
101. Texas Longhorn Cavern	Texas	
102. Cobb Cavern	Texas	
103. Wonder Cave	Texas	

104. Natural Bridge Caverns   Texas

105. Cascade Caverns   Texas

106. Century Caverns   Texas

107. Caverns of Sonora   Texas

108.
    Bethlehem Cave   South Dakota

    Jewel Cave Nat'l Park   South Dakota

    Nameless Cave   South Dakota

    Rushmore Cave   South Dakota

    Sitting Bull Crystal
    Caverns   South Dakota

    Stage Barn Crystal
    Cave   South Dakota

    Thunder Head Falls   South Dakota

    Wild Cat Cave   South Dakota

    Wind Cave Nat'l Park   South Dakota

    Wonderland Cave   South Dakota

109. White Sulphur Spring
    Cave   Wyoming

110. Cave of the Winds   Colorado

111. Mesa Verde Nat'l Park   Colorado

112. Bandelier Nat'l
    Monument   Colorado

113. Desert Ice Box   New Mexico

114. Carlsbad Caverns   New Mexico

115. Gila Cliff Dwelling
    Nat'l Monument   New Mexico

116. Colossal Cave   Arizona

117. Tonto Nat'l
    Monument   Arizona

118. Montezuma Castle
    Nat'l Monument   Arizona

119. Sunset Crater Nat'l
    Monument   Arizona

120. Canyon de Chelly
    Nat'l Monument   Arizona

121. Navajo Nat'l
    Monument   Arizona

122. Grand Canyon
    Caverns   Arizona

123. Lehman Caves Nat'l
    Monument   Nevada

124. Timpanogos Cave   Utah

125. Minnetonka Cave   Idaho

126. Crystal Ice Caves   Idaho

127. Craters of the Moon
    Nat'l Monument   Idaho

128. Shoshone Indian Ice
    Caves   Idaho

129. Lewis & Clark Cavern
    State Park   Montana

130. Lava River Caves State
    Park   Oregon

131. Lavacicle Cave   Oregon

132. Sea Lion Caves   Oregon

133. Oregon Caves Nat'l
    Monument   Oregon

134. Lava Beds Nat'l
    Monument   California

135. Lake Shasta Caverns   California

136. Subway Cave   California

137. Mercer Caverns   California

138. Moaning Cave   California

139. Pinnacles Nat'l
    Monument   California

140. Boyden Cave   California

141. Crystal Cave   California

142. Mitchell Caverns State
    Reserve   California

Another major cave region is in the limestone belt of the Appalachian Mountains, from Pennsylvania to Alabama. Most of the caves are in the large synclinal limestone valleys that occur mostly on the east side of the mountain complex. The most famous region is the Shenandoah Valley of Virginia, where at least 1,000 caves are known to exist, with a score of 50 of the most famous and accessible ones commercially operated. Many of the caves are on the flanks of the synclinal ridges. The most famous and spectacular cave in this group is Luray Caverns, unusual because it underlies a haystack - like hill, which is a remnant limestone mass. The Black Hills of South Dakota constitute another interesting cave region. In a limestone belt surrounding the Black Hills dome are extensive caverns, many of which have never been explored. At least seven are open to the public; the most famous, Wind Cave, has unusual box work formations. Also in the western United States is Cave of the Winds, near Manitou Colorado. It has been formed in a small bed of limestone perched on top of an isolated hill of crystalline rock.

In Europe and Asia there are extensive karst and cavernous areas. Many of the European caves are more explored and better mapped than those in the United States. Northern Italy and Yugoslavia are renowned for their large caverns and sinkholes. Holloch Cave in Muotatal, Switzerland is the longest in Europe being 48.7 miles (78.0 km) long. The Pyrenees contain the deepest caves in the world. Gouffre de la Pierre St. Martin in the Basses region of France goes over 2,000 feet (610 meters) below its entrance. In western Germany and eastern Belgium are caverns that are well-known because of the remnants of ancient man found in them. These caves are too numerous to plot on the world map. In the Urals, the Caucasus, and other parts of the southern USSR are found extensive caverns and karst. Other noted cavern areas are in central England, North Africa, southern China and Indochina, the islands of the East Indies and in parts of southern and southeastern Australia. New Zealand has several caverns noted for glowworms.

On pages 265-268 is a list of 45 of the largest and better known world caves and caverns. Where the length is known figures are given in miles and kilometers.

REFERENCES

The following list includes the major references that have been referred to in compiling these maps and text but does not begin to cover all the literature consulted in compiling this section. As space precludes a complete listing, an attempt has been made to include books and articles that are representative, especially those which afford broad reviews of the topics covered and also contain well-selected bibliographies.

GENERAL GEOGRAPHY

The Associated Press Almanac, Hammond Almanac, Inc., Maplewood, N. J., 1975, p. 15.

The Atlas of Britain and Northern Ireland, (1963), Clarendon Press, Oxford.

Atlas of Canada (1957), Dept. of Mines and Technical Surveys Geographic Branch, Ottawa.

The World Almanac and Book of Facts 1976, Newspaper Enterprise, Assoc., Inc., New York, 1976.

Encyclopedia Americana, America Corporation, New York, 1968.

Atwood, Wallace W. (1940). *The Physiographic Provinces of North America,* Ginn and Company, New York.

Bauer, H. A. (1933). "A World Map of Tides," *The Geographical Review,* Vol. 23, pp. 259-270.

Becker, R. (1936). Dunung und Wind des Atlantischen Ozeans im Bereich des Meteorologischen Aguators. *Ann. d. Hydrographie U. mar. Metrologie,* Zweites Koppen-Heft, 1-4.

Bird, J. B. (1967). *The Physiography of Artic Canada,* The John Hopkins Press, Baltimore, pp. 35-45.

Bird, J. B. (1972). *The Natural Landscapes of Canada,* John Wiley and Sons, Inc., New York.

Birot, J. B. (1967). "Le Cycle d'Erosion pour Les Differents Climats," *Curso de Altos Estudos Geographicos,* Vol. 1, University of Brazil.

Black, R. F. (1954). "Permafrost--A Revies," *Bulletin of the Geologic Society of America,* pp. 65, 839-855.

Board, C. R. J. Chorley, P. Haggett, and D. S. Stoddart (1969). *Progress in Geography,* Vol. 1, N. Y., St. Martin's Press, pp. 159-222.

Bruns, E. (1953). *Handbuck der Wellen der Meere und Ozeans.* Berlin.

Bryan, K. (1946). "Cryopedology--The Study of Frozen Ground and Intense Frost Action with Suggestions of Nomenclature," *American Journal of Science,* Vol. 244, pp. 622-642.

Budel, J. (1948). "Das System der klimatischen Geomorphologie (Beitrage zur Geomorphologie der Klimatzonen und Verzeitklimate V), *Verhandlungen Deutschen Geographentag,* Munchen, Vol. 27, pp. 65-100.

Budel, J. (1963). "Klima: Genetische Geomorphologie," *Geogr. Rundschau,* Vol. 15, pp. 269-286.

Budel, J. (1969). "Das System der Klima-genetischen Geomorphologie," *Erdkunde,* Vol. 23, No. 3.

Butzer, K. W. (1964). *Environment and Archaeology: An Introduction to Pleistocene Geography,* Aldine Publishing Company, Chicago.

Butzer, K. W. (1976). *Geomorphology From the Earth,* Harper and Row, Publishers, New York, 463 pp.

Chapman, L., and D. Putnam (1967). *The Physiography of Arctic Canada,* John Hopkins Press, Baltimore.

Chorley, R. J. (1969). *Water, Earth and Man: A Synthesis of Hydrology, Geomorphology and Socioeconomic Geography,* Methuen and Co., Ltd., London.

Clark, T. H., and C. W. Stearn (1960). *The Geological Evolution of North America,* The Ronald Press Co., New York.

Cooke, R. U., and A. Warren (1973). *Geomorphology of Deserts,* University of California Press, Berkeley and Los Angeles.

Cotton, C. A. (1941). *Landscape, As Developed by the Processes of Normal Erosion,* The University Press, Cambridge.

Cotton, C. A. (1949). *Geomorphology: An Introduction to the Study of Landforms,* (5th Ed.), John Wiley and Sons, Inc., New York.

Cressey, G. B. (1963). *Asia's Lands and Peoples,* McGraw-Hill Book Co., Inc., New York.

Davis, W. M. (1901). *Physical Geography,* Ginn and Co., Boston, 339 pp.

Defant, A. (1961. *Physical Oceanography,* Heinemann, London.

Doty, M. S. (1957). "Rocky Intertidal Surfaces," *Geol. Soc. Am., Memoir 67,* Vol. 1, pp. 535-585.

Dury, G. H. (1966). *Essays in Geomorphology,* Heinemann, London.

Emmons, W. H., G. A. Thiel, C. R. Stauffer, and I. S. Allison (1955). *Geology,* McGraw-Hill Book Co., Inc., New York and London.

Fairbridge, R. W. (1968). *The Encyclopedia of Geomorphology,* Reinhold Book Corp., New York.

Fiziko-Geografischeskiy Atlas Mira *(Physical Geographic Atlas of the World)* (1964). Academy of Sciences USSR and the Main Administration of Geodesy and Cartography, State Geological Committee USSR, Moscow. Translated in *Soviet Geography: Review and Translation,* American Geographical Society, New York, May-June 1965.

Gentille, J., and R. W. Fairbridge (1951). *Physiographic Diagram of Australia,* C. S. Hammond and Company, Maplewood, New Jersey.

Grabau, A. W. (1920). *A Comprehensive Geology,* Part I, General Geology, D. C. Heath and Company, Publishers, New York.

Gregory, H. E. (1915). *Geology of Today: A Popular Introduction in Simple Language,* J. B. Lippincott Co., Philadelphia.

Gresswell, R. K. (1967). *Physical Geography,* Longmans Inc., London.

Helle, J. R. (1958). Surf Statistics for the Coasts of the United States, *Tech. Memo. Beach Erosion Board, U. S.,* No. 108.

Hinds, N. E. A. (1943). *Geomorphology, the Evolution of Landscape,* Prentice-Hall, Inc., New York.

Holmes, A. (1945). *Principles of Physical Geology,* The Ronald Press Company, New York.

Hunt, G. B. (1967). *Physiography of the United States,* W. H. Freeman and Company, San Francisco, 480 pp.

Hunt, C. B. (1974). *Natural Regions of the United States and Canada,* W. H. Freeman and Company, San Francisco, 725 pp.

Jenness, J. L. (1949). "Permafrost in Canada," *Artic,* Vol. 2, pp. 13-27.

King, C. A. M. (1966). *Techniques in Geomorphology,* Edward Arnold, Ltd., London.

King, L. C. (1962). *The Morphology of the Earth,* Oliver and Boyd, Edinburgh and London.

Lobeck, A. K. (1939). *Geomorphology,* McGraw-Hill Company, Inc., New York.

Lobeck, A. K. (1945). *Physiographic Diagram of Asia,* C. S. Hammond and Company, Maplewood, New Jersey.

Lobeck, A. K. (1946). *Physiographic Diagram of Africa,* C. S. Hammond and Company, Maplewood, New Jersey.

Lobeck, A. K. (1950). *Physiographic Diagram of North America,* C. S. Hammond and Company, Maplewood, New Jersey.

Lobeck, A. K. (1951). *Physiographic Diagram of Europe,* C. S. Hammond and Company, Maplewood, New Jersey.

Lobeck, A. K. (1958). *Things Maps Don't Tell Us,* The MacMillan Company, New York.

Lowman, P. D. Jr. (1968). *Space Panorama,* Weltflugbild Reinhold A. Müller Feldmeilen, Zurich.

Marmer, H. A. (1926). *The Tide,* D. Appleton and Co., New York.

Marmer, H. A. (1935). "The Variety of Tides," Smithsonian Institute Annual Report, Washington.

McGinnes, W. C., *Deserts of the World,* Univ. of Arizona Press, Tucson, 788 pp.

Meigs, P. (1953). "World distribution of arid and semi-arid homoclimates" in *Reviews of Research on Arid Zone Hydrology,* (UNESCO, Paris), pp. 203-9.

Mellor, R. E. (1965). *Geography of the USSR,* MacMillan and Co., London, St. Martin's Press, New York.

Mitchell, J. B., (ed.), (1962). *Great Britain: Geographical Essays,* Cambridge University Press, Cambridge, England.

Mountjoy, A. B. and C. Embleton (1967). *Africa: A New Geographical Survey,* F. A. Praeger Pub., Inc.

Morski Atlas (Oceanic Atlas) (1950), edited by I. S. Isakov, Naval General Staff, Moscow.

Murphy, R. E. (1967). "A Spatial Classification of Landforms Based on Both Genetic and Empirical Factors--A Revision," *Annals, Association of American Geographers,* Vol. 57, No. 1, pp. 185-186.

Murphy, R. E. (1968). "Landforms of the World," "Annals Map Supplement No. 9," *Annals, Association of American Geographers,* Vol. 58, No. 1.

U. S. Geological Survey (1970). *The National Atlas of the United States of America,* U. S. Geological Survey, Washington, D. C., U. S. Geol. Survey.

Peltier, L. C. (1950). "The Geographic Cycle in Periglacial Regions as it is Related to Climatic Geomorphology," *Annals, Association of American Geographers,* Vol. 40, pp. 214-236.

Peltier, L. C. (1962). "Area Sampling for Terrain Analysis," *Professional Geographer,* Vol. 14, No. 2, pp. 24-28.

Pergamon World Atlas (1968). Pergamon Press, New York (PWN--Poland, Polish Scientific Publishers, Warsaw).

Powers, W. E. (1966). *Physical Geography,* Appleton-Century-Crofts, New York.

Pitty, A. F. (1971). *Introduction to Geomorphology,* Barnes and Noble, Inc., New York, 526 pp.

Putnam, D. F. (1952). *Canadian Regions: A Geography of Canada,* Thomas Y. Crowell Co., New York.

Raisz, E. (1949-1950). *Landform Map of Canada,* Institute of Geographical Exploration, Harvard University, Cambridge, Massachusetts.

Raisz, E. (1957). *Landforms of the United States,* Scale about 1/4,000,000 (Sixth rev. ed.).

Raisz, E. (1966). *Landform Map of Alaska* (2nd ed.). Originally prepared for the Environmental Protection Section. Office of the Quartermaster General, Hydrography by R. L. Williams.

Rees, H. (1964). *Australia, New Zealand, and the Pacific Islands,* (2nd ed.), McDonald and Evans, London.

Schott, W. (1952). "Zur Klimaschichtung der Tiefseedlimente in aquatorialen Atlantischen Ozean," *Geol. Rdsch.,* Vol. 40, H. 1.

Schou, A. (1949). *Atlas of Denmark,* Part 1, *The Landscapes,* H. Hagerup, Kobenhaven.

Schubart, L., and W. Mockel (1949). Dunung im Atlantischen Ozean. Deutsche Hydrographische Zeitchrift, Vol. 2, pp. 280-285.

Shelton, J. S. (1966). *Geology Illustrated,* W. H. Freeman and Company, San Francisco.

Short, N. M., P. D. Lowman, Jr., S. C. Freden and W. A. Finch, Jr. (1976). *Mission to the Earth: Landsat Views the World,* National Aeronautics and Space Administration, Washington, D. C., 459 pp.

Skylab Explores the Earth (1977). NASA SP-380, National Aeronautics and Space Administration, Washington, D. C., 517 pp.

Smith, G. H. (1935). *Physiographic Diagram of South America,* C. S. Hammond and Company, Maplewood, New Jersey.

Somme, A. (1960). *A Geography of Norden: Denmark, Finland, Norway, Sweden,* J. W. Coppelens Forlag, Oslo.

Sparks, B. W. (1960). *Geomorphology,* Longmans, Green and Co., Ltd., London.

Spencer, J. E., and W. L. Thomas (1971). *Asia, East by South: A Cultural Geography,* John Wiley and Sons, New York.

Stamp, D. L. (1962). *Asia: A Regional and Economic Geography,* 11th Ed., Dutton, New York, Methuen and Co., Ltd., London.

Steers, J. A., ed. (1964). *Field Studies in the British Isles,* Thomas Nelson and Sons.

Strahler, A. (1963). *The Earth Sciences,* Harper and Row, New York.

Strahler, A. (1969). *Physical Geography,* 3rd Ed., John Wiley and Sons, Inc., New York.

Strahler, A. (1970). *Introduction to Physical Geography* (2nd ed.), John Wiley and Sons, Inc., New York.

Strakhof, N. M. (1967). *Principles of Lithogenesis,* Vol. 1 (translated by J. Paul Fitzsimmons), Oliver and Boyd, Edinburgh and London.

Suess, E. (1908). *The Face of the Earth,* Vol. 2, Clarendon Press, Oxford.

Suslov, S. P. (1961). *Physical Geography of Asiatic Russia,* W. H. Freeman and Company, San Francisco.

Sverdrup, H. V., M. W. Johnson, and R. H. Fleming (1942). *The Oceans, Their Physics, Chemistry and General Biology,* Prentice-Hall, Inc., New York.

Taber, S. (1943). "Perennially Frozen Ground in Alaska: Its Origin and History," *Bulletin, Geological Society of America,* Vol. 54, pp. 1433-1548.

Tarr, R. S. and O. D. Von Engeln (1962). *New Physical Geography,* MacMillan and Company, New York.

Taylor, G. (1943). *Australia,* E. P. Dutton and Co., New York.

Thornbury, W. D. (1965). *Regional Geomorphology of the United States,* John Wiley and Sons, Inc., New York.

Thornbury, W. D. (1969). *Principles of Geomorphology,* 2nd Ed., John Wiley and Sons, Inc., New York.

Thornthwaite, C. W., "An Approach Toward a Rational Classification of Climate," *Geographical Review.*

Trewartha, G. T., A. H. Robinson, and E. H. Hammond (1967). *Physical Elements of Geography,* 5th Ed., McGraw-Hill Book Company, New York.

Tricart, J., and A. Cailleux (1965). *Introduction a la Geomorphologie Climatigue,* Societe d'Editions d'Enseignement Superieur, Paris.

UNESCO (1952). Arid Zone Programme I Reviews of Research on Arid Zone Hydrology, Paris.

Van Mieghem, J., and P. van Oye (1965). *Biogeography and Zoology in Antarctica,* Dr. W. Junk N. Y., The Hague.

Van Riper, J. E. (1962). *Man's Physical World,* McGraw-Hill Book Co., Inc., New York.

Von Engeln, O. D. (1960). *Geomorphology,* The MacMillan Company, New York.

Walton, K. (1969). *The Arid Zones,* Hutchinson University Library, London.

Williams, H. (1958). *Landscapes of Alaska,* University of California Press, Berkeley.

Woldstedt, P. (1964-1965). *Das Eiszeithlter; Grundlinieneiner Geologie des Quartars,* 3 vols, Ferdinand Enee, Stuttgart.

Wooldridge, S. D., and S. W. Morgan (1937). *Outlines of Geomorphology,* Longmans, Green and Co., New York.

Wooldridge, S. W., and R. S. Morgan (1959). *Outline of Geomorphology: The Physical Basis of Geography* (2nd ed.), Longmans, Green and Co., New York.

Worcester, P. G. (1945). *Geomorphology,* D. Van Nostrand Co., Inc., New York.

The World Almanac and Book of Facts, The Albuquerque Tribune, 1976.

Wright, H. E., Jr., and David G. Frey (1965). *The Quaternary of the United States,* Princeton University Press, Princeton, New Jersey.

Wyckoff, J. (1966). *Rock, Time, and Landforms,* Harper and Row, New York.

Zubov, N. M. (1947). *Dinamicheskaya Okeanologiya, Hydrometeorological Publishing House,* Moscow.

STRUCTURE AND TECTONICS

Barazangi, M., and J. Dorman (1969). "World Seismicity Maps Compiled from ESSA, Coast and Geodetic Survey, Epicenter Data, 1961-1967," *Bulletin of the Seismological Society of America,* Vol. 59, No. 1.

Billings, M. P. (1942). *Structural Geology,* Prentice-Hall, Inc., New York.

Bullard, F. M. (1962). *Volcanoes,* University of Texas Press, Austin.

Chhibber, H. L. (1934). *The Geology of Burma,* Macmillan and Co., London.

Coats, R. R. (1950). "Volcanic activity in the Aleutian Arc," *U. S. Geological Survey Bulletin,* 974-B, pp. 35-49.

Craig, E. H. C. (1912). *Oil Finding, an Introduction to the Geological Study of Petroleum,* Longmans, Green and Co., London, pp. 103-108.

DeGotyer, E. L. (1919). "The West Point, Texas, Salt Dome, Freestone County," *Journal of Geology,* Vol. 27, pp. 647-663.

Dewey, J. F. (1972). "Plate Tectonics," *Scientific American,* pp. 56-68.

Dietz, R. S.; and Holden, J. C. (1970). Reconstruction of Pangaea: Break-up and Dispersion of Continents, Permian to Present," *Journal of Geophysical Research,* Vol. 75, No. 26, pp. 4939-4956.

Dietz, R. S.; and Holden, J. C. (1970). "The Breakup of Pangaea," Scientific American, pp. 30-41.

Dillon, L. S. (1974). "Neovolcanism: A Proposed Replacement for the Concepts of Plate Tectonics and Continental Drift," Fig. 21, p. 212 in C. F. Kahle, *Plate Tectonics Assessments and Re-Assessments,* The American Association of Petroleum Geologists, Tulsa, Oklahoma, 514 pp.

Fisher, W. B. (1968), *The Cambridge History of Iran,* Vol. 1 (The Land of Iran), Cambridge University Press, Cambridge, England.

Gorshkov, G. S. (1970). *Volcanism and the Upper Mantle; Investigations in the Kurile Island Arc,* Plenum Publishing Corporation, New York, 385 pp.

Guest, John (1971). *The Earth and Its Satellite,* David McKay Company, Inc., New York.

Harrison, J. V. (1941). "Coastal Makran," *Geographical Journal,* Vol. XCVII, No. 1, pp. 1-17.

Harrison, T. S. (1927). "Colorado-Utah Salt Domes," *Bulletin, American Association of Petroleum Geologists,* Vol. II, pp. 111-113.

Hodgson, J. H. (1964). *Earthquakes and Earth Structures,* Prentice-Hall, Inc., Englewood Cliffs, New Jersey.

Holmes, A. (1965). *Principles of Physical Geology* (2nd ed.), The Ronald Press Company, New York.

Iddings, J. P. (1914). *The Problem of Volcanism,* Yale University Press, New Haven, Connecticut.

Lalicker, C. G. (1949). *Principles of Petroleum Geology,* Appleton-Century-Crofts, Inc., New York.

Leet, L. D. (1942). "Mechanics of Earthquakes Where There is No Surface Faulting," *Bulletin, Seismological Society of America,* Vol. 32, pp. 93-96.

Leet, L. D., and S. Judson (1965). *Physical Geology* (3rd ed.), Prentice-Hall, Inc., Englewood Cliffs, New Jersey.

Levorsen, A. I. (1954). *Geology of Petroleum,* W. H. Freeman and Co., San Francisco, p. 21.

MacDonald, G. A. (1972). *Volcanoes,* Prentice-Hall, Inc., Englewood Cliffs, N. J., Fig. 14-1, pp. 429-450.

Mattson, P. H. (1977). *West Indies Island Arcs,* Dowden, Hutchinson and Ross, Inc., Stroudsburg, Pa., 382 pp.

Murray, G. E. (1961). *Geology of the Atlantic and Gulf Coastal Province of North America,* Harper and Bros., New York.

Neck, N. H. (1936). *Earthquakes,* Princeton University Press, Princeton, New Jersey.

Neumann van Pedang, M.; Richards, A. F.; Machad, F.; Bravo, T.; Baker, P. E.; and R. W. LeMaitre (1967). *Catalog of the Active Volcanoes of the World including Solfatara Fields,* Part 21, "Atlantic Ocean," Int'l. Assoc. of Volcanology, Naples, 128 pp.

Nichols, R. L. (1938). "Groved Lave," *Journal of Geology,* Vol. 46, pp. 604-614.

Nichols, R. L. (1939). "Sneeze-ups," *Journal of Geology,* Vol. 47, pp. 421-425.

Powers, S. (1920). "The Butler Salt Doome, Freestone County, Texas," *American Journal of Science,* No. 1,99, pp. 127-142.

Riggs, E. A. (1960). *Major Basins and Structural Features of United States* (Map), C. S. Hammond and Company, Maplewood, New Jersey, 1960.

Rittman, A. (1962). *Volcanoes and Their Activity,* John Wiley and Sons, Inc., New York.

Rittman, A. (1936). *Vulkane und ihre Tatigkeit,* Ferdinand Enke Verlag, Stuttgart, pp. 162-163.

Runcorn, S. K. (1962). *Continental Drift,* Academic Press, Inc.

Sanders, C. W. (1939). "Emba Salt Dome Region, USSR, and Some Comparisons with Some Other Salt Dome Regions," *Bulletin, American Association of Petroleum Geologists,* Vol. 23, pp. 492-516.

Short, N. M. (1966). "Shock Processes in Geology," *Journal of Geological Education,* Vol. 14, pp. 149-166.

Snead, R. E. (1964). "Active Mud Volcanoes of Baluchistan, West Pakistan," *The Geographical Review,* Vol. LIV, No. 4, pp. 546-560.

Snead, R. E. (1967). "Recent Morphological Changes Along the Coast of West Pakistan," *Annals,* Association of American Geographers, Vol. 57, No. 3, pp. 550-556.

Snead, R. E. (1970). *Physical Geography of the Makrán Coastal Plain of Iran,* National Technical Information Service, U. S. Department of Commerce, 5825 Port Royal Rd. Springfield, Va. 22151.

Taylor, G. (1943). *Australia,* F. P. Dutton and Co., Inc., New York, P.50.

"Tectonic Map of the United States: Exclusive of Alaska and Hawaii" (1962). U. S. Geological Survey and the American Association of Petroleum Geologists (Scale 1:2,500,000).

Twidale, C. R. (1971). *Structural Landforms,* The MIT Press, Cambridge.

Umbgrove, J. H. F. (1947). *The Pulse of the Earth,* Martinus Nihoff, The Hague.

Wadia, D. N. (1953). *Geology of India,* Macmillan and Company, Ltd., London.

Waring, G. A. (1965). "Thermal Springs of the United States and Other Countries of the World: A Summary," *Geological Survey Professional Paper 492,* U. S. Government Printing, Printing Office, Washington, D. C.

Weeks, L. G. (November 1952). "Factors of Sedimentary Basin Development that Control Oil Occurrence." *Bulletin of the American Association for Petroleum Geology,* Vol. 36, Fig. 2, p. 2076.

Wilson, J. Tuzo (1972). *Continents Adrift,* W. H. Freeman and Company.

Woollard, G. P. (1958). "Areas of Tectonic Activity of the United States as Indicated by Earthquake Epicenters," *American Geophysical Union Transactions,* Vol. 39, No. 6, pp. 1135-1160.

OCEANOGRAPHY AND HYDROLOGY

Credner, G. R. (1878). "Die Deltas, Ihre Morphologie, Geographische Verbreitung und Entstehungs-Bedingungen," *Petermann's Geograpahische Mitteilungen,* Eng. Bd. 12, No. 56.

Crossette, G. (1975). "Waterfalls Nature's Extravaganzas," *Explorer's Journal,* Vol. 53, No. 4, pp. 146-153.

De Martonne, E. (1927). "Regions of Interior Basin Drainage," *The Geographical Review,* Vol. 17, pp. 397-414.

Gilmer, F. W. (1818). "On the Geological Formation of Natural Bridge of Virginia," *American Philosophical Society Transaction,* Vol. 1, pp. 187-192.

Iseri, K. I., and Langbein, W. B. (1974). *Large Rivers of the United States,* Geological Survey Circular 686, U. S. Government Printing Office, Washington, D. C., pp. 8-10.

Reclamation Project Data (Supplement) United States Department of the Interior, Bureau of Reclamation, U. S. Government Printing Office, Washington, D. C. 1966 (Revised May 1975).

Russell, R. J. (1967). *River Plains and Sea Coasts,* University of California Press, Berkeley and Los Angeles.

Shirley, Martha Lou, and James Ragsdale (1966). *Deltas in the Geologic Framework,* Houston Geological Society, Houston, Texas.

COASTAL FEATURES

Amiran, D. H. K. and A. W. Wilson (1973). *Coastal Deserts: Their Natural and Human Environments,* The University of Arizona Press, Tucson, 207 pp.

Bascom, W. (1964). Waves and Beaches: *The Dynamics of the Ocean Surface,* Anchor Books, Garden City, New York.

Bird, E. C. F. (1964). *Coastal Landforms: An Introduction to Coastal Geomorphology with Australian Examples,* Publishing Center, The Australian National University, Canberra.

Bird, E. C. F. (1968). *Coasts,* Australian National University Press, Canberra.

Case, G. O. (1921). *Coast Sand Dunes, Sand Spits, and Sand Wastes,* St. Bride's Press, Ltd., London.

Coleman, J. M. (1966). *Recent Coastal Sedimentation: Central Louisiana Coast* (Technical Report No. 29), Coastal Studies Institute, Louisiana State University, Baton Rough.

Cooper, W. S. (1958). "Coastal Sand Dunes of Oregon and Washington," *Geological Society of America Memoir 72,* 169 pp.

Daly, R. A. (1915). "The Glacial-Control Theory of Coral Reefs," *American Academy of Arts and Sciences Proceedings,* Vol. 51, No. 4.

Davias, J. L. (1973). *Geographical Variation in Coastal Development,* Hafner Pub. Co., New York, 1973, 204 pp.

Davis, W. M. (1928). "The Coral Reef Problem," *American Geographic Society, Special Publication 9.*

DeVirville, A. D. (1940. "Les Zones De Vegetation Sur Le Littoral Atlantique," *Societe de Biogeographie Memoires,* Vol. 7.

Dietrich, G. (1963). *General Oceanography,* Wiley Interscience, New York.

Dolan, R., and others. *Classification of the Coastal Environments of the World, Part I. The Americas. Part II. Africa.* Tech. Reports 1 and 3, Office of Naval Research, Geography Programs, 1972 and 1973.

El- Ashry, M. T. (1977). *Air Photography and Coastal Problems,* Dowden, Hutchinson and Ross, Inc., Stroudsburg, Pa., 425 pp.

Emery, K. O., J. I. Tracey, Jr., and H. S. Ladd (1954). "Geology of Bikini and Nearby Atolls," *Geological Survey Professional Paper 260-A,* U. S. Government Printing Office, Washington, D. C.

Gresswell, R. K. (1962). *The Physical Geography of Beaches and Coastlines,* Hulton Educational Publications, Ltd., London.

Guilcher, A. (translated by B. W. Sparks and R. H. W. Kneese) (1958). Coastal and Submarine Morphology, John Wiley and Sons, Inc., New York (Methuen and Co., Ltd., London).

Inman, D. L., and W. R. Nordstrom (1971). "On the Tectonic and Morphologic Classification of Coasts," *Journal of Geology,* Vol. 79, pp. 1-21.

Issac, W. E. (1937). "South African Coastal Waters in Relation to Ocean Curren ts," *Geographic Review,* Vol. 27, pp. 651-664.

Johnson, D. W. (1919). *Shore Processes and Shoreline Development,* John Wiley and Sons, Inc., New York.

Johnson, D. W. (1925). *The New England-Acadian Shoreline,* John Wiley and Sons, Inc., New York.

King, C. A. M. (1961). *Beaches and Coasts,* Edward Arnold, Ltd., London.

Kuenen, P. H. (1955). "Sea Level and Crustal Warping," *Geol. Am. Spec. Papers,* 62, (Crust of Eathe), pp. 193-204.

Ladd, H. S. (1950). "Recent Reefs," *Bulletin, American Association of Petroleum Geologists,* Vol. 34, pp. 203-214.

McGill, J. R. (1958). "Coastal Landforms of the World" (Map at scale 1:25,000,000), accompanying Putnam, W. C., D. I. Axelrod, H. P. Bailey, and J. T. McGill, *Natural Coastal Environments of the World,* University of California, Los Angeles.

Mitchell, C. (1971). *Isles of the Caribbees,* National Geographic Society, Washington, D. D.

Putnam, W. C., D. I. Axelrod, H. P. Bailey and J. T. McGill (1960). *Natural Coastal Environments of the World,* University of California, Los Angeles.

Russell, R. J. (1963). "Recent Recession of Tropical Cliffy Coasts," *Science,* Vol. 139, No. 3549, pp. 9-14.

Russell, R. J. (1967). *River Plains and Sea Coasts,* University of California Press, Berkeley.

Shepard, F. P. and H. R. Wanless (1971). *Our Changing Coastlines,* McGraw-Hill Book Company, New York 579 pp.

Steers, J. A. (1962). *The Sea Coast,* Collins, London.

Schwartz, M. L. (1972). *Spits and Bars,* Dowden, Hutchinson and Ross, Inc., Stroudsburg, Pa., 452 pp.

U. S. Geological Survey (1970). *Natural Atlas of the United States of America,* U. S. Government Printing Office, Washington, D. C.

Valentin, H. (1952). "Die Kusten der Erde," Peter. Geog. Mitt., 246, 118 pp.

Wells, J. W. (1957). "Coral Reefs," in J. W. Hegspeth (ed.), "Treatise on Marine Ecology and Palecology," *Geological Society of America, Memoir 67.*

West, R. C. (1956). "Mangrove Swamps of the Pacific Coast of Colombia," *Annals Association of American Geographers,* Vol. 46, No. 1.

Williams, W. W. (1960). *Coastal Changes,* Routledge and Kegan Paul, London.

Womersley, H. B. S. (1954). "The Species of Macrocystis With Special Reference to Those on Southern Australian Coasts," *University of California Publications in Botany,* Vol. 27, No. 2, pp. 109-132.

GLACIATION SECTION

Atlas of Canada (1957). Department of Mines and Technical Surveys, Geographic Branch, Ottawa.

Alden, W. C. (1918). "The Quaternary Geology of Southern Wisconsin, *"U.S. Geological Survey, Prof. Paper, No. 106,* 356 pp.

Antevs, E. (1928). *The Last Glaciation* (American Geographical Society Research Series No. 17).

Aranow, S. (1959). "Drumlins and Related Streamline Features in the Warwick-Tokio Area, North Dakota, *"American Journal of Science,* Vol. 257, No. 3, pp. 191-203.

Atwood, W.A. (1940). *The Physiographic Provinces of North America,* Ginn and Company, New York.

Bird, J.B. (1967). *The Physiography of Arctic Canada,* The John Hopkins Press, Baltimore.

Butzer, K. W. (1964). *Environment and Archaeology; An Introduction to Pleistocene Geography,* Aldine Pub. Co., Chicago.

Butzer, K. W. (1976). *Geomorphology From The Earth,* Harper and Row, Publishers, New York.

Charlesworth, J. K. (1957). *The Quaternary Era with special reference to its glaciation,* 2 vols., MacMillan Company, London.

Charnley, F. E. (1959). "Some Observations on the Glaciers of Mount Kenya, "Journal of Glaciology, Vol. 3, No. 26, pp. 483-492.

Clark, T. H., and C. W. Stearn (1960). *The Geological Evolution of North America,* The Ronald Press Co., New York.

Costin, A. B., and others (1964). "Snow Action on Mount Twynam, Snow Mountains, Australia, *"Journal of Glaciology,* Vol. 5, No. 38, pp. 219-227.

Cotton, C. A. (1941). *Landscape, As Developed by the Processes of Normal Erosion,* The University Press, Cambridge.

Daly, R. A. (1934). *The Changing World of the Ice Age,* Yale University Press, New Haven.

Daly, R. W. (1963). *The Changing World of the Ice Age,* Hafner Publishing Company, New York.

Davis, W. M. (1901). *Physical Geography,* Ginn & Co., Boston, 339 pp.

Embleton, C., and C. A. M. King (1968). *Glacial and Periglacial Geomophology,* St. Martin's Press, New York.

Fairchild, H. L. (1896). "Kame Areas in Western New York," *Journal Of Geol.,* Vol. 4, pp. 129-159.

Fairchild, H. L. (1927). *Geologic Romance of the Finger Lakes* Smithsonian Institute Annual Report, M Section), Smithsonian Institute, Washington, D. C.

Field, W. O. and Associates (1965). *Atlas of Mountain Glaciers* in teh Northern Hemisphere, U. S. Army Quartermaster Research and Engineering Command, Natick, Massachusetts.

Flint, R. F., *et al.* (1945). "Glacial Map of North America," *Geological Society of America, Special Paper 60, Part I,* Glacial map (in color); Part II, explanatory notes, 37 pp.

Flint, R. F. 1953). *Glacial Geology and the Pleistocene Epoch,* John Wiley and Sons, Inc., New York.

Flint, R. F. (1957). *Glacial and Pleistocene Geology,* John Wiley and Sons, Inc., New York.

Flint, R. F. (1971). *Glacial and Quaternary Geology,* John Wiley and Sons, Inc., New York.

Garner, H. F. (1974). *The Origin of Landscapes, A Synthesis of Geomorphology,* Oxford University Press, New York.

Gregory, J. W. (1913). *The Nature and Origin of Fiords,* John Murray, New York.

Gunn, B. M. (1964). "Flow Rates and Secondary Structures of the Fox and Franz Joseph Glaciers, New Zealand," *Journal of Glaciology,* Vol. 5, No. 38, pp. 173-190.

Harrisson, A. E. (1952). "Glacial Activity in the Western United States, "*Journal of Glaciology,* Vol. 2, No. 19, pp. 666-668.

Hershey, O. H. (1897). "Eskers Indicating Stages of Glacial Recession in the Kamson Epoch in Northern Illinois," *Am. Geol.,* Vol. I, pp. 197-209, pp. 237-253.

Holmes, A. (1945). *Principles of Physical Geology,* The Ronald Press, Co., New York.

Hoppe, G. (1959). "Glacial Morphology and Inland Ice Recession in Northern Sweden," *Geografiska Annaler,* Vol. 41, pp. 193-212.

Hough, J. L. (1958), *Geology of the Great Lakes,* University of Illinois Press, Urbana.

Humphries, D. W. (1959). "Preliminary Notes on the Glaciology of Kilimanjaro," *Journal of Glaciology,* Vol. 3, No. 26, pp. 13-27.

Jenness, J. L. (1949). "Permafrost in Canada," *Arctic,* Vol. 2, pp. 13-27.

Kick, W. (1960). "The First Glaciologists in Central Asia," *Journal of Glaciology,* Vol. 3, No. 28, pp. 687-692.

King, C. A. M., and J. D. Ives (1955). "Glaciological observations on some of the outlet glaciers of Southwest Vatnajokull, Iceland, *Journal of Glaciology,* Vol. 2, No. 18, pp. 563-569.

Krinsley, D. B. (1965). "Pleistocene Geology of the Southwest Yukon Territory, Canada," *Journal of Glaciology,* Vol. 5, No. 40, pp. 385-397.

Kupsch, W. O. (1955). "Drumlins with Jointed Boulders Near Dollard Saskatchewan," *Bull. Geol. So. Am.,* Vol. 66, pp. 327-338.

Lemke, R. W. (1958). "Narrow Linear Drumlinoids Near Velva, North Dakota," *Am. J. Sci.,* Vol. 256, pp. 270-283.

Leverett, F., and R. B. Taylor (1915). "The Pleistocene of Indiana and Michigan and the History of the Great Lakes," *U. S. Geological Survey Monograph,* No. 53, 529 pp.

Lewis, W. V. (1949). "An Esker in Process of Formation: Boverbreen, Jotunheimen, 1947, "*Journal of Glaciology,* Vol. 1, No. 6, pp. 314-319.

Lewis, W. V. (1960). "The Problem of Cirque Erosion," in (Lewis, W. V., editor), "Norwegian Cirque Glaciers," (Vol. 4), Royal Geographical Society Res. Ser., London, John Murray.

Liboutry, L. (1953). "Snow and Ice in the Monte Fitz Roy Region (Patagonia)," *Journal of Glaciology,* Vol. 2, No. 14, pp. 255-261.

Lobeck, A. K. (1939). *Geomorphology,* McGraw-Hill Company, Inc., New York.

Lobeck, A. K. (1958). *Things Maps Don't Tell Us,* The MacMillan Company, New York.

Mercer, J. H. (1967). *Southern Hemisphere Glacial Atlas,* Technical Report 67-76-ES, Earth Sciences Laboratory, U. S. Army Natick, Massachusetts.

Miller, M. M. (1963). "The Vaughan Lewis Glacier, Juneau Icefield, Alaska," *Journal of Glaciology,* Vol. 4, No. 36, pp. 666-667.

Mosley, A. (1937). "The Ponds, Lakes, and Streams of the Kergiz Steppes, *Scottish Geographic Magazine,* Vol. 53, pp. 6-9.

Niewiarowski, W. (1963). "Types of Kames occurring within the area of the last glaciation in Poland compared with kames known from other regions," *Report of 6th Conference of the International Association for Quaternary Research (Warsaw, 1961),* Lodz, pp. 475-485.

Niewiarowski, W. (1965). "Conditions of occurrence and distribution of kame landscapes in the Peribalticum. " *Geographia Polonica,* Vol. 6, pp. 7-18.

Odell, N. E. (1937). "Franz Josef Fjord and the Mystery Lake District, N. E. Greenland," *Scottish Geographic Magazine,* Vol. 53, pp. 207-322.

Odell, N. E. (1960). "The Mountains and Glaciers of New Zealand," *Journal of Glaciology,* Vol. 3, No. 28, pp. 739-742.

Odell, N. E. (1963). "The Kolahoi Northern Glacier, Kashmir," *Journal of Glaciology,* Vol. 2, No. 1, pp. 15-26.

Orme, A. R. (1969). "The Quaternary Glaciation of Ireland," *Geographica Viewpoint,* Vol. 2, No. 1, pp. 15-26.

Ostrem, G. (1961). "A New Approach to End Moraine Chronology," *Geografiska Annaler,* Vol. 43, pp. 418-419.

Paige, R. A. (1965). "Advance of Walsh Glacier," *Journal of Glaciology,* Vol. 5, No. 42, pp. 876-878.

Pewe, T. L. (1969). *The Periglacial Environment, Past and Present,* McGill-Queens University Press, Montreal.

Powers, W. E. (1966). *Physical Geography,* Appleton-Century-Crofts, New York.

Raisz, E. (1949-1950). *Landform Map of Canada,* Institute of Geographical Exploration, Harvard Univ., Cambridge.

Rankama, K. (1965). *The Quaternary,* Vol. 1, John Wiley and Sons, Inc., New York.

Schou, A. (1949). *Atlas of Denmark,* Part I, *The Landscapes,* H. Hagerup, San Francisco.

Shelton, J. S. (1966). *Geology Illustrated,* W. H. Freeman and Co. San Francisco.

Smith, H. T. V., and L. L. Ray (1941). "Southernmost Glaciated Peak in the United States." *Science,* Vol. 93, p. 209.

Somme, A. (1960). *A Geography of Norden:* Denmark, Finland, Norway, Sweden, J. W. Coppelens Forlag, Oslo.

Suggate, R. P. (1950). "Franz Josef and Other Glaciers of the Southern Alps, New Zealand," *Journal of Glaciology,* Vol. 1, No. 8, pp. 422-429.

Suslov, S. P. (1961). *Physical Geography of Asiatic Russia,* W. H. Freeman and Co., San Francisco.

Synge, F. M. (1956). "The Glaciation of North-East Scotland," *The Scottish Geographical Magazine,* Vol 72, No. 3, pp. 129-143.

Taylor, F. B. (1912). "The Glacial and Postglacial Lakes of the Great Lakes Region," *Annual Report,* 1912, Smithsonian Institution, Washington, D. C.

Temple, P. (1965). "Some Aspects of Cirque Distribution in the West-Central Lake District," *Geographiska Annaler,* Vol. 47A, pp. 185-193.

Thornbury, W. D. (1965). *Regional Geomorphology of the United States,* John Wiley and Sons, Inc., New York.

Thornbury, W. D. (1969). *Principles of Geomorphology,* John Wiley and Sons, Inc., New York.

Totten, S. M. (1969). "Overridden Recessional Moraines of North-Central Ohio," *Geological Society of America Bulletin,* Vol. 80, No. 10, pp. 1931-1946.

Tricart, J. (1965). "Observations on the Quaternary Firn Line in Peru," *Journal of Glaciology,* Vol. 5, No. 42, pp. 857-863.

Vanni, M. (1963). "Variations of the Italian Glaciers in 1961," *Journal of Glaciology,* Vol. 4, No. 34, pp. 467-470.

Vernon, P. (1966). "Drumlins and Pleistocene Ice Flow Over the Ards Peninsula, Strangford Lough Area, County Down, Ireland," *Journal of Glaciology,* Vol. 6, No. 45, pp. 401-409.

West, R. G. (1968). *Pleistocene Geology and Biology,* Longmans, Green and Co., Ltd., London.

White, S. E. (1954). "The Firn Field of the Volcano Popocateptl, Mexico," *Journal of Glaciology,* Vol. 2, No. 16, pp. 389-392.

Whittow, J. B. (1959). "The Glaciers of Mount Baker, Ruwenzori," *Geomorphology,* Longman, Green and Co., New York.

Wooldridge, S. D., and S. W. Morgan (1937). *Outlines of Geomorphology,* Longman, Green and Co., New York.

Wright, W. B. (1937). *The Quaternary Ice Age* (2nd ed.), MacMillan And Company, London.

Wright, H. E., Jr., and David G. Frey (1965). *The Quaternary of the United States,* Princeton University Press, Princeton, New Jersey.

Zeuner, F. E. (1945). *The Pleistocene Period, Its Climate, Chronology and Faunal Successions,* Bernard Quaritch, London.

WIND CREATED LANDFORM

Amiran, D. H. K., and A. W. Wilson (1973). *Coastal Deserts, Their Natural and Human Environments,* Univ. of Arizona Press, Tucson.

Bagnold, R. A. (1941). *The Physics of Blown Sand and Desert Dunes,* Methuen and Co., Ltd., London.

Bulla, B. (1954). *Altalanos Ferme's Zeti Fololrodjz,* Budapest.

Cooke, R. U., and A. Warren (1973). *Geomorphology in Deserts,* Univ. of California Press, Berkeley.

Fisk, H. N. (1951). "Loess and Quaternary Geology of the Lower Mississippi Valley," *Journal of Geology,* Vol. 59, No. 4, pp. 333-356.

Flint, R. F. (1971). *Glacial and Quaternary Geology,* John Wiley and Sons, pp. 262-265.

Hoyt, J. B. (1967). *Man and Earth,* 2nd Edition, p. 83.

Jaeger, E. C., et al. (1955). *The California Deserts,* Stanford University Press, Stanford.

Jennings, J. D. (1968). "A Revised Map of the Desert Dunes of Australia," *Australian Geographer,* Vol. 10, pp. 408-409.

King, L. C. (1962). *The Morphology of the Earth,* Oliver and Boyd, Edinburgh and London.

Kriger, N. I. (1965). "Loess, Its Characteristics and Relation to Geographical Environment," USSR Academy of Science Committee for Study of Quaternary Period, Moscow, *Navka,* p. 296.

Leopold, A. S. and Editors of Life (1962). *The Desert,* Time, New York.

Lindsay, John F. (1973). "Reversing Barchan Dunes in Lower Victoria Valley, Antarctica," *Geol. Soc. of Am. Bull.,* Vol. 84, pp. 1799-1806.

Lobeck, A. K. (1939). *Geomorphology,* McGraw-Hill Company, Inc., New York, pp. 392-395.

Lobeck, A. K. (1960). *Things Maps Don't Tell Us,* The MacMillan Company, New York.

Lugn, A. L. (1962). *The Origin and Sources of Loess,* University of Nebraska Studies, Lincoln, Nebraska, Plate I.

McGinnies, W. G., and B. J. Goldman (1969). *Arid Lands in Perspective,* University of Arizona Press, Tucson.

Obruchev, V. A. (1964). *Loess of Northern China,* V. V. Popov, editor, Israel Program for Scientific Translations, Jerusalem.

Reed, E. C., and V. H. Dreezen (1965). "Revision of the Classification of the Pleistocene Deposits of Nebraska," *University of Nebraska, Conservation and Survey Division,* Lincoln, p. 2.

Schultz, C. B., and J. C. Frye (1968). *Loess and Related Eolian Deposits of The World,* Proceedings VIIth Congress, INQUA, 1965; University of Nebraska Press, Lincoln, p. 355.

Smalley, I. J. (1975). *Loess: Lithology and Genesis,* Dowden, Hutchinson and Ross, Inc., Stroudsburg, Pa., 429 pp.

Smith, H. T. U. (1961 and 1964). *Periglacial Aeolian Phenomena in the United States,* Report of 6th Conference International Association for Quaternary Research, Warsaw Lodz.

(1965). "Dune Morphology and Chronology in Central and Western Nebraska," *Journal of Geology,* Vol. 73, pp. 557-578.

Thorp, James, et al. (1962). [Map of] Pleistocene Eolian Deposits of the United States, Alaska and Parts of Canada. Scale 1:2,500,000 (In colors). *Geol. Soc. America,* New York.

WATER CREATED LANDFORM

Bretz, J. H. (1956). "Caves of Missouri," *Missouri Geological Survey and Water Research,* Vol. 39, 2nd Series.

Dicken, S. N. (1935). "Kentucky Karst Landscapes," *Journal of Geology,* Vol. 43, No. 7, pp. 708-728.

Folsom, F. (1956). *Exploring American Caves (Their History, Geology, Lore and Location: A Spelunker's Guide),* Crown Publishers, Inc., New York.

Hack, J. T., and L. H. Durloo, Jr. (1962). "Geology of Luray Caverns, Virginia," *Virginia Division of the Mineral Research Report Investigation,* Vol. 3.

Herak, M. (1972). *Karst: Important Karst Regions of the Northern Hemisphere,* Elsevier Publishing Co., New York.

Horberg, L. (1949). "Geomorphic History of the Carlsbad Caverns Area, New Mexico," *Journal of Geology,* Vol. 57, No. 5, pp. 464-476.

Jennings, J. N. (1971). *Karst,* The M.I.T. Press, Cambridge, Massachusetts and London, England.

Jennings, J. N. (1973). *Karst,* The M.I.T. Press, Cambridge, Massachusetts, 257 pp.

Lehmann, H. (1960). "International Contributions to Karst Phenomena," *Zeitschrift fur Gemorphologie,* Supplementband 2, 197 pp.

Malott, C. A., and R. R. Shrock (1930). "Origin and Development of Natural Bridge, Virginia," *American Journal of Science,* Vol. 219, pp. 257-273.

Malott, C. A., (1945). "Significant Features of the Indiana Karst," *Proceedings of the Indiana Academy of Sciences,* vol. 54, pp. 8-24.

Meyerhoff, H. A., (1938). "The Texture of Karst Topography in Cuba and Puerto Rico," *Journal of Gecmorphology,* Vol. 1, No. 4, pp. 279-295.

Powell, R. L., (1961). "Caves of Indiana," *Indiana State Geological Survey Circular,* Vol.8.

Silar, J., (1965). "Development of Tower Karst of China and North Viet Nam," *National Speological Society Bulletin,* Vol. 26, pp. 35-46.

Sweeting, N. M. (1958). "The Karstlands of Jamaica," *Geographical Journal,* Vol. CXXIV, Part 2, pp. 184-199.

Sweeting, N. M. (1973). *Karst Landforms,* Columbia University Press, New York.

White, W. B., (1965). "World's Longest Caves," *National Speological Society News,* Vol. 23, pp. 68-69.

Woodward, H. P. (1936). "Natural Bridge and Natural Tunnel, Virginia," *Journal of Geology,* Vol. 44, pp. 604-616.

Wray, D. A., (1922). "The Karstlands of Western Jugoslavia," *Geology Magazine,* Vol. 59, pp. 59-62.

Section Three
OCEANOGRAPHIC and
HYDROGRAPHIC FEATURES
Pages 105-154

Section Six
WIND CREATED LANDFORMS
Pages 241-252